D0148389

Tourism, Security and Safety

From Theory to Practice

Tourism, Security and Safety

From Theory to Practice

First edition

Yoel Mansfeld
Abraham Pizam

ELSEVIER

AMSTERDAM • BOSTON • HEIDELBERG • LONDON
NEW YORK • OXFORD • PARIS • SAN DIEGO
SAN FRANCISCO • SINGAPORE • SYDNEY • TOKYO
Butterworth-Heinemann is an imprint of Elsevier

Elsevier Butterworth–Heinemann
30 Corporate Drive, Suite 400, Burlington, MA 01803, USA
Linacre House, Jordan Hill, Oxford OX2 8DP, UK

♾ Recognizing the importance of preserving what has been written, Elsevier prints its books on acid-free paper whenever possible.

Library of Congress Cataloging-in-Publication Data
Mansfeld, Y. (Yoel)
Tourism, security, and safety : from theory to practice / Yoel Mansfeld and Abraham Pizam.
p. cm.
1. Tourism. 2. Tourism—Safety measures. I. Pizam, Abraham. II. Title.
G155.A1M264 2006
910′.68′4—dc22 2005020378

British Library Cataloguing-in-Publication Data
A catalogue record for this book is available from the British Library.

ISBN 13: 978-0-7506-7898-8
ISBN 10: 0-7506-7898-4

For information on all Elsevier Butterworth–Heinemann publications visit our Website at www.books.elsevier.com

Printed in the United States of America
05 06 07 08 09 10 10 9 8 7 6 5 4 3 2 1

Contents

Contributors

Alistair Anderson
Aberdeen Business School
Robert Gordon University
Garthdee, Aberdeen
AB 24 7QE UK

Alex Kobina Armoo
Department of Hospitality and Tourism Management
Baltimore International College
17 Commerce Street
Baltimore, MD 21202-3230 USA

Eli Avraham
Department of Communications
University of Haifa
Haifa 31905 Israel

David Beirman
Israel Tourism Office
395 New South Head Rd
Double Bay
NSW 2028 Australia

Tracy Berno
School of Food and Hospitality
CPIT
PO Box 540
Christchurch 8015 New Zealand

Zelia Breda
Departamento de Economia, Gestao e Engenharia Industrial (DEGEI)
Universidade de Aveiro
Campus Universitario de Santiago
3810-193 Aveiro Portugal

Steven Buccola
Department of Agricultural and Resource Economics
Oregon State University
Corvallis, OR 97331 USA

Nevenka Cavlek
University of Zagreb
Graduate School of Economics & Business
Department of Tourism
Trg J.F. Kennedyja 6
10000 Zagreb Croatia

Carlos Costa
Departamento de Economia, Gestao e Engenharia Industrial (DEGEI)
Universidade de Aveiro
Campus Universitario de Santiago
3810-193 Aveiro Portugal

Aliza Fleischer
Department of Agricultural Economics and Management
The Hebrew University of Jerusalem
P.O. Box 12, Rehovot 76-100 Israel

Martin Fluker
School of Hospitality, Tourism & Marketing
Victoria University
PO Box 14428
Melbourne Victoria 8001 Australia

Dee Wood Harper, Jr.
Department of Criminal Justice
Loyola University New Orleans
6363 Saint Charles Avenue
Stallings Hall, Box 14
New Orleans, LA 70118 USA

Judy Holcomb
Rosen College of Hospitality Management
University of Central Florida,
9907 Universal Blvd
Orlando, FL 32819-9357 USA

Wilson Irvine
Aberdeen Business School
Robert Gordon University
Garthdee, Aberdeen
AB 24 7QE U.K.

Aviad Israeli
Department of Hotel and Tourism Management
The School of Management
Ben Gurion University of the Negev
P.O.Box 653
Beer Sheva, 84105 Israel

Brian King
School of Hospitality, Tourism & Marketing,
Victoria University,
PO Box 14428 MC,
Melbourne, Victoria, Australia 8001

Yoel Mansfeld
Department of Geography & Environmental Studies & Center for Tourism, Pilgrimage & Recreation Research
University of Haifa
Haifa 31905 Israel

Jerome McElroy
Department of Business and Economics
Saint Mary's College
Notre Dame, IN 46556 USA

Damian Morgan
School of Business
Monash University
Gippsland Campus
Churchill Victoria 3842 Australia

Gianna Moscardo
School of Business
James Cook University
Townsville QLD 4811 Australia

Abraham Pizam
Rosen College of Hospitality Management
University of Central Florida
9907 Universal Blvd
Orlando, FL 32819-9357 USA

Arie Reichel
Department of Hospitality and Tourism Management
The School of Management
Ben Gurion University of the Negev
P.O.Box 653
Beer Sheva, 84105
Israel

Greg Stafford
Holiday Inn-on-the-Hill
415 New Jersey Ave., NW
Washington, DC 20001 USA

Peter Tarlow
Tourism & More Inc.
1218 Merry Oaks,
College Station, TX 77840-2609 USA

Matthew Taverner
Begawan Giri Estate
P.O. Box 54
Ubud 80571
Gianyar, Bali, Indonesia

Geoffrey Wall
Faculty of Environmental Studies
University of Waterloo
Waterloo, Ontario N2L 3G1 Canada

Barbara Woods
School of Business
James Cook University
Townsville QLD 4811 Australia

Larry Yu
Department of Hospitality and Tourism Management
School of Business
The George Washington University
600 21st Street, NW
Washington, DC 20052 USA

Prologue

Tourism and security incidents are inevitably interwoven phenomena. When security incidents such as wars, terrorism, crime, and civil unrest take place at or close to areas where tourism forms an important land use, the tourism industry, tourists, and the local community are always affected. Although this form of interrelationship has been manifested in various security related situations around the world since the beginning of modern tourism, it has gained substantial global interest only in the aftermath of the September 11, 2001 attacks on US targets by al Qaida. Ever since this tragic event, tourism academics and practitioners alike have realized that security situations no longer negatively affect only a specific location, be it a given destination, a region, or an entire country. Apparently, these incidents cause an imbalance in global tourism systems, forcing this highly important yet fragile global industry to operate under high uncertainty and risk levels.

The coincident local and global fragility of the tourism industry as a result of security situations and the damage incurred by a crisis calls for long-term monitoring and studying of these interrelations in order to mitigate the negative impacts. The first studies into the relationship between tourism and security incidents emerged in the beginning of the 1990s. These studies were triggered by a variety of security incidents, for example in the Middle East (mainly wars and terror); in Central America (mainly civil unrest); and in North America, South America and Africa (mainly in the form of high rates of crime against tourists). These events, and their repercussions, triggered academic interest in this field in the form of a number of academic conferences, academic articles, and an edited book entitled *Tourism, Crime and International Security Issues.* That book, which was published in 1995, was our first collaborative effort to put security and tourism in the forefront of academic literature yet serve as relevant and practically oriented tourism research. Ten years later our second book is an attempt to pursue these goals a few steps further. The goals of this volume are twofold: First, the book attempts to develop the first two building blocks of a *tourism security theory,* which is based on the knowledge generated by the research and academic community as well as by the accumulated experiences of numerous destinations that were afflicted by security crises. Second, it provides a multidimensional discussion on managing security crises in affected tourist destinations. In this respect, this book represents a shift from case studies documenting the characteristics of impacts of security incidents on tourism, tourists, and host communities, to an academic discourse on how to handle and mitigate the consequences of such events.

Indeed, in line with these goals, the book takes the reader from theory into the practical aspects. It starts with an in-depth theoretical chapter that synthesizes the scientific knowledge developed so far in this field of research. Subsequently, both theoretical and practical matters are discussed under four themed sections. The book ends with a summary that draws on the recommended research agenda to further explore security and tourism relationships and thus expand and deepen both their theoretical foundations and the practical strategies to mitigate their negative impacts.

We hope that this book will be meaningful and beneficial to students and researchers engaged in the study of the multidisciplinary fields of tourism and hospitality management. Concurrently we wish that tourism and hospitality practitioners will find this book a good and helpful reference and a practical guide for successful operations of tourism enterprises before, during, and after security crises.

Special thanks go to our valued contributors to this book who worked diligently to produce a quality product on time. We thank them for their attitude, patience, and teamwork. Yoel Mansfeld would like to convey his special thanks to all his colleagues at the Department of Tourism Management at the Management School of the University of Waikato, Hamilton, New Zealand, who hosted him during his sabbatical and thus facilitated the work on this book. We would also like to thank the publishers of *Annals of Tourism Research; Journal of Hospitality and Tourism Management; International Journal of Hospitality Management; Cornell Hotel and Restaurant Administration Quarterly;* and *Applied Economics* for allowing us to republish exceptionally high quality articles on tourism and security in the form of chapters in our book. Lastly, we would like to convey our sincere gratitude to two special ladies, Mrs. Olga Sagi and Mrs. Genoveba Breitstein, who, for the third time, collaborated with us on the technical production of this book.

1

Toward a Theory of Tourism Security

Abraham Pizam and Yoel Mansfeld

Learning Objectives

- To understand the process of theory building in the field of tourism security.
- To understand the importance of theory building as part of developing appropriate strategies to control the negative impacts security incidents have on the tourism system.
- To become acquainted with the fundamentals of tourism security theory.
- To become familiar with the nature of security incidents.
- To understand the array of impact security incidents have on tourists, the tourism industry, and the host community.
- To become aware of future research directions needed in order to refine or redefine tourism security theory.

Introduction

Bailey (1982, p. 39) defines theory as "an attempt to explain a particular phenomenon." In his opinion, to be valid a theory must: "predict and explain a phenomenon and be testable, at least ultimately" (Bailey, 1982, p. 40). Therefore, the objective of this chapter is to *start* the process of crafting a "tourism security" theory by constructing its first two building blocks, namely the statement of concepts and propositions.

It is expected that by the time this theory is completed it should be able to answer the following questions:

- Why incidents of security such as crime, terrorism, wars, riots, and civil unrest exist at tourist destinations;
- What are the motives of the perpetrators/offenders;

- What are the impacts of such incidents on the tourists, the tourism industry, the destination, and the community at large;
- How do the tourism industry, the tourists, the destination, the media, and the community react to the crises caused by such incidents;
- What effective recovery methods can be undertaken by the public and private sectors at the destination;
- What methods of prevention or reduction of such incidents can be used by the destination in order to avoid or minimize the impacts of future security crises?

A successful theory will provide explanations and predictions of the phenomenon of "tourism security" by relating some of its components (i.e., the variables of crimes, terrorism, war, riots) to some other phenomena (e.g., the variables of tourism demand, offenders' motivation, victims' behavior, opportunity, location, etc.). Though the ultimate aim of such a theory would be to state the relationship between these phenomena in *causal* terms (e.g., increased rates of crimes against tourists *causes* a decrease in tourist visitation), in the immediate future this theory would at best suggest only the direction of the hypothesized relationships (e.g., a negative direction indicates that the higher the crime rates at a tourist destination, the lower the tourist arrivals, whereas a positive direction indicates that the more uniformed police officers are visible at the tourist destination, the more secure tourists feel about the destination).

As suggested by Bailey (1982, p. 40) the first two steps in theory construction are the statement of concepts and the writing of one or more propositions. "*Concepts* are . . . mental images or perceptions . . . they may be impossible to observe directly," such as fear of being robbed in the case of tourism security, "or they may have referents that are readily observable," such as a gun or knife in the case of tourism security. "On the other hand, many concepts contain several categories, values, or subconcepts, often falling along a recognizable dimension or continuum," such as the number of tourist related robberies in a given year. "Concepts that can take on more than one value along a continuum are called variables" (Bailey, 1982, p. 40).

Propositions are "statements about one or more concepts or variables." Subtypes of propositions include "hypotheses, empirical generalizations, axioms, postulates, and theorems." Hypotheses are propositions that are "stated in a testable form and predict a particular relationship between two (or more) variables. In other words, if we think that a relationship exists, we first state it as a hypothesis and then test the hypothesis in the field" (Bailey, 1982, p. 40). "In contrast to a hypothesis, an empirical generalization is a relationship that represents an exercise in induction. Rather than hypothesizing that a relationship exists and then testing this hypothesis, an empirical generalization is a statement of relationship that is constructed by first observing the existence of a relationship (in one or a few instances) and then generalizing to say that the observed relationship holds in all cases (or most cases)" (Bailey, 1982, pp. 41–42).

For the purpose of constructing a tourism security theory, empirical generalizations rather than hypotheses were developed in this chapter because by now researchers have had the opportunity to observe numerous tourism security crises throughout the world and have examined their impacts on the tourists, the destinations, and the tourism industry. Furthermore, in the aftermath of these incidents, researchers also managed to study and scrutinize the effectiveness of various recovery and prevention methods that were put in place by the affected destinations. This enables suggesting a group of empirical generalizations that will ultimately lead to the crafting of a comprehensive theory that will predict and explain the tourism security phenomenon.

The following sections will list and define the various concepts of this theory and put forward a set of propositions stated in the form of empirical generalizations.

Tourism and Security: Concepts and Their Respective Variables

To formulate and construct the basis for a theory of tourism security it is necessary, first, to define the major concepts that are derived from the relationship between tourism and security incidents. Once these concepts and their respective variables are defined they will lay the foundations for the theoretical development of over eighty propositions, stated in the form of empirical generalizations.

In the next section we define the relevant concepts and their corresponding variables, grouped by common subjects.

In recent years the theoretical discourse on the relationship between tourism and security has been conducted around three main groups of concepts and their derived variables. These groups are:

- Group A: Concepts relating to the nature of tourism-related security incidents and crises (including types, causes, mode of operation, motives, targets, etc.);
- Group B: Concepts relating to the impacts of security incidents and crises on the tourism industry, the tourists, and host communities;
- Group C: Concepts relating to the short-, medium-, and long-term reactions of all tourism stakeholders to existing and potential security incidents and crises.

A Typology of Tourism Security Concepts and Variables

Based on the grouping of concepts proposed above, we propose the following concepts and their corresponding variables.

Group A: The Nature of Tourism-Related Security Incidents and Crises

Types of Security Incidents

The first and perhaps the most fundamental concept to be discussed under this group is the type of security incident that affects tourism. This concept's centrality stems from its substantial influence on its potential impact on tourism. So far, the literature dealing with these interrelations identified four major types of security incidents that triggered some form of negative impact on the host communities, the tourism industry, and the tourists themselves. The four possible generators of a given security situation that might harmfully impinge on the tourism system are: crime-related incidents, terrorism, war, and civil/political unrest.

Crime-related incidents can be in the form of:

- Larceny;
- Theft;
- Robbery;
- Rape;
- Murder;
- Piracy; and
- Kidnapping.

These crime-related incidents may take place in various scenarios, such as crimes committed by local residents against tourists; crimes committed by tourists against local residents; crimes committed by tourists against other tourists; and organized crime against tourism enterprises.

Terrorism can take the form of:

- Domestic terrorism;
- International terrorism; and
- Cross-border terrorism.

The relationship between tourism and terrorism can be manifested in three possible scenarios: Terrorism that is aimed at civil targets yet sometimes victimizes tourists as well; terrorism that is directed at economic targets that are functionally related to tourism; and finally, terrorism that targets tourism and/or tourists since both are regarded as "soft targets" with relatively high-impact media coverage.

Wars, either full-scale or limited to a given region, have also had major impacts on tourist demand, both for the involved countries as well as on global tourist flows. The outbreak of wars, unlike terror activities, tends to have a negative tourism impact on larger areas and for a longer period of time. Historically, the types of wars that have been found to have an impact on tourism are:

- Cross-border wars;
- Trans-border wars;
- Wars of attrition; and
- Civil wars.

Civil and/or political unrest can be in the form of:

- Coup d'état;
- Violent demonstrations;
- Uprising; and
- Riots.

The above incidents have caused major declines in tourism demand in various parts of the world. Whether it is a coup d'état in Fiji, violent demonstrations against the Group of Seven nations (G7) in several different venues, the uprising of the Palestinians in the West Bank and Gaza, or riots in the Chiapas region of Mexico, such incidents paralyzed or severely impacted the local tourism industry as a result of trip cancellation behavior and a shift of bookings to safer alternative destinations.

Frequency of Security Incidents

Empirical evidence so far shows that the higher the frequency of such incidents and the more media coverage they obtain, the greater the negative impact on tourist demand. A high frequency of security incidents causes changes in tourists' booking and cancellation behavior, selective spatial behavior in the affected destination, and other tourism demand characteristics. The frequency of security incidents is usually measured by the following variables:

- Number of security incidents in a given period of time; and
- Scaled frequency pattern within a given period of time.

Motives and Targets of Security Incidents

As previously indicated, to predict the impacts that security incidents have on tourism, it is imperative to understand the motives behind such incidents. An in-depth study of these motives could provide valuable information on potential targets. A greater understanding of this cause and effect relationship can lead to more effective contingency and mitigation plans for affected destinations. Thus, the variables most often used to detect goals and targets are:

- Types of (declared or undeclared) motives;
  - Political;
  - Religious;
  - Social;
  - Economic;
  - Hostility to tourists;
  - Publicity seeking; and
  - Destruction of an area's economy.
- Types of (declared or undeclared) targets;
  - Tourists on the way to and from their travel destinations;
  - Tourists vacationing in a given travel destination;
  - Tourism and hospitality installations and facilities;
  - Strategic and non-strategic transportation facilities serving tourists; and
  - Public and private services and businesses also serving tourists.

Severity of Security Incidents

The evidence so far shows that the impact of security incidents on tourism, tourists, and hosts is directly correlated with the severity of the incidents. Although it is difficult to objectively define the levels of severity of security incidents, we propose the following variables that can be used as measurement scales:

- Extent of overall damage to tourism properties caused by security incidents;
- Extent of damage to private sector tourism properties caused by security incidents;
- Extent of damage to public sector tourism properties caused by security incidents; and
- Extent of damage to life caused by security incidents.

Location

Understanding the geographical dimension of security incidents is of great importance when handling security related tourism crises. Host governments and the tourism industry will do their utmost to ensure that the impacts of security incidents will be confined to the location where the security incident actually took place, and will not spill over to other locations. Mapping the relationship between the location where the security incident occurred and the tourist destination may result in three main situations. The first is when the security location and the tourist destination overlap. The second is when there is a geographical proximity between these two locations. The third situation is when those two locations are far apart. It is assumed that the closer the two locations, the more severe would be the impact of the security incident on the tourism industry. However, this statement sometimes tends to oversimplify the relationship between location and severity of impact on the tourism industry. For example, in some cases, terrorist attacks in major city centers such as Madrid, London, or Paris only marginally affected tourists' demand to these cities and only for a short period of time. However, in other cases the impact was extremely severe and long lasting, such as happened in New York City in the aftermath of 9/11, and in Tel Aviv following frequent suicide bombing of local buses. This lack of coherent reaction suggests that locational factors are only one part of the anatomy of security incidents. The most relevant variables used to examine the geographical dimension of security situations are:

- Geographical range of impact;
- Geographical distribution of affected areas;
- On- vs off-the-premises of tourist enterprises;
- High vs low crime areas;
- Physical characteristics of the urban environment;
- Physical characteristics of the tourist installations; and
- Location of potentially crime-generating tourist activities.

Group B: Impacts of Security Incidents

The accumulated evidence throughout the world shows that the impacts of security incidents on the tourism industry, the destination, the local community, and the tourists are, in most cases, negative and multifaceted. Consequently, impact concepts are grouped here into six subgroups depicting different facets of the impact of security situations on tourism. Each subgroup of concepts also includes commonly used variables that measure these impacts.

Impact on the Destination Itself

When a tourism crisis occurs in a given destination, one of the first actions taken by local decision makers is to assess the damage. This assessment is needed in order to help in formulating contingency plans and policies to handle and mitigate the damage to the local tourism industry in the wake of the security incidents. Some of the more common variables used for the purpose of assessing the impact of security incidents on a macro-level destination performance are:

- Tourist overall arrivals in a given period;
- Tourist segmented arrivals in any given period;
- Tourist overall receipts in any given period;
- Tourist segmented receipts in any given period;
- Duration of impact (crisis); and
- Destination life cycle.

Impact on Tourists' Behavior

In most cases, security incidents cause changes in tourists' perception of risk, and thus are always translated into travel decisions. These could be in the form of cancellations of booked trips, avoiding booking trips to affected destinations, or, for those already in the affected destination, moving to a safer place or evacuating the destination and returning home. Such decisions are based on a variety of considerations and circumstances that will be discussed later on. However, at this stage it is important to note that measuring tourists' behavior following a major change in the security level of a given destination is imperative in order to formulate crisis-management plans.

The most frequent variables used in pursuit of understanding tourists' reaction to changing security situations are:

- Intention to travel to affected destination;
- Actual cancellations;
- Actual bookings;
- Actual avoidance of unsafe destinations;
- Risk-taking tendency of various tourist segments;
- Change in use of risk-related travel information prior to destination choice;
- Perceived vulnerability to specific types of crimes;
- Characteristics of tourist image projection;
- Familiarity with safe and unsafe areas within a given destination; and
- Involvement in illicit activities.

Impact on the Tourism Industry

In the case of leisure tourists on organized trips, the tourists' travel behavior is facilitated by two stakeholders in the tourism system—tour operators in the generating markets and tour operators in the receiving destination. Both share a common objective of mitigating the almost inevitable damage resulting from a change in the security climate of a given destination. Since these stakeholders do not normally coordinate their reactions to security-oriented crises, each has to perform an individual assessment of the other side's actions taken to mitigate the damage.

The most common variables used to characterize the behavior of the tourism industry in the wake of evolving security situations are:

- Evacuation of tourists by tour operators;
- Local investors' behavior;
- Transnationals' investing behavior;

- Human resource restructuring behavior;
- Inclusion/exclusion of destination in tour operators' brochures;
- Cost of doing or ceasing doing business;
- Cash flow assessment;
- Profitability;
- Projection of destination image by tour operators and travel agents; and
- Extent of economic interest in tourism business at the destination.

Impact on Host Governments

In many communities, tourism serves as an important contributor to the local, regional, and national economies, and in some cases is a major contributor to foreign currency earnings. Because security incidents might have a major negative impact on these economies, such incidents are normally a cause of major concern for local, regional, and national governments. Such concern might change governments' policies towards the future of this sector, its relative role in the economy, and the level of involvement governments wish to exert once they realize the fragility and the potential instability of this economic sector. Host governments in affected destinations usually monitor and assess the impact of security incidents on a dynamic basis. In addition these governments: (a) initiate the implementation of new and/or improved security measures aimed at preventing and/or diminishing the occurrence of future security incidents taking place in tourist areas; (b) assist in the process of damage control when the security situation deteriorates; and (c) provide ad hoc financial assistance to cope with all the major negative ramifications of security induced tourism crises.

To evaluate the impact of security situations on tourism from the governmental perspective the following variables may be used:

- Changes in level of security measures in affected destinations;
- Changes in short-, medium-, and long-term government policies towards tourism;
- Extent of governmental direct/indirect operational involvement in tourism;
- Extent of governmental direct/indirect financial involvement in tourism; and
- Extent of governmental direct/indirect marketing involvement in tourism.

Impact on Governments of Generating Markets

Potential tourists' travel behavior is influenced to a certain extent by the risk assessment conducted and published by some of their respective governments. These governments issue frequent bulletins that assess the risk involved in traveling to affected destinations. Many travelers tend to highly value the accuracy of these assessments as they lack the ability to make their own judgment as to the real risks involved. Affected destinations have learned the hard way that it is extremely important to understand the serious consequences of such warnings and to try to influence governments in the generating countries to make them objective and unbiased as well as update them periodically.

The most common variables used to evaluate the impact of governmental warnings are:

- Availability of travel advisories in given generating markets;
- Level of exposure to travel advisories in generating markets;

- Position on travel advisories' risk scale; and
- Frequency of travel advisory updates.

Media Behavior

Security incidents are regarded by the media as important news generators. Thus, when they take place, the media becomes preoccupied in providing its customers with the most vivid and explicit information and analyses of these incidents. Thus, potential tourists in the generating markets are saturated with up-to-date and real-life information which consciously or unconsciously establishes a perceived high-risk image of the affected destinations. In some cases it was evident that the information and assessment provided by the media about the severity of the incidents were biased and the media exaggerated the real risk involved in traveling to the affected areas.

In order to establish empirical evidence on this possible bias and in pursuit of objective assessment of media behavior in times of security induced tourism crisis, the following variables may be useful:

- Extent of coverage of the incident;
- Types of media coverage;
- Forms of media coverage (informative vs interpretive);
- Relative coverage of security situations by media platforms;
- Level of biased information;
- Level of biased interpretation of security situations;
- The impact of media warnings; and
- Extent of media messages directly aimed at potential tourists.

Group C: Reaction to Tourism Crises by All Tourism Stakeholders

The concepts gathered under Group C represent the expected and actual efforts made by the various stakeholders in the tourism system in response to security incidents that either:

- Might affect tourist destinations in the future;
- Are currently affecting tourist destinations causing a crisis situation; or
- Affected tourist destinations in the past.

For many tourist destinations around the world, security incidents and security crises are not, unfortunately, a matter of a past episode but rather a stage in a perpetual cycle of crises and recoveries. Therefore, the concepts illustrated below, as well as their derived variables, are of a dynamic nature. Although these groups of concepts and variables are discussed separately it is important to emphasize that there are many cross-functional relations between them.

Destination Behavior

In times of security oriented tourism crises, affected destinations play a key role in fighting for their economic and social survival. The key questions to be investigated when evaluating destinations' behavior in times of security crises are: (a) to what extent are destinations proactive or reactive as the crisis emerges; and (b) are destinations involved in a concerted multistakeholder (the tourism industry, local community, and the local/regional governments) effort to mitigate the consequences?

Following are some of the most frequent variables used to measure and evaluate the performance of each of the destination stakeholders separately and jointly as a concerted destination effort to mitigate the damage incurred:

- Extent of publicity and public relations activities;
- Availability of contingency and crisis plans;
- Availability of marketing campaigns;
- Level of implementation of contingency and crisis plans;
- Level of cooperation among stakeholders on planning and implementation of crisis management operations;
- Characteristics of marketing campaigns;
- Availability of tourist education programs;
- Availability of image enhancement programs; and
- Availability of crisis management funding.

Image and Perception Management

When security incidents take place and the security situation in tourist destinations deteriorates, the result does not always lead to a long-term detrimental effect on the local tourism industry. However, when the situation involves global media coverage, the information it conveys creates a strong negative image among potential tourists. If this negative image is translated by would-be travelers into unacceptable risk levels, potential tourists would most likely cancel their bookings or choose to book alternative and more secure destinations. It is, therefore, in the interest of the tourism industry and host governments to try to balance the negative images by conveying their own more accurate, less biased, and marketing-oriented messages. However, in order to choose the right strategy in pursuit of a better perception management, affected destinations have to detect and analyze the perceived images and their interpretation by their potential markets.

The following variables are to be used in order to unveil the characteristics, image, and risk perception of security affected destinations:

- Nature of perceived destination image following security incidents;
- Levels of perceived risk;
- Effect of mass media on destination image;
- Effect of travel trade on destination image;
- Effect of friends and relatives on destination image;
- Effect of risk-taking tendency on destination image; and
- Effect of risk takers' experience on destination image.

Risk and Crisis Management (Prevention/Reduction/Mitigation) Techniques

Past experience shows that forward-thinking destinations that were concerned about being affected by security incidents dealt with the situation in one or both of the following ways, either (1) before an incident took place, or (2) when an incident occurred and caused some sort of tourism crisis. In the first case, destinations prepared themselves by developing contingency plans as part of a proactive risk-management policy. In the second case, when an incident occurred, these destinations pulled their crisis-management plans out of the drawer and implemented them. Obviously, the better a destination was prepared the more effective was its response to the security crisis. However, for both cases the effectiveness of its prevention, reduction, and/or mitigation plans was a function of the cooperation between all tourism stakeholders in the affected destinations and between these stakeholders and those in the generating markets.

Common variables evaluating the extent of operational cooperation and success of crisis management plans are:

- Availability of risk related information to tourists and potential tourists;
- Availability of integrated contingency marketing plans for each crisis stage;
- Availability of media and image-management plans;
- Availability of attractive incentives for domestic tourists;
- Level of labor cost reduction in private enterprises;
- Level of dissemination of positive communication;
- Development, operation, and updating of travel advisories among generating markets and host destinations;
- Presence of law enforcement or the military in tourist zones;
- Level of technologically based means of protection in and around tourism installations;
- Availability of dedicated tourist police units;
- Level of dedicated tourism policing;
- Level of visibility of security measures;
- Availability of rewards for information leading to arrests of offenders;
- Facilitation of tourist victims' testimony in criminal cases;
- Training of tourism employees in security matters;
- Public–private cooperation in security provisions;
- Availability of tourism and security education programs;
- Adoption of CPTED (Crime Prevention Through Environmental Design) principles in the design of tourism physical plants;
- Designating crime against tourists a major criminal offense;
- Maintaining a database of crimes against tourists;
- Educating local citizens;
- Creating and maintaining safe roads; and
- Partnership between the leaders of the local community and governments.

Recovery Methods

Past experience has shown that those destinations that conducted well-coordinated efforts to regain tourists' trust when the crisis was over managed to increase tourist demand and recovered in a relatively short time. Recovery efforts involved various actions taken by different tourism stakeholders. But in all cases these actions were successful only when they were backed by sufficient financial resources.

Measuring the effectiveness of recovery methods uses the following most common variables:

- The effect of price reduction strategies;
- Availability of funds for marketing recovery plans;

- Ability to develop new market segments;
- Availability of new and innovative promotional campaigns;
- Availability of destination-specific marketing strategies;
- Effectiveness of marketing campaigns by the private sector;
- Availability of comprehensive marketing campaigns by Destination Management Organizations (DMOs), Non-Governmental Organizations (NGOs), and governments;
- Scheduling of special events;
- Availability of incentives to tourists;
- Availability of financial assistance from governmental agencies;
- Level of local community involvement in recovery oriented efforts;
- Level of positive public relations campaigns to improve public opinion among the media, tourists, and locals; and
- Level of disseminating positive information to existing and potential tourists.

Empirical Generalizations

The next stage in the development of our tourism security theory was the formation of some one hundred empirical generalizations listed in the form of statements. These statements represent a summary of the current best practices in the field of tourism security. They are organized in groups and subsections employing the same typology used previously. Under no circumstances should this list be considered as exhaustive. As with many other theories that are in a state of development, the following list is a work in progress and as such it is expected that with time new statements will be added, while others may be modified or dropped.

Group A: The Nature of Tourism-Related Security Incidents and Situations

Types of Security Incidents

- When multiple types of security incidents occur over a short period of time at the same destination, the negative impact on tourism demand is more severe than when a single type of incident occurs.
- Destinations that accommodate large numbers of tourists tend to develop more enduring and chronic security problems, mostly in the way of crime.
- Crime incidents at tourist destinations are enduring and more difficult to eradicate, while terror and war related situations tend to be more sporadic and of shorter duration.
- International terrorism is the most destructive type of short-term security incident that impacts tourism destinations.
- Full-scale wars are the most destructive medium- to long-term security incidents, since they often include overwhelming obliteration of tourism infrastructure and natural tourism assets. The consequences of such security incidents on the tourism industry could be in the form of irreversible damage or extremely long and highly expensive rehabilitation processes.

- Civil unrest can cause major damage to the local, regional, or national tourism industry especially when the insurgent groups are part and parcel of the cultural tourism product (i.e., their culture is of interest to tourists and exposure of their culture is part of the national, regional, and/or local tourist product).

Frequency of Security Incidents

- All else being equal (AEBE), security incidents occurring more frequently will have a more intense, widespread, and lengthy effect on tourism demand than those occurring less frequently. This is mainly due to high media coverage of these frequent incidents.
- Frequent and severe security incidents (i.e., loss of life and property) have a more detrimental impact on tourism demand to affected destinations than do less frequent and severe incidents.

Motives and Targets of Security Incidents

- AEBE, the political and religious motives of the perpetrators of crimes or terrorism at tourist destinations have the most intense, widespread, and lengthy effects on tourist arrivals and can ultimately lead to the demise of the tourist destination. Economic and social motives have the second strongest effect followed by personal motives, which have the lowest effect.
- At the level of street culture, committing petty crimes against tourists is seen as a way of outsmarting the naïve tourist.
- Some criminals in economically deprived areas possess a Robin Hood sense of entitlement that justifies in their mind robbing the rich (i.e., the tourists) and giving to the poor (i.e., themselves).
- Attacks against mass tourist destinations are particularly desirable to terrorists because:
  - Tourist destinations are easy (soft) targets;
  - Tourist destinations are symbols of national and cultural identity and a strike against them is a strike against a nation and/or its culture;
  - The tourism economy at the destination is intertwined with the regional, state, and countrywide economies and its destruction can cause catastrophic damages to these economies;
  - They result in large number of fatalities; and
  - They create instantaneous mass publicity.
- In security affected destinations, the tourists' look, behavior, and lack of awareness of high-risk crime areas make them more vulnerable to street crimes than local residents.
- In security affected destinations, tourists are much more vulnerable to property crimes (i.e., robbery and larceny) than residents. This is mainly as a result of carrying more money and valuables than local residents.
- In security affected destinations, tourists are more likely than residents to be victims of violent crimes such as murder, rape, and major assault.

Severity of Security Incidents

- Tourist destinations are subject to differential severity levels of security incidents.
- In the short term, the more severe the security incident, the more severe is its negative impact on the local tourism industry and on tourism demand. Acts that cause

mass destruction of life and property—such as war and terrorism—have a more devastating effect on tourist arrivals than acts that cause some loss of life (i.e., murder), which in turn will have a more negative impact than those that cause only bodily harm (i.e., assault and rape). Lastly, acts that cause only loss of property might have only a minimal or negligible impact on tourism arrivals.

- In many situations, the media portrays the severity of security incidents occurring at tourist destinations more harshly than the actual reality. Likewise, would-be tourists in their own communities perceive the level of severity of these incidents to be higher and more serious than those tourists who are already onsite. However, in some cases the situation is reversed and the level of severity perceived by onsite tourists is higher than that of would-be travelers. The nature of these differences is controlled by various factors such as the level of severity of the incident; its duration; its location relative to distinctive tourist areas; the way it was portrayed in the media; the level of exposure of potential tourists to the conveyed images; and finally, to the level of exposure to the actual security situations by onsite tourists.

Location of Incident

- The decline in tourist visitation following safety and security incidents is not restricted to the local community in which the incident occurs. It usually spreads quickly to other regions within and outside the country affected. This spillover effect is generated either by the tourists' lack of geographical knowledge, which distorts their geographical image of the conflict area, or by a biased media coverage which does not supply detailed geographical information on the affected area.
- In cases of very severe security incidents (i.e., terrorism, war) there would be no significant difference in tourist arrivals between acts conducted on or off the premises of tourism enterprises.
- In cases of less severe security incidents (i.e., crimes), acts conducted on the premises of tourism businesses will have a greater effect on tourist visitation than those conducted off the premises.
- Crimes against tourists tend to occur more in geographical areas that have a higher level of conventional crimes.
- Tourist locations are more conducive to crime (hot spots) due to their inherent activities and hedonistic orientation.
- Most crimes that occur in tourist destinations tend to be on the perimeter and in areas with low pedestrian traffic and no apparent police presence.
- AEBE, some of the physical characteristics of tourist plants (i.e., dimly lit parking lots, motels with external corridors) may be a contributor to crime.
- Countries or regions that possess a significant narcoeconomy acquire a tarnished image that creates the impression of an unsafe tourist destination.

Group B: Impacts of Security Incidents

Impacts on the Destination Itself

- All forms of security incidents that occur at tourist destinations—be they war, terrorism, political upheaval, or crimes—negatively affect their image and can cause a decline in tourist arrivals. This phenomenon is more evident in long-term trends and more specifically related to long-lasting security situations. Declines in tourist arrivals lead to diminishing tourist receipts and may result in a full-fledged economic recession in destinations that specialize in tourism.

- The longer a security crisis lasts, the higher is its aggregated negative impact on a tourist destination.
- The decline in tourist arrivals following one or several security incidents can last anywhere from a few weeks to indefinitely. The factor that most significantly affects the duration of the decline in tourist arrivals is the frequency of the incidents rather than their severity.
- Instability of the tourism industry in security affected destinations forces many investors in other sectors to pull out when the tourism sector is the most significant contributor to the local economy.
- Different tourist market segments possess different levels of sensitivity to security situations. Some tourists tend to be less concerned with security threats and will continue to travel to affected destinations, while others may either avoid them altogether or postpone their trip until the security situation improves.
- Following security incidents there is usually a change in the risk-taking profile of visitors to affected destinations. More security-sensitive segments of the market are replaced by segments that are highly price sensitive and more risk takers. These segments have less spending power; thus often they do not generate expected levels of income and cash flow.
- The profitability of businesses that partially rely on tourism is negatively affected by security incidents. Thus, frequent security incidents may make them insolvent and hence, negatively affect the quality of life of the entire host community.
- The high cost involved in providing security for both the private and public sectors increases the cost of providing services to tourists and makes the affected destinations far less competitive.
- With the exception of very severe security incidents committed against local residents at tourist destinations (i.e., mass terrorism or war), acts committed against tourists have a stronger effect on tourism demand than those committed against local residents, political figures, famous personalities, or businesspersons.

Impact on Tourists' Behavior

- Personal security is a major concern for tourists. Thus, most tourists will seek safe and secure destinations and avoid those that have been plagued by all sorts of violent incidents.
- The perceived risk of traveling to a security-affected destination is shaped by:
  - The objective facts on the ground;
  - Mass media;
  - The travel trade (i.e., travel agents and tour operators);
  - Personal information sources (i.e., friends and relatives); and
  - The subjective acceptable risk threshold of the individual traveler.
- Following a security incident, the general public and would-be tourists' perceptions about its severity and impact are more negative than the facts or real circumstances.
- Leisure tourists are more prone to taking risks while on vacation than local residents, and less likely to observe safety precautions. This is due to lack of understanding and

awareness of local risks and as a result of common belief that while on vacation nothing bad could happen to them.

- Tourists present lucrative targets to criminals because:
  - They tend to carry much portable wealth;
  - They ignore normal precautions;
  - They are unfamiliar with their surroundings;
  - They are less likely to report crimes;
  - They cannot correctly identify their assailants; and
  - They do not return as witnesses at trial.
- In most cases, the likelihood of prosecuting offenders who victimize tourists is relatively low because the victims/witnesses have returned home and, unless they were seriously injured or experienced a large but recoverable loss, they are not likely to return to press charges.
- Many tourist robberies go unreported to the police because of guilt feelings and the embarrassment of having had a desire for illicit activities that led to victimization.
- In pursuit of risk-free travel, potential tourists use a variety of security-related information sources to facilitate their destination-choice behavior.

Impact on the Tourism Industry

- Peace, safety, and security are the primary conditions for successful tourism development.
- Tour operators are severely affected by unexpected security incidents. This is due to their large investments in purchasing tourist products that might perish following a security incident. Moreover, the cost of finding alternative solutions for tourists who have already booked their trips to affected destinations is high and might even lead to business failure.
- In times of security crises, government-regulated tour operators are the first to react and will either evacuate their guests, exclude the affected destination from their travel brochures, stop operation in destinations already included in their products, or temporarily relocate their traveling clients.
- Security incidents at tourist destinations result in diminishing tourist arrivals. Consequently, affected destinations lose both professional employees and entrepreneurs, who are essential for the successful operation of the tourism industry.
- In destinations that have been affected by security incidents the quality of tourist installations and services may become degraded, since many employees are made redundant and funds for regular maintenance are not available. The long-term implication is a need to reinvest large sums of money on rehabilitation of the affected infrastructure and superstructures once the security situation is over.
- Many transnational tourism companies—mainly international hotel and restaurant chains—tend to cease their operation in security affected destinations, causing major damages to their marketing infrastructure by removing affected destinations from their global distribution networks.
- Following severe or frequent security incidents many tourism companies in affected destinations have to restructure their human resources, which means making many employees redundant and operating their services on a very tight budget. This has major bearing on the quality of service provided and on the level of satisfaction of those tourists who are willing to take the risk and visit these destinations.

- Airlines and cruise lines tend to cut short or discontinue their service to affected destinations due to a reduction in demand and an increase in insurance premiums, which causes a severe decrease in the profitability level, or even losses. Thus, the accessibility of affected destination deteriorates substantially.

Impact on Host Governments

- Frequent and recurring security incidents force host governments to usually choose between two possible courses of action. The first option is for governments to reduce their involvement in tourism development and tourism promotion due to the high risk involved and the vulnerability of this industry. Alternatively, they will make substantial investments in improving security measures and tourism promotion to help the private sector overcome the tourism crisis.
- Some host governments will coordinate their contingency plans with all stakeholders in the local, regional, and/or national tourism system in order to mitigate the negative effects of security incidents.

Impacts on Governments of Generating Markets

- Governments in major generating markets normally develop and publish travel advisories to help their citizens in assessing the risks involved in traveling to security affected destinations. One of their motivations for doing this is to reduce the need for possible evacuation of their citizens from affected destinations and/or assist them in a foreign territory.
- One of the consequences of the publishing of travel advisories is that governments in the generating markets often determine the choice space for international tourists seeking tourist destinations. This is done as a result of high insurance premiums levied by insurance companies on security affected destinations. These high rates are based on risk assessments made by these governments through their travel advisories. Once insurance companies raise their premiums, the overall costs of travel to an affected destination increase. Moreover, if a government's risk assessment of a given destination is an overwhelming "Don't go," insurance companies will not issue even an expensive policy, causing potential tourists to drop the idea of travel to such destinations altogether.

Media Behavior

- Intensive mass media coverage of security incidents contributes to the decline of tourist visitation in affected destinations.
- In cases of recurrent and frequent security incidents, the media's continuous coverage and interpretation of the conflict deepen the fixation of a long-term negative image of such affected destinations.
- Unless the media reports on destinations that are traditionally in the forefront of public interest, its level of coverage of security incidents weakens with time. This time-decay function creates a negative image of an affected destination in the short term, but the negative image fades as time goes by.
- Electronic live media coverage causes the most detrimental effect on tourists' perception of risk and on the image of affected destinations.

- In many cases the news media tends to distort the actual security situation on the ground and to exaggerate the risk involved in traveling to affected destinations.
- The media, though not always objective, tends to report not just on the occurrence of security incidents and their actual impacts on tourism, but also takes upon itself the role of interpreter and assessor of the risk involved in traveling to the affected destinations.

Group C: Reaction to Tourism Crises by all Tourism Stakeholders

Destination Behavior

- Following security incidents, the tendency of DMOs is to assume that the life span of the ensuing crisis will be short and that they can count on tourists' short memories about such incidents.
- The majority of DMOs does not possess contingency plans to handle security-induced tourism crises.
- The involvement of most DMOs in security crises focuses on co-funding and coordinating marketing and PR campaigns to regenerate tourist demand once the security crisis is over.
- DMOs and the tourism industry in security affected destinations provide only reactive situational information when security incidents take place, and sometimes when the security crisis is over.
- Most DMOs in destinations affected by security incidents do not engage their private and public tourism sectors in proactive security information dissemination to tourists and tourist gatekeepers (travel agents, tour operators, meeting planners, the mass media, and governmental travel advisories).
- Many DMOs and private sector tourism enterprises are reluctant to put in place tourist security education programs because of concerns over the potentially negative effect of advertising the existence of tourist security threats.
- Following a security incident the first response of the public and private sector at the affected destination is to claim that the magnitude of the incident is exaggerated by the media and/or other entities outside the area.

Image and Perception Management

- Only a small number of destinations practice image management following a security incident for the purpose of diminishing the negative image created by the media.
- When DMOs, governments, and/or the tourism industry of the host destinations do not issue frequent and up-to-date security-related information during security crises, potential and existing tourists base their perceptions of risk on the images conveyed by external and often exaggerated media sources.
- Risk reduction strategies, such as dissemination of positive communication, can influence the risk perception of not only potential tourists who are engaged in a destination-choice process, but also to those who:
  - Have booked but not already taken their trips;
  - Are in the process of visiting the destination; and
  - Are returning from a recent trip.

Risk and Crisis Management (Prevention/Reduction/Mitigation) Techniques

- Tourism security crises are for the most part unavoidable, since they are generated in many cases by exogenous factors that are beyond the control of the tourism industry or the tourist destination.
- Travel advisories issued by foreign governments are perceived by affected destinations as influential and therefore result in some actions being taken by the tourism industry and/or governmental tourism agencies.
- Improving tourist security by housing tourists in gated all-inclusive resorts has a negative effect on small tourism businesses, which feel shut out and reinforce the perceptions of inequality between tourists and local residents.
- Many tourism practitioners feel that too many visible security measures will cause visitors to wonder if they should be afraid and thus even speaking about these subjects could frighten customers.
- AEBE, tourist destinations that have created special police units aimed at preventing and reducing crimes against tourists have managed to reduce their tourist crime rates and/or have lower tourist crime rates than their counterparts.
- Tourism police units that are effective in the reduction and/or prevention of crimes against tourists normally conduct the following activities:
  - Train their officers in tourism issues;
  - Are visible, accessible, and friendly to tourists;
  - Work closely with the community and tourism industry representatives;
  - Advise and train tourism industry employees in crime prevention techniques;
  - Assist the tourism industry by conducting background checks for employees;
  - Facilitate tourist victims' testimony in criminal cases;
  - Encourage tourism enterprises to adopt crime prevention/reduction practices (i.e., installation of electronic room locks, surveillance cameras, room safety deposit boxes, employing full-time security officers, etc.);
  - Develop and implement tourist education programs aimed at reducing the risk of being victimized; and
  - Increase the presence of uniformed officers in tourist zones.
- Tourist destinations that offer rewards for information leading to the arrest and conviction of those who commit serious crimes against tourists have better records of conviction, and in turn lead to lower rates of crimes against tourists.
- AEBE, tourism enterprises that incorporate the principles of Crime Prevention Through Environmental Design (CPTED) tend to have lower rates of crimes committed against tourists than their counterparts.
- AEBE, tourism enterprises that consider the function of security important to the success of their business and allocate a significant portion of their financial and human resources to it, tend to have lower rates of crimes committed against tourists than do their counterparts.
- AEBE, tourist destinations that make their tourists aware of the possibility of becoming victimized by criminals and instruct them in crime prevention methods tend to have lower rates of crimes committed against tourists than do their counterparts.
- Crime prevention/reduction methods—such as security hardware and security policies—that are used by tourism enterprises have varying levels of effectiveness.

- Destinations that collect and maintain tourism crime data at the property and destination levels are better able to evaluate the effectiveness of crime prevention/reduction techniques than their counterparts.
- AEBE, destinations that have made crime against tourists a major criminal offense have lower crime rates committed against tourists than do their counterparts.
- AEBE, destinations that have educated their citizens on the serious impacts that crime against tourists can have on their communities and engaged them in a local neighborhood watch tend to have lower tourism crime rates than do their counterparts.
- AEBE, destinations that have erected special highway signs that provide visitors with directions for travel on safe and well-patrolled routes and provide them with appropriate maps tend to have lower rates of car-related tourist crimes than do their counterparts.
- Tourist destinations that established a partnership between law enforcement agencies, tourism enterprises, the community at large, and the tourists themselves have lower crime rates than do their counterparts.
- Tourist destinations that established a partnership between the leaders of the local community and the national and local governments have been more successful in the prevention/reduction of acts of riot and political unrest at tourist destinations than their counterparts.

Recovery Methods

- In the aftermath of a decline in tourist visitation that is caused by a security incident, most tourist enterprises will try to reduce their operational costs by laying off a proportion of their employees.
- Following terrorist incidents, the tourism industry in the affected destination will seek to compensate for declines in leisure travel by appealing to local, regional, and national governments to promote and encourage government-related business travel.
- In the aftermath of terrorist incidents, tourist enterprises seek to compensate for international tourists' declines by reducing their prices to appeal to domestic tourists.
- To recover from a decline in tourist arrivals caused by security incidents, the public and/or private sectors of the affected destination will in most cases undertake:
  - Intensive marketing campaigns to convince the general public that things are back to normal; and
  - The scheduling of special events to attract local residents and out-of-town tourists.
- To recover from the decline in tourist visitations caused by a security incident, the private sector at the affected destination will in most cases reduce prices and offer a variety of incentives (e.g., package deals) to bring the tourists back to the destination.
- AEBE, following serious incidents of tourism security, tourist businesses that obtained financial assistance (i.e., grants, tax holidays, subsidized loans, etc.) from their local, state, and/or national governments manage to recover faster from declines in tourist arrivals caused by these incidents than do their counterparts.
- AEBE, tourist destinations that secure governmental grants for financing the promotion of their destination following serious incidents of tourism security (i.e.,

terrorist attacks or warfare) manage to recover faster from declines in visitations caused by these incidents than do their counterparts.

- Following terrorist incidents DMOs in cooperation with other NGOs and governmental agencies will undertake comprehensive marketing campaigns, the purpose of which is to convince local, regional, national, and international travel markets to return to the affected destination.
- Some marketing campaigns initiated after a terrorist attack aim to encourage area residents within a short drive of the affected destination to visit their region or hometown by appealing to their sense of local patriotism.
- Destinations that are able to effectively recover from the aftermath of terrorist incidents use a combination of the following strategies:
  - Establish a mechanism for sharing information and coordination of publicity and PR activities aimed at creating positive public opinion among the media, local community, and customers;
  - Engage their local community in the effort of recovery;
  - Reassure and calm their existing and potential clients by providing them with current and updated information aimed at persuading them that the destination is open for business as usual; and
  - Secure funding for the development and implementation of a recovery marketing plan.
- Following one or a series of security incidents, to compensate for the resultant decline in tourist arrivals, tourism enterprises will employ one or a combination of the following strategies:
  - Reduce labor costs;
  - Decrease prices for their services and goods;
  - Initiate new promotional campaigns;
  - Develop new products:
  - Identify and develop new market segments;
  - Postpone major expenditures on maintenance and renovation; and
  - Request financial assistance from governmental agencies.
- AEBE, tourist destinations and/or tourist enterprises that possess and implement crisis and contingency plans manage to recover better and faster than do their counterparts.
- Destinations that employ a destination specific marketing strategy—a tactic employed by a local destination to dissociate itself from a larger tourist destination that has an undesirable security image—are more successful in avoiding the decline in tourist arrivals caused by a series of security incidents such as repeated terrorist attacks or warfare than are their counterparts.

Summary

The aim of this chapter was to take the initial steps towards developing a tourism security theory. This challenging task involved the creation of the first two fundamental building blocks of the theory. The chapter started with a construction of tourism and security concepts and their corresponding variables as the first building block. Subsequently, as the second block, it assembled a wide array of empirical generalizations that represent the current best practices in the field of tourism security.

With time, these building blocks will be further refined as more and more empirical studies will confirm or refute the proposed empirical generalizations and thus lead to an accepted and tested tourism security theory. Indeed what is needed now in order to improve the paradigmatic basis for a confirmed tourism security theory is a research agenda that develops scientific knowledge in two distinctive directions.

The first research direction is to conduct a set of studies examining the relationship between tourism and security on a destination-specific basis. The aim of this direction is to further deepen the understanding of causes and effects in tourism and security relations. This can be achieved through an inductive research approach that moves from specific observations and measures, to detecting patterns and regularities, formulating empirical generalizations or hypotheses that should be tested and confirmed, and finally leading to general conclusions in the form of a theory.

The second research direction is to encourage the conduct of comparative (i.e., local, regional, national, international) studies to test the level of universalism of the proposed tourism security theory. Achieving this goal is imperative if the tourism stakeholders wish to develop and adopt effective strategies and methods for the prevention, mitigation, and reduction of security incidents at tourist destinations.

Concept Definitions

Theory An attempt to explain and predict a particular phenomenon.

Proposition A statement about one or more concepts or variables. Subtypes of propositions include hypotheses, empirical generalizations, axioms, postulates, and theorems.

Hypothesis Proposition that is stated in a testable form and predicts a particular relationship between two (or more) variables.

Empirical generalization A statement of relationship that is constructed by first observing the existence of a relationship (in one or a few instances) and then generalizing to say that the observed relationship holds in all or most cases.

Security incidents An act of violence or threat of violence, such as crimes, terrorism, wars, and civil or political unrest committed at a tourist destination against tourists or local residents.

Travel advisory A statement issued by a government in a generating market intended to advise its citizens about possible risks involved in traveling to security affected destinations.

Destination-specific marketing strategy A tactic employed by a local destination to disassociate itself from a larger tourist destination that has an undesirable security image.

Review Questions

1. Why do we need a theory on tourism security?
2. Discuss the means by which it is possible to increase the prosecution rate of offenders who victimized tourists.
3. What are the differences between the impacts of terrorism and the impacts of crime on affected tourist destinations?

4. What is the typical reaction of destinations that suffer from a tourism security crisis?
5. Which has a greater impact on tourism demand, frequency of security incidents or severity of security incidents?
6. Discuss the methods employed by tourism enterprises to survive during serious tourism crises.
7. What, if any, impacts do governments in the generating markets have on recovery from security crises?

References

Following is a list of recommended literature on security and tourism issues that, together with the chapters in this volume as well as many other sources, formed the basis for the above theoretical chapter.

Aziz, H. (1995). Understanding attacks on tourists in Egypt. *Tourism Management,* 16, 91–95.

Bach, S., and Pizam, A. (1996). Crimes in hotels. *Hospitality Research Journal,* 20(2), 59–76.

Bailey, K. D. (1982). *Methods of Social Research,* 2nd ed. New York: Free Press.

Barker, M., Page, S. J., and Meyer, D. (2002). Modeling tourism crime: The 2000 America's Cup. *Annals of Tourism Research,* 29(3), 762–782.

Barker, M., Page, S. J., and Meyer, D. (2003). Urban visitor perceptions of safety during a special event. *Journal of Travel Research,* 41, 355–361.

Beirman, D. (2003). *Restoring Tourism Destinations in Crisis—A Strategic Marketing Approach.* Wallingford, UK: CABI International.

Brunt, P., and Hambly, Z. (1999). Tourism and crime: A research agenda. *Crime Prevention and Community Safety: An International Journal,* 1(2), 25–36.

Brunt, P., Mawby, R., and Hambly, Z. (2000). Tourist victimization and the fear of crime on holiday. *Tourism Management,* 21(4), 417–424.

Caric, A. (1999). Tourism-related crime: Forms, causes and prevention. *Turizam,* 47(1).

Cassedy, K. (1992). Preparedness in the face of crisis: An examination of crisis management planning in the travel and tourism industry. *World Travel and Tourism Review,* 2, 8–10.

Cavlek, N. (2002). Tour operators and destination safety. *Annals of Tourism Research,* 29(2), 478–496.

Chesney-Lind, M., and Lind, I. Y. (1986). Visitors as victims: Crimes against tourists in Hawaii. *Annals of Tourism Research,* 13, 167–191.

Cohen, E. (1987). The tourist as victim and protégé of law enforcing agencies. *Leisure Studies,* 6, 181–198.

Cohen, E. (1997). Tourism-related crime: Towards a sociology of crime and tourism. *Visions in Leisure and Business,* 16(1), 4–14.

Cohen, L., and Felson, M. (1979). Social change and crime rate trends: A routine activity approach. *American Sociological Review,* 44, 588–608.

Crotts, J. C. (1996). Theoretical perspectives on tourist criminal victimization. *The Journal of Tourism Studies,* 7(1), 2–9.

de Albuquerque, K., and McElroy, J. (1999). Tourism and crime in the Caribbean. *Annals of Tourism Research,* 26(4), 968–984.

de Albuquerque, K., and McElroy, J. (2001). Tourist harassment: Barbados survey results. *Annals of Tourism Research,* 28(2), 477–492.

Dimanche, F., and Lepetic, A. (1999). New Orleans tourism and crime: A case study. *Journal of Travel Research,* 38(August), 19–23.

Elliot, L., and Ryan, C. (1993). The impact of crime on Corsican tourism: A descriptive assessment. *World Travel and Tourism Review,* 3, 287–293.

Ferreira, S. L. A., and Harmse, A. C. (2000). Crime and tourism in South Africa: International tourists' perceptions and risk. *South African Geographical Journal,* 82(2), 8–85.

Fujii, E. T., and Mak, J. (1979). The impact of alternative regional development strategies on crime rates: Tourism vs agriculture in Hawaii. *Annals of Regional Science,* 13(3), 42–56.

Fujii, E. T., and Mak, J. (1980). Tourism and crime: Implications for regional development policy. *Regional Studies,* 14, 27–36.

Fuchs, G., and Reichel, A. (2004). Cultural differences in tourist destination risk perception: An exploratory study. *Tourism,* 52(4), 7–20.

Gartner, W. C., and Shen, J. (1992). The impact of Tiananmen Square in China's tourism image. *Journal of Travel Research,* 30(4), 47–52.

George, R. (2001). The impact of crime on international tourist numbers to Cape Town. *Crime Prevention and Community Safety: An International Journal,* 3(3), 19–29.

George, R. (2003a). Tourists' fear of crime while on holiday in Cape Town. *Crime Prevention and Community Safety: An International Journal,* 5(1), 13–25.

George, R. (2003b). Tourist's perceptions of safety and security while visiting Cape Town. *Tourism Management,* 24(5), 575–585.

Goodrich, J. N. (1991). An American study of tourism marketing: Impact of the Persian Gulf War. *Journal of Travel Research,* 30(2), 37–41.

Goodrich, J. N. (2002). September 11, 2001 attack on America: A record of the immediate impacts and reactions in the USA travel and tourism industry. *Tourism Management,* 23, 573–580.

Hall, C. M., Timothy, D. J., and Duval, D. T. (eds.). (2004). *Safety and Security in Tourism: Relationships, management, and marketing.* Binghamton, NY: Haworth Press.

Hall, C., Selwood, J., and McKewon, E. (1995). Hedonists, ladies and larrikins: Crime, prostitution and the 1987 America's Cup. *Visions in Leisure and Business,* 14(3), 28–51.

Harper, D. (2001). Comparing tourists' crime victimization. *Annals of Tourism Research,* 28(4), 1053–1056.

Jenkins, O. H. (1999). Understanding and measuring tourist destination images. *International Journal of Tourism Research,* 1, 1–15.

Jud, G. D. (1975). Tourism and crime in Mexico. *Social Sciences Quarterly,* 56, 324–330.

Kathrada, M., Burger, C. J., and Dohnal, M. (1999). Holistic tourism—Crime modeling. *Tourism Management,* 20, 115–122.

Kelly, I. (1993). Tourist destination crime rates: An examination of Cairns and the Gold Coast, Australia. *The Journal of Tourism Studies,* 4(2), 2–11.

Lepp, A., and Gibson, H. (2003). Tourist roles, perceived risk and international tourism. *Annals of Tourism Research,* 30(3), 606–624.

Leong, C. (2001). Improving safety and security at tourism destinations. *Journal of Travel & Tourism Marketing,* 10(1), 129–135.

Levantis, T., and Gani, A. (2000). Tourism demand and the nuisance of crime. *International Journal of Social Economics,* 27(7/8/9/10), 959–967.

Lin, V. L., and Loeb, P. D. (1977). Tourism and crime in Mexico: Some comments. *Social Science Quarterly,* 58, 164–167.

Loeb, P. D., and Lin, V. L. (1981). The economics of tourism and crime: A specific error approach. *Resource Management and Optimization,* 1(4), 315–331.

Mansfeld, Y. (2000). Crime, in J. Jafari (ed.), *Encyclopedia of Tourism.* London: Routledge, p. 118.

Mansfeld, Y. (2000). Risk, in J. Jafari (ed.), *Encyclopedia of Tourism.* London: Routledge, p. 508.

Mansfeld, Y. (1999). Cycles of war, terror and peace: Determinants and management of crisis and recovery of the Israeli tourism industry. *Journal of Travel Research,* 38(1), 30–36.

Mawby, R. I. (2000). Tourists' perceptions of security: The risk-fear paradox. *Tourism Economics,* 6(2), 109–121.

Mawby, R. I., Brunt, P., and Hambly, Z. (1999). Victimisation on holiday: A British survey. *International Review of Victimology,* 6, 201–211.

Mawby, R. I., Brunt, P., and Hambly, Z. (2000). Fear of crime among British holidaymakers. *British Journal of Criminology,* 40, 468–479.

McPheters, L. R., and Stronge, W. B. (1974). Crime as an environmental externality of tourism. *Florida. Land Economics,* 50(2), 288–292.

Milman, A., and Bach, S. (1999). The impact of security devices on tourists' perceived safety: The central Florida example. *Journal of Hospitality and Tourism Research,* 23(4), 371–386.

Muehsam, M. J., and Tarlow, P. E. (1995). Involving the police in tourism. *Tourism Management,* 16(1), 9–14.

Nicholls, L. L. (1976). Tourism and crime. *Annals of Tourism Research,* 3(4), 176–182.

Olsen, M. D., and Pizam, A. (1998). *Think-Tank Findings on Safety and Security (Orlando, USA).* Paris: International Hotel and Restaurant Association.

Olsen, M. D., and Pizam, A. (1999). *Think-Tank on Safety and Security: Key Findings (Stockholm, Sweden).* Paris: International Hotel and Restaurant Association.

Page, S. J. (2002). Tourist health and safety. *Travel and Tourism Analyst,* October, 1–31.

Pinhey, T. K., and Iverson, T. J. (1994). Safety concerns of Japanese visitors to Guam. *Journal of Travel & Tourism Marketing,* 3(2), 87–94.

Pizam, A., Jeong, G. H., Reichel, A., Van Boemmel, H., Lusson, J. M., Steynberg, L., State-Costache, O., Volo, S., Kroesbacher, C., Kucerova, J., and Montmany, N. (2004). The relationship between risk-taking, sensation-seeking, and the tourist behavior of young adults: A cross-cultural study. *Journal of Travel Research.* 42(3), 251–260.

Pizam, A. (2002). Editorial: Tourism and terrorism. *International Journal of Hospitality Management,* 21, 1–3.

Pizam, A., and Smith, G. (2000). Tourism and terrorism: A quantitative analysis of major terrorist acts and their impact on tourism destinations. *Tourism Economics,* 6(2), 123–138.

Pizam, A. (1999). A comprehensive approach to classifying acts of crime and violence at tourism destinations. *Journal of Travel Research,* 38(3), 5–12.

Pizam, A., Tarlow, P. E., and Bloom, J. (1997). Making tourists feel safe: Whose responsibility is it? *Journal of Travel Research,* 36(1), 23–28.

Pizam, A., and Mansfeld, Y. (eds.). (1996). *Tourism, Crime and Security Issues.* Chichester, UK: John Wiley & Sons.

Pizam, A. (1982). Tourism and crime: Is there a relationship? *Journal of Travel Research,* 20(3), 8–20.

Prideaux, B., and Dunn, A. (1995). Tourism and crime: How can the tourism industry respond? The Gold Coast experience. *Australian Journal of Hospitality Management,* 2(1), 7–15.

Richter, L. K., and Waugh, W. L. (1986). Terrorism and tourism as logical companions. *Tourism Management,* 7, 230–238.

Roehl, W. S., and Fesenmaier, D. R. (1992). Risk perceptions and pleasure travel: An exploratory analysis. *Journal of Travel Research,* 30(4), 17–26.

Ryan, C. (1991). *Tourism, Terrorism and Violence: The Risks of Wider World Travel. Conflict Studies, No. 244.* London: Research Institute for the Study of Conflict and Terrorism, p. 6.

Ryan, C. (1993). Crime, violence, terrorism, and tourism: An accidental or intrinsic relationship? *Tourism Management,* 14, 173–183.

Santana, G. (2001). Globalization, safety and national security, in S. Wahab and C. Cooper (eds.), *Tourism in the Age of Globalization.* London: Routledge, pp. 213–241.

Santana, G. (2003). Crisis management and tourism: Beyond the rhetoric. *Journal of Travel and Tourism Marketing,* 15(4), 299–321.

Sharpley, R., and Sharpley, J. (1995). Travel advice—Security or politics? in *Security and Risks in Travel and Tourism. Proceedings of the Talk at the Top Conference,* Mid-Sweden University, Östersund, Sweden, 168–182.

Sherman, L.W., Gartin, P., and Beurger, M. (1989). Hot spots of predatory crime: Routine activities and the criminology of place. *Criminology,* 27(1), 27–56.

Smith, G. (1999). Towards a United States policy on traveler safety and security: 1980–2000. *Journal of Travel Research,* 38(1), 62–65.

Sönmez, S. F., and Graefe, A. R. (1998a). Determining future travel behavior from past travel experiences and perceptions of risk and safety. *Journal of Travel Research,* 37(2), 171–177.

Sönmez, S. F., and Graefe, A. R. (1998b). Influence of terrorism risk on foreign tourism decisions. *Annals of Tourism Research,* 25(1), 112–144.

Sönmez, S. F., Apostolopoulos, Y., and Tarlow, P. (1999). Tourism in crisis: Managing the effects of terrorism. *Journal of Travel Research,* 38(1), 13–21.

Sonmez, S. F. (1994). An exploratory analysis of the influence of personal factors on international vacation decisions within the context of terrorism and/or political instability risk. Unpublished Ph.D. dissertation, Pennsylvania: Pennsylvania State University.

Sonmez, S. F. (1998). Tourism, terrorism, and political instability. *Annals of Tourism Research,* 25(2), 416–456.

Steene, A. (1999). Risk management within tourism and travel. *Turizam,* 47(1), 13–18.

Tarlow, P. E., and Santana, G. (2002). Providing safety for tourists: A study of a selected sample of tourist destinations in the United States and Brazil. *Journal of Travel Research,* 40, 424–431.

Tsaur, S. H., Tzeng, G. H., and Wanf, K. C. (1997). Evaluating tourist risks from fuzzy perspectives. *Annals of Tourism Research,* 24(4), 796–812.

Walmsley, D. J., Boskovic, R. M., and Pigram, J. J. (1983). Tourism and crime: An Australian perspective. *Journal of Leisure Research,* 13(2), 136–155.

Wilks, J., and Page, S. J. (eds.) (2003). *Managing Tourist Health and Safety in the New Millennium.* Oxford, UK: Pergamon.

World Tourism Organization (1991). *Recommended Measures for Tourism Safety.* Madrid: World Tourism Organization.

World Tourism Organization (1995). *Best Practice Manual on Traveler Safety*. Madrid: World Tourism Organization.
World Tourism Organization (1997). *Tourist, Safety and Security: Practical Measures for Destinations,* 2nd ed. Madrid: World Tourism Organization.
World Tourism Organization (2003). *Safety and Security in Tourism: Partnerships and Practical Guidelines for Destinations*. Madrid: World Tourism Organization.

Tourism, Terrorism, and Civil Unrest Issues

Yoel Mansfeld and Abraham Pizam

Terrorism and civil unrest have been the cause of numerous tourism crises since the beginning of modern tourism in the late 1950s. While civil unrest is more confined to specific destinations and the frequency is relatively low, terrorism has evolved into a major global concern for the tourism industry, tourists, and for hosting communities. In all cases the major concern is when these events take place at tourist destinations or in very close proximity, thus damaging the image, infrastructure, and competitiveness of the destination.

The relationship between civil unrest and terrorism on the one hand, and tourism on the other, has had various manifestations. In some cases, as often happened with the IRA terror activity in the UK or with the ETA separatist group in Spain, terrorist groups use threats to hit tourism facilities and/or tourists as a means to achieve their goals. If such threats make their way through to reach the media, they could negatively affect a given destination without even reaching the stage of actual on-the-ground operation. Occasionally, when civil unrest and terrorism do take place, the declared targets are simply economic or strategic ones, yet hitting them does have negative impacts on tourism. This was the case in the September 11 attack in 2001 in the United States and the civil unrest events of the Chiapas in Mexico in 1994. In both cases, tourism was badly affected, access to various tourism attractions was temporarily denied due to security and safety reasons, and local communities faced various levels of short-term economic hardship.

The most detrimental effect that civil unrest, especially terrorism, has on tourism is when the violent activity directly targets it. Some examples are Bolivia in 2003; Spain with the ETA group since 1995; in Egypt on several occasions, but mainly in the attack on tourists in Luxor 1997; and on the Hilton Taba Hotel in the north tip of the Sinai peninsula in 2004. In all these instances, the image of such destinations is severely damaged and the recovery process takes much longer. Moreover, when security events such as terror and civil unrest are directly aimed at tourism, both public and private tourism sectors need to exert the most demanding concerted efforts to reactivate tourist flows.

In this first section of the book, the above relationships and their impact on tourism are discussed on both a theoretical and a case-study level. The first chapter, by Peter Tarlow, provides a theoretical discussion on the roots and characteristics of the relationship between terrorism and tourism, and examines the reasons why terrorism often selects tourism targets to pursue its political-ideological interests. This chapter also deals with how the tourism industry has been transformed since September 11, 2001, in order to adjust itself to the reality that at least in the foreseeable future it will continue to be a prime target for terrorist organizations on a global scale. His main conclusions are that:

- The tourism industry must undergo a paradigm shift and accept the fact that terrorism will regard tourism as a target in the future and thus the industry must prepare for it.
- Tourists will place more and more emphasis on choosing safe destinations in their destination-choice process.
- In its paradigm shift the tourism industry has to incorporate safety, security, and economic viability considerations into one entity.
- The provision of good security has to be regarded as part of the overall service quality offered by the tourism industry.

The second chapter, by Aliza Fleischer and Steven Buccola, focuses on the economic ramifications of war and terrorism using the Arab-Israeli conflict and its reflection on Israel's hotel sector from the late 1980s to 1990. Using a supply and demand model of the Israeli lodging sector, the chapter evaluates the impact of hostility on management's price response to demand shifts. It also assesses the international and domestic markets' reaction to security incidents in terms of price elasticity and level of sensitivity to regional warfare and terror. By doing so, this chapter shines a very interesting light on the complexity of handling the negative effects of violence on a specific tourism sector that caters to both domestic and international markets. Fleischer and Buccola's main conclusions are that:

- The use of a supply and demand model of a nation's accommodation sector is an efficient research tool to evaluate management price response to demand shifts as a result of security incidents.
- A price discrimination model might be a better vehicle for assessing the impacts of security incidents than a supply and demand model. This is mainly true in cases where such a strategy is used to differentiate between international and domestic markets.

Finally, the impact of civil unrest on tourism destinations is dealt with in a chapter by Brian King and Tracy Berno, who discuss the handling of two tourism crises in the wake of the coups d'état in 1987 and 2000 in Fiji, and the ensuing recovery efforts of the tourism sector. The comparative analysis of the two civil unrest events, and their impact on the tourism industry, characterizes in a comprehensive and multidimensional manner the nature of the impact that political and social unrest events have on tourism. Although the case study depicts these impacts on Fiji, it can be seen as an insightful reflection on many other developing countries that developed high levels of economic dependency on foreign tourists' willingness to spend their holidays in their resorts and tourism destinations. Moreover, the analysis of the recovery processes, and the measures taken by the various Fiji tourism stakeholders confronting the evolving tourism crises, sheds light on the relative importance of crisis communication and travel advisories provided by the generating countries to secure the safety of their traveling citizens. King and Berno's main conclusions are that:

- When well-structured and carefully planned strategies are employed to confront tourism crises they provide tangible positive results.
- Concerted efforts by both private and public tourism stakeholders yield positive results and boost recovery from security induced tourism crises.
- When the tourist product is attractive enough and there exists a strong captive market for the affected destination, recovery is almost guaranteed and can be attained within a short period of time.

2

A Social Theory of Terrorism and Tourism

Peter E. Tarlow

Learning Objectives

- To understand the long history of the relationship between terrorism and tourism.
- To understand that warfare has radically changed in the latter part of the last century and now in the twenty-first century.
- To understand that the roles of tourism, nostalgia, and terrorism intermix.
- To understand that classical European thought has bled into other parts of the world and has become an inspiration for terrorism.
- To understand the major difference between tourism crimes and acts of terrorism against tourism centers.
- To understand that the tourism industry is going through a major paradigm shift in its adjustment to a post September 11 world.

Toward the end of the twentieth century the postmodernist Francis Fukuyama published his now famous thesis called "The End of History" (Fukuyama, 1989). Fukuyama predicted that with the end of the historic battle called the Cold War, a new "posthistory" would develop, that democracy had triumphed, and that war as we had understood it had come to an end. Fukuyama predicted that while this posthistory would also be a time of nostalgia, it would not be a time of tranquility. Instead he predicted that boredom fueled by lack of major conflict would "continue to fuel competition and conflict even in the posthistorical world for some time to come" (Fukuyama, 1989). Fukuyama's vision of nostalgia turning to posthistorical quasi-wars may first have been manifested in the ongoing Balkan conflicts (Ottolenghi, May 9, 2004). Ottolenghi writes about the current slaughter of Serbians by Albanians: "(reality) was more like two ethnic groups animated by a centuries-old reciprocal hatred, with good and evil cutting across both communities" (Ottolenghi, May 9, 2004). This postmodern sense of conflict provides a major insight into our understanding of how postmodern *kleinkriege* (small wars and incidents that punctuate history) will impact the world's most postmodern industry, travel and tourism.

Writing as a Hegelian, Fukuyama's vision has proved to be all too true in the first years of the twenty-first century. The September 11, 2001, attacks against New York's World Trade Center and the Pentagon in Washington, DC, may mark the dividing point between the first and second stage of post-history. In this new historical period, nations no longer fight against nations, but rather against amorphous terrorism cells. In this post-history, wars are being fought on a new battlefield—the centers of civilian leisure life. Examples of this war against civilians abound. From Saudi Arabia to Peru, from Spain to Japan, from Israel to the United States, civilians are no longer victims of what the military often calls friendly fire or collateral damage. Rather they now are the specific intended targets. For example, Muslim extremist groups regularly attack civilians with the hope of not only hurting the infidels' economy and political structure, but also as a means of demonstrating that the victims (especially Jews and Christians) are the enemy, spreading terror throughout the victims' society, and lifting the terrorists' spirits.

In this new world, today's battles are fought and won as much on the television screen as anywhere else. Fukuyama's prediction of a "bellicose nostalgia" has great implications for the travel and tourism industry. The travel and tourism industry has suffered from terrorism attacks ever since the 1970s. These pre-post-history attacks, however, were never defining moments. Now in the early years of the twenty-first century tourism has become a major target of those who seek not merely to conquer but also of those who seek destruction for its own sake. The September 11 mega-attack and its aftermath plunged the world of travel and tourism into a new paradigm in which the industry may be forced to fight for its very survival.

This chapter introduces the social-historical aspects of terrorism as it impacts the travel and tourism industry. The chapter focuses on understanding terrorism through the prism of classical sociological thought. The first part of the chapter seeks to explain how late nineteenth century and twentieth century sociological authors provide a paradigm for understanding the reasons that terrorists have often targeted the tourism industry. We begin our journey by reviewing some of the basic ideas of European theorists who created, although perhaps unwillingly, the academic foundations of modern terrorism. The chapter then turns to the role of nostalgia in understanding terrorism. Using the work of the Russian-American writer Svetlana Boym as a basis, this chapter connects nostalgia in travel to acts of terrorism against travel. The chapter then examines the social-historic connection of overt freedom as one of the principles that inspired the cult of death. By comparing the works of Hegel, Marx, Camus, modernism, and pre-modernism to those of modern Middle Eastern figures such as the school of Sayyid Qutb, the reader will gain a greater understanding into these classical authors' influence on modern militant Islam. Finally, this chapter addresses philosophical questions such as, is tourism terrorism a form of crime, and as such, should it be treated as a problem with a potential for compromise or should it be seen as ontological in nature and thus exist outside of the realm of compromise? What is the role of fear in terrorism and how can the tourism industry work to overcome fear? Are some countries, cities, or attractions more likely targets than others? If so, what makes these locales more vulnerable?

Historical Literary Background

Mass travel and tourism* as a sociological and economic phenomenon is symbolic of the modern world. The mass travel and tourism industry, as it is known today, is no more than 50 or 60 years old. Prior to the 1950s travel was restricted to the very wealthy or those, such as soldiers, who had to travel. Travel was dangerous and uncomfortable. It is for this reason that the English word *travel* is derived from the French word for work (*travail*), which in turn is derived from the Latin word for pitchfork (*tripalium*). Modern travel is a result of a confluence of multiple social factors, among them economic surplus and relief from ennui. Future historians reviewing the current fin de siècle period may well call our period "The Age of Attention Deficit Disorder (ADD)." This first decade of the twenty-first century also approaches what latter historians may call the age of *a-history*. Being a time of collective amnesia, historical revisionism, political correctness, and a sensationalist media, an impartial observer may wonder if most current events either take place in a vacuum or have no historical basis. Certainly the media's portrayal of terrorism would fit this description. The modern media has portrayed terrorism as an almost recent phenomenon. In reality, terrorism and suicidal acts aimed at the murder of the innocent can be traced to prior to the French revolution.

One of the earliest forms of terrorism may have occurred during the fifteenth century. The Peasants' Revolt of Germany is such an example of early terrorism in which the innocent were murdered as a form of social pressure. Led by radical theologians and second ranked knights and nobles the so-called revolt occurred during a period of rapid social change in which thousands of people lost their social bearings. Its Osma Ben Laden was the radical cleric Thomas Muentzier. Peters has written of Muentzier: "Muentzier left a trail of devastation across the middle of the Germanies that only ceased when a coalition of the nobility and knights brought him to a final apocalyptic that ended with an uncompromising pursuit and massacre of the insurgents, followed by the ingenious torture and execution of their captured leaders" (Peters, 2002, p. 48). The Peasants' Revolt may have influenced the nineteenth century German philosopher Hegel in producing his countertheory of humanity. The rationalists who preceded the French revolution rejected the idea that some people are inherently evil and instead developed the position that humanity was perfectible through science and knowledge. The Enlightenment philosopher believed that history might be brought to its successful conclusion through social science and understanding.

Philosophically opposed to this rational train of thought, Hegel proposed his dialectic that for every thesis there is an antithesis. Thus, if humans are perfectible they must also have a dark side or that, as humanity becomes ever more rational, there will also surface the irrational side to humanity. From a Hegelian perspective, the realistic or irrational counterpoint to the French philosophes argued that at least some humans seek to dominate others both politically and economically.

At approximately the same time the idea of combining murder with suicide, the glorification of death appeared throughout European literary thought. Examples abound from Victor Hugo's play *Hernani* to the Russian author Dostoevsky's famous character Ivan who states that in a world without God, all is permitted. In

* The travel and tourism indutry goes by many names, such as "travel," "hospitalipy," "visitor," etc. For purposes of this book all terms have been reduced to simelr "tourism industry."

England, Joseph Conrad reflects many of these same principles in his short story, "The Secret Agent." Even the Robin Hood legends carry a hint of terrorism. In these legends self-appointed heroes take from the rich to give to the poor and fight with the evil sheriff. The Robin Hood tale assumes that the rich are evil and that the poor are good. It also assumes the right of wealth redistribution that leisure and service are by their very nature evil.

Under such assumptions, it is not hard to make the leap that tourists are enslaving the poor workers and that not to be at work is to be evil. This same concept is shown sociologically by Veblen in his work *The Theory of the Leisure Class*. In the most Robin Hood of styles Veblen writes, "the term 'leisure' as here used does not connote indolence or quiescence. What it connotes is a non-productive consumption of time. Time is consumptive non-productively (1) from a sense of the unworthiness of productive work, and (2) as an evidence of pecuniary ability to afford a life of idleness" (Veblen, 1963, p. 46). From the perspective of the terrorist, the non-pilgrim tourist is merely a user of idleness, one who loves money and hates productivity, and an industry based on this type of consumption will necessarily deprive its consumers' future paradise for this world's pleasures.

As Paul Berman notes in his book *Terror and Liberalism*, the theoretical model used by nineteenth century Russians (based on the ideas of the French) is the following. Humanity can go from freedom, to the freedom to murder, to the freedom to commit the ultimate murder, that of oneself, while murdering others. In such a world, the cult of death became the philosophical underpinning for the irrational movements that ended in mass murder. Ironically, history notes some of the greatest massacres in places where the West sought to bring its humanistic civilization (such as India, the Congo, South Africa, or Algeria). The ultimate expression of this rationalized irrationality of course is the Nazi Germany and the Holocaust. For all intents and purposes the idea that humanity was perfectible is replaced after World War I with a new political theory, that of totalitarianism. Totalitarianism replaced the individual with the needs of the state, and then replaced the value of the human life with the needs of the state's life. Death now became elegant. Franco's *Falange* party expressed it best with its ironic motto *¡Viva la muerte!* (Long live death!). The French writer Camus expressed the idea of anomie in his novel *The Rebel*. Fascist Europe, including Stalin, who was in theory part of the left but acted as if he were a fascist, was unable to reinvigorate the continent's economy but was able to deliver on one promise; it succeeded in delivering death to the masses.

The West has exported to the Middle East a great many concepts and ideas, one of which is the philosophy of death and fascism. While terrorism exists in many parts of the world, the form of extreme Muslim terrorism may serve as a paradigm or model of modern terrorism in general. The Middle East is home to great literary traditions. Nevertheless, some of the ideas that have influenced its greatest writers are Western imports refitted to meet local Middle Eastern needs. Islamic terrorism should not be confused with Islam. There are a great many differences between the two. For example, Islam is a religious philosophy; Militant Islamism is a political philosophy based on active nostalgia that seeks world conquest through terror. Based on nostalgia, Militant Islamism plays a major role in the relationship between tourism and terrorism.

The Russian-American author, Svetlana Boym, in her major study of nostalgia (2001), notes that although Swiss physicians first coined the word to describe an ailment affecting Swiss soldiers, nostalgia has now become a necessary concept in our understanding of terrorism. Terrorism, on some levels, like messianism, began as an idea that explains a physical desire to return home and became a spiritual desire to return to a postmodern world. Today, nostalgia functions as the ideological underpinning of terrorism, a seeking to return not to the past but to the future. Viewed in these terms, nostalgia is another manifestation of Hegelian theory. Understood in this manner, scientific progress leads to a counterreaction by some groups that seek to return to the past (or what they perceive to be the past). In a like manner, tourism helps to unite past with future turning it into a constant present. From the perspective of tourism and terrorism, Boym's paradigm defining and delineating restorative versus reflexive nostalgia is extremely helpful. The reader will note in Table 1 that terrorism is based on a misguided form of restorative nostalgia. This is a form of nostalgia that seeks a mystical past as the road map for a desired future. Restorative nostalgia is postmodern in thought and often reflects in political terms what postmodern historically built villages reflect in cultural tourism.

Boym's ideas are supported by the work of Roland Robertson. Robertson predicted that in our post-historical interconnected age fits of nostalgia would develop both within nations and individuals. In Robertson's words, "national societies are increasingly exposed internally to problems of heterogeneity and diversity and at the same time are experiencing both external and internal pressures to reconstruct their collective identities along pluralistic lines" (Robertson, 1990, p. 57).

As Hegel might have predicted, restorative nostalgia, like terrorism, when mixed with the political will to destroy the present so as to advance to the past, is the search for the particular over the universal (Boym, 2001, p. 11). The Mujahad, the Baathist parties of Iraq and Syria, and Iranian Khomeineism are all examples of this desire to advance to the past. The clearest expression of this form of Militant Islamism may be the Saddam Hussein legacy. During his reign, over 1,000,000 people were reported to have died in the Iran–Iraq war. Militant Islamism often seeks enemies who demonstrate that diverse peoples can live together in relative harmony. Thus, among the Muslim nations, secular Turkey exemplifies a nation where Islam, Christianity, Judaism, and Secularism have been able to coexist (see Table 2).

Table 1 Characteristics of Restorative and Reflexive Nostalgia

Type of Nostalgia	*Restorative*	*Reflexive*
Stress	Action of going home	The longing
Push for homecoming	Quickens it	Delays it
Way it thinks of itself	Truth and tradition	Faces modernity
Dealing with absolutes	Protects absolute truth	Questions absolute truth
Politics	National revivals	How do we inhabit two places at the same time
Emphasis on	Symbols	Details
Memory	National and linear	Social and varied
Plots	Restore national origins and conspiracy theories. A paranoiac reconstruction of "home" based on rational delusions (p. 41)	The past is dealt with irony and humor while mourning is mixed with a sense of play and points to the future.

Source: Boym (2001, p. xviii).

Table 2 Key Differences between Militant Islam (MI) and Religious Islam (RI)

Issue	*Militant Islamism/European Totalitarianism*	*Religious Islam*
Operates in the sphere of	Public life	Spiritual/private life
Led by	Engineers	Religious scholars
Outside influences	Importation of Western ideas such as Friday as a Sabbath (not part of totalitarianism)	Does not seek to emulate the West. Friday is day of assembly
Locus of law	In place	In the person
Attitudes toward the West	Seeks to confront West/ pluralistic democracies	Seeks to develop its own relationship to God and humanity's place in the universe
Population	Demographics as a weapon. Murder of children	Love of children
Key nations as models	Iran, Saddam's Iraq	Turkey
Type of philosophy	Political (totalitarian)	Religious (seeks converts)
View on nostalgia	Nostalgia based on pseudo-history	Adaptation to modernity
Main recruitment source	People trained at secular universities	People with high levels of anxiety or fear
Logic	Uses Western logic	Appeals to the emotional and mystical
View of existing system	Seeks to change it	Seeks to destroy it

Source: Compiled from Peters (2002).

These differences are well noted in the collective works of Ralph Peters. Peters divides terrorists into two specific categories, the practical terrorist and the apocalyptic terrorist. While both forms of terrorism practice a restorative rather than reflexive form of nostalgia, Peters notes that there are great differences between these two forms of restorative nostalgia and these differences cast a dark shadow over tourism.

Practical terrorists seek to restore or create new–old political realities. Its practitioners do not seek to destroy a society, but to eliminate what they consider to be a political evil. Practical terrorists usually manifest limited goals. For example, anti-abortion terrorists do not wish to destroy the United States but rather to right (from their viewpoint) a political mistake. Apocalyptic terrorists are different. They do not so much seek to change as to destroy; their political agendum is one of total annihilation. "The religious apocalyptic terrorist is a captive of his own rage, disappointments, and fantasies" (Peters, 2002, p. 23; see Table 3).

Crime vs Terrorism

All too often crime and terrorism and their impact are confused. This confusion is understandable in light of their overlapping at times. For example, Victor Davis Hanson has noted "Unfortunately to the sniper's innocent targets in Virginia and Maryland(emdash)or any others who will die by an unhinged Al Qaeda wannabe(emdash) it makes very little difference if they were victims of terror or terroristic (sic) behavior" (Hanson, 2004, p. 11). They ought not to be. Although both are

Table 3 Differences between Practical and Apocalyptic Terrorists

	Practical Terrorist	*Apocalyptic Terrorist*
Mental state	Hopes to change a policy through violence	Victim of self-rage and anger
Time frame for success	This world oriented	Next world oriented
Attitude toward religion	Tends toward secularization	Tends toward religious mystical experiences
Suicide	Rarely suicidal, not a key goal	Highly suicidal; suicide is a means to a greater end
Goal	Recreation of state or policy	Annihilation of the state, or people
Willingness to use WMDs	Limited use of chemical	Biological or nuclear
Value of human life	Low	Nonexistent

Source: Peters (2002).

social illnesses that are highly destructive to a tourism industry, to misdiagnose the malady and to mistreat it can create an even worse situation. On the whole, crime lives off a tourism industry in a parasitic relationship. Criminals rarely seek publicity and tend to work on a one-on-one basis, be that basis an aggregate of many or simply of two individuals. From the tourism criminal's standpoint, a tourism industry in decline will result in lost opportunities. Thus, on the macro level the criminal will want a successful tourism industry, even though on the micro level he/she may be doing great harm. The opposite is true for the terrorist. Terrorists are not out for profit, but for a cause. As such they seek the macro destruction of the tourism industry as a way to hurt or destroy a particular nation's economy. Terrorists rarely seek profit, but instead determine success by the quantity of dead bodies and by the loss of economic opportunities. Table 4 provides some nine different categories and how criminal acts and terrorism differ from each other.

Crime and Terrorism Basics

Table 4 demonstrates there is not only a major difference between criminal acts and acts of terrorism, but also that there are different remedies to deal with these two very separate social illnesses. It is incumbent upon tourism scholars and practitioners to distinguish between these two social maladies and develop responses that are appropriate to the threat. Criminal acts require well-trained police forces. The emphasis is both on displacement (i.e., hardening the target so that the victimizer goes somewhere else) and on active prevention. Dealing with crime requires clear guidelines in which police, security professionals, legal professionals (such as district attorneys), and the tourism industry work together. Terrorism, on the other hand, is more warlike in nature. Its goal is destruction through victimization. It seeks publicity and mass economic destruction with the added bonus of loss of life. Elliot M. Krammer has noted, "What is news(worthy) to the media is a function of numbers (of people) killed or hurt" (Kramer, 2003, p. 82). Assuming that Kramer is correct, then the media encourages mass murder by making that story its number one news item. Crime coexists with a modern tourism

Table 4 Key Differences between Acts of Tourism Crime and Terrorism

	Crime	*Terrorism*
Goal	Usually economic or social gain	To gain publicity and sometimes sympathy for a cause
Usual type of victim	Person may be known to the perpetrator or selected because he/she may yield economic gain	Killing is a random act and appears to be more in line with a stochastic model. Numbers may or may not be important
Defenses in use	Often reactive, reports taken	Some proactive devices such as radar detectors
Political ideology	Usually none	Robin Hood model
Publicity	Usually local and rarely makes the international news	Almost always is broadcast around the world
Most common forms in tourism industry are:	Crimes of distraction Robbery Sexual assault	Domestic terrorism International terrorism Bombings Potential for biochemical warfare
Statistical accuracy	Often very low, in many cases the travel and tourism industry does every-thing possible to hide the information	Almost impossible to hide. Numbers are reported with great accuracy and repeated often
Length of negative effects on the local tourism industry	In most cases, it is short term	In most cases, it is long term unless replaced by new positive image

Source: Tarlow (2001, pp. 134–135).

industry. Terrorism seeks to destroy modernity and thus seeks tourism's demise. Thus, tourism industries that face a terrorism problem require a very different set of responses. Terrorism is war and as such cannot be rooted out with police action. Perface nonhese is this exeplified better thn in the issue of gesm warface. Milles, Eagleberg and Broad have noted that "the wold's response to the growing dangers of germ warface has faller far short of what is needed (miller at al., 2001, p. 315).

Crime and terrorism are two thirds of what may be called the triangle of violence. The third leg of this triangle is Meetings cum Demonstrations or MCDs. The MCD phenomenon first took the tourism industry by surprise at the Seattle, Washington, U.S. meetings of the World Trade Organization. An MCD is a planned riot for political purposes. While MCDs rapidly generate into a multitude of criminal acts, their participants often exhibit the same form of nostalgia that is typical of terrorism (restorative) and demonstrate a practical form of terrorism similar to acts of violence by violent groups with a specific cause. The effects of an MCD can be devastating for a local tourism industry. Not only do visitors stay away from the locale and a great deal of damage may be done to the locale's property, but often reputations can be destroyed for long periods of time.

Table 5 delineates some of the differences between domestic terrorism, protest terrorism (MCDs), and international terrorism.

Table 5 Differences between Domestic Terrorism, Protest Terrorism (Meetings cum Demonstrations or MCDs) and International Terrorism

	Domestic Crime	*MCDs*	*Acts of Terrorism*
Viewed as	Crime	Politics	War
Goal	Overthrow government or policy	Change policy	Conquest
Preparation time	Very little or none	Great deal of time	Very little or none
Targets	Government buildings	Meetings	Economic or transportation centers. Tourism most at risk here of a direct attack
Effects on tourism	Major short-term effect. Can become a part of dark tourism	Major effect during short and medium term memory	Can have long-term effects, especially if it is repeated

Source: Tarlow (2002, p. 136).

A Social Theory of Terrorism and Tourism

The Italian social commentator, Giovani Sartori, has brought the relationship between modernity and internationalization to the forefront. Sartori suggests that living in the age of television has turned *homo sapiens* into *homo videns*. By connecting the idea of *homo videns* with that of the global village, Sartori reminds us that in a postmodern world any place is every place, that fear can now be transmitted at the speed of light, and that careful manipulation of the media is essential in postmodern warfare. Terrorism seeks worldwide publicity and tourism centers act as magnets for such publicity. If postmodernity is centered around the concept of the global village, then tourism is the social glue that unites this village. In the world of images, terrorism is transformed from the horrendous event into the iconic event.

The tourism industry is the world's largest peacetime industry. For example, in the United State, it ranks as one of the top three industries in every state of the nation. Tourism is also a major export item. Each year nations around the world see tourism as a major source of revenue and as a means to earn foreign currency. It is then logical that anyone seeking to destroy a national economy would seek as one of its targets that nation's tourism industry. For these and some of the other reasons mentioned below, tourism is a major battleground for terrorism and needs to be a concern for anyone interested in his or her nation's homeland. Benjamin Netanyahu has noted "The societies targeted (by terrorists) have included Britain, Italy, France, Holland, Spain, Germany, Japan, Argentina, Israel, and most recently the United States itself. No country is immune and few are spared (Netanyahu, 2001, p. 7). Over the last 30 years tourism providers and sites have been targets for terrorists. From the viewpoint of the terrorist, some of these attacks have been very successful. A review of terrorism over the last 30 years reveals that 1) there does not seem to be a relationship between a locale's base population size and the act of terrorism. That is, terrorism has occurred in all sorts of rural and urban settings and 2) acts of terrorism tend to increase until decisive passive restraints plus active pursuit of terrorists occur.

Ever since the period of the 1960s and plane hijackings, terrorism has found ways to attack the travel and tourism industry. Here are the themes of terrorism during the last 40 years.

The 1960s and 1970s was the time of plane hijackings. In the 1960s, these hijackings were often from the U.S. to Cuba and were not aimed at harming passengers. In the 1970s the model was modified and plane hijackings were now part of terrorism. The purpose of 1970s actions (often committed by radical Muslim Mid-Eastern groups) was loss of life plus property. During the 1970s the Munich Olympic Games became the site of one of the most notorious terrorist attacks and set a new standard for terrorism at major world events. The 1970s also saw attacks at airports, where passengers were murdered for the sake of murder. The 1980s also saw plane bombings such as bombs aboard both TWA (U.S.) and Air India planes. The 1990s and the first part of the early years of the twentieth-first century have seen terrorism increase in the following areas: attacks against ground transportation, bus and train bombings, attacks against major buildings, attempts at mass murder through terrorism, and attacks against hotels.

Here is a partial list of places where terrorism has been launched against the tourism industry within the last 10 years.

- Bali
- Casablanca
- Israel
- Kenya
- Los Angeles
- Mexico
- Morocco
- Peru
- The Philippines.

These locales have nothing more in common than a successful tourism industry. Students of tourism and its professionals have wondered what attracts terrorism to tourism. Below are some of the reasons for this interaction. Tourism is interconnected with transportation centers.

In all these cases the careful observer will note that terrorism has:

- Sought to increase both economic and human damage
- Become more severe in its attacks
- Sought new travel and tourism targets
- Moved from the fringe to the mainstream.

Terrorists, having learned the lesson of mass media, have also tended now to seek urban over rural areas. The use of urban areas is a good way to gain instant publicity and assure the greatest numbers of casualties. Attacks in cities mean that not only are there direct victims, but also those who witness the attack become secondary (psychologically hurt) victims. Cities being centers of economic activity, urban attacks have the advantage of greater economic destruction. Cities are also easy places into which a terrorist can fade and often find a safe house among the terrorist group's sympathizers. This analysis would lead one to believe that rural tourism centers are thereby safer. Tourism, however, no matter where it is located, acts as if it were an urban center. For example, attacks on tourism centers almost instantaneously produce major news coverage (an example is the Bali bombings). In a like manner, tourism is an economic engine and often assembles large numbers of people even in rural fields (for example, outdoor music festivals such as Woodstock).

Traditionally, many tourism professionals have avoided addressing issues of tourism security and tourism safety altogether. There has been a common feeling among these professionals that visitors will wonder if too much security indicates that they should be afraid and that even speaking about these subjects will frighten customers. Thus, especially in the years prior to 2001, the industry often took the position that the less said about tourism security and safety the better. The terrorist attacks in New York City, Washington, DC, Bali, and Kenya have destroyed more than many thousands of lives and hundreds of millions of dollars in property value. The attacks also forced the travel industry to deal with a major travel paradigm shift. This shift in travelers' mindsets had been occurring prior to September 11; however, these terrorist attacks, and the possibility of new attacks, have given the travel and tourism a major wake-up call.

Unfortunately, too many in the travel and tourism industry were the last people to realize that the shift was already under way. Put in its simplest of terms, travelers no longer fear tourism security but demand it. In the old travel industry paradigm, security was a dark secret. Industry leaders rarely spoke about threats to tourists in public fearing that such openness would scare away visitors. The common belief was that security was a necessary evil that one had to have, but that security added nothing to the business's bottom line. For this reason, tourism and travel security were rarely publicized, never mentioned in marketing campaigns, and underfunded, and its practitioners were often underpaid. The old paradigm led to poor security at airports, hotels, restaurants, and attractions. Security professionals who spoke of acts of terrorism, biochemical attacks, and crime were seen as alarmist and asked to rephrase their warnings in ways that would be acceptable for public consumption.

The public, however, was beginning to change its views. Even prior to the September 11 attacks, there were multiple signs that the public was concerned about and demanded good tourism security. Throughout the travel and tourism industry, anecdotal evidence began to appear that tourism's customers were choosing locations and venues precisely because they were safe and secure. Studies and scholarly articles noted that security personnel were beginning to obtain some of the tourism security training that was needed.

Toward a Theory of Terrorism and Tourism

The Italian social-commentator, Giovani Sartori has brought the relationship between modernity and internationalization to the forefront. Sartori suggests that living in the age of television has turned *homo sapiens* into *homo videns*. By connecting the idea of *homo videns* with that of the global village, Sartori reminds us that in a postmodern world any place is every place, that fear can now be transmitted at the speed of life, and that a careful manipulation of the media is essential in postmodern warfare. Terrorism seeks worldwide publicity and tourism centers act as magnets for such publicity. If postmodernity is centered around the concept of global village than tourism is the social glue that unites this village. In the world of images, terrorism is transformed from the horrendous event into the iconic event.

Terrorism is much older than modern mass tourism. Terrorism has been drawn to tourism for a number of reasons. Because tourism is a modern to postmodern industry, there are those who see it as an evil in and of itself, in addition to the economic damage that a terrorist attack can cause. Here are a number of reasons why terrorism has often attacked tourism areas.

- Tourism officials have often been afraid to take creative steps to stop terrorism fearing that such hardening of their targets will frighten customers away. The end result has been that tourism often provides very soft (easy) targets.
- Tourism is big business and terrorism seeks to destroy economies.
- Tourism is interrelated with multiple other industries; thus an attack on the tourism industry may also wipe out a number of secondary industries.
- Tourism is highly media oriented and terrorism seeks publicity.
- Tourism must deal with people who have no history, thus there are often no databases, and it is easy for terrorists to simply blend into the crowd.
- Tourism must deal with a constant flow of new people, thus terrorists are rarely suspected.
- Tourism is a nation's parlor that it is the keeper of a nation's self-image, icons, and history. Tourism centers are the living museum of a nation's cultural riches.
- Terrorists tend to seek targets that offer at least three out of the following four possibilities and these same possibilities often exist in the world of tourism:
    - Potential for mass casualties;
    - Potential for mass publicity;
    - Potential to do great economic damage; and
    - Potential to destroy an icon.

A study of tourism and terrorism permits the following postulates to be developed.

- *Terrorism is the opposite of a predictable peace.* If Hegel is correct that social movements often result in counter social movements, then an industry based on relaxation, freedom, and open borders may well produce people who see the world in opposite colors. Thus, those who hate freedom of travel and peaceful coexistence and who foster xenophobia are bound to see tourism as a destructive agent that ought to be destroyed.
- *Tourism is a competitor with terrorism for the nostalgia.* Much of tourism is based on a nostalgic view of the past. In the world of tourism, reality is often presented in the least realistic terms and truths are often sacrificed on the altar of marketing. Terrorists have learned from the tourism industry. Although their story is different and the purpose is destruction for the sake of destruction, both industries appeal to the non-factual realities. In a strange sense both are at opposite ends of the social poles held together by the gravitational pull of fantasy portrayed as reality.
- *Tourism and terrorism view women from opposite perspectives.* Terrorism may use women to gain power, but in most cases women are never given power. Seen from a Freudian viewpoint, those cultures that repress women and keep men in a psychologically infantile state have a higher probability of spawning terrorism. Tourism is an industry filled with working adult women and often these women hold the most powerful and prestigious of jobs. From the terrorist perspective then, tourism undercuts the idea that women are incapable, and forces men to deal with women on a one-to-one human basis rather than as infants demanding service from their mothers.

- *Tourism and terrorism know no borders.* Tourism is a universal phenomenon; it seeks to obliterate xenophobia and it brings peoples together. Tourism is a celebration of the distinct and the other. Such a philosophy is counter to that of terrorists who seek the establishment of apocalyptic restorative nostalgic states that may never have existed.
- *In a world in which the media believes in balanced news, terrorist strikes on innocent civilians and tourists are a "logical act of war."* Much of the Western media seeks to delineate causality for terrorism attacks. Based on the classical liberal idea that all people are intrinsically good, the media seeks to understand the frustrations and anger expressed by terrorism. This lack of moral clarity means that terrorist attacks, especially against the innocent, make a great amount of sense. Terrorist attacks against tourism locations not only spread the terrorist's message but at the same time encourage other terrorist actions, cause a lessening of personal interchanges, and in the end undercut the economy upon which the media depends.

Tourism is an integral part of globalization. Tourism not only mixes people and cultures but is a major source for the dissemination of information. Using the words of Ralph Peters we can see that tourism helps to bring about an "onslaught of information—a plague of ideas, good and bad, immune to quarantine or ready cures, under whose assault those societies, states, and even civilizations without acquired resistance to information disorders will shatter irreparably" (Peters, 2002, p. 140). Terrorism fears the onslaught of a free flow of human interaction and information for which tourism stands. Tourism requires globalization, while terrorism cannot live with globalization; tourism promotes cultural and information interchange, but terrorism dies from such interchanges; tourism requires caring and human kindness, and terrorism promotes the principle that only one's own are true human beings. Under the terrorist rubric, tourism can be seen as the instrument by which much of the world will be infected and as such to attack tourism is not only seen as fair play but as a positive step. If terrorism must depend on the beliefs of the ignorant, then non-group tourism undercuts such beliefs by allowing people to interact on the micro- and personal levels.

Tourism seeks stability and tranquility, while terrorism seeks to undercut that stability. Few people, other than the extreme allocentric tourist, seek out war zones as places to visit. Tourism seeks peace. All forms of violence, be it violence in the form of crime or terrorism, tend to destroy a tourism industry. It then follows that groups that seek to control information, humble women, maintain men in an infantile state, and turn their societies into what they believe was rather than is, will use any form of violence to undercut and then destroy a local tourism industry.

Summary

The postulates given above means that the industry must go through a major paradigm shift if it is to survive. In order for this paradigm shift to occur, the travel and tourism industry will have to take the following into consideration.

The fact that terrorism will target the tourism industry and that the industry cannot afford to ignore the threat is widely accepted. The targeted list may include:

- Airlines
- Cruise ships

- Buses;
- Restaurants and outdoor cafes
- Major events, or sporting or cultural institutions
- Places where people congregate
- Wherever people are carefree and happy.

The realization that travelers and tourists, for the most part, will seek out places where there is a sense of security and safety. Although there is a small minority of travelers who seek out the dangerous, most visitors want to know what the industry is doing to protect them, and how well prepared a local industry is in case a security or safety issue should occur.

The merging of security and safety concerns into tourism surety concerns everyone. Classically security has been defined as the protection of a person, site, or reputation against a person or thing that seeks to do harm. Safety is often defined as protecting people against unintended consequences of an involuntary nature. In the case of the travel and tourism industry, both a safety and a security mishap can destroy not only a vacation but also the industry. It is for this reason that the twenty-first century paradigm will combine these two fields as "tourism surety." Tourism surety is more than merely the point where safety, security, and economic viability meet.

Surety refers to a lowering of the probability that a negative event will occur. Surety does not promise perfection, but rather improvement and takes into account that to live is to risk. Because few industry people work according to strict academic guidelines, the remainder of this chapter will use the terms surety, security, and safety interchangeably.

This paradigm change will not come about easily. Many CEOs of attractions, hotels, and regional, state (provincial), and national tourism offices along with other tourism professionals will continue to see security as adding nothing to their bottom line. Terrorism is often perceived to be a unique occurrence and/or something that will happen to the other person or at the other's location. Other tourism officials may misdiagnose terrorism as a form of crime and thus believe it to be a manifestation of psychological, economic, or political frustration, rather than a tool used by highly trained and calculating professionals. If the above social ills were the cause of terrorism then countries such as those in sub-Saharan Africa and Bangladesh would be the world's great terrorism exporters. In a like manner, no people had more right to be frustrated than Jews living in Nazi occupied Europe, yet there are no recorded incidents of actions taken against innocent civilian population centers.

Poor economic conditions, lack of educational opportunities, and/or political frustration may well be the cause of criminal acts, but rarely if ever are they the cause of terrorism.

The problem becomes a question of how the tourism industry balances security and profitability. This matter is not easy, as much terrorism tends to take security out of the hands of tourism officials and place it into the hands of government officials. An example of this phenomenon can be seen in a *Houston Chronicle* headline. The headline reads: "Tighter security squeezing out tourists" (*Houston Chronicle*, 7/a, April 22, 2004). Quoting Secretary of State Powell, the *Chronicle* reports his having said in reference to foreign student visas, "People aren't going to take that (tighter security regulations) for very long, and when the word gets out they will start going elsewhere. Despite the fact that this article mixes foreign students seeking advanced academic degrees with leisure tourists, the article does make several important points. The article highlights the following theoretical concepts:

- There is a difference between security measures and hassle measures.
- Tourism requires both good security and good service. Half the equation is not enough.
- Tourism security is different from other forms of security. It depends to a much greater extent on the quality of personnel training. Machines and passive measures are not a substitute for a well-trained employee.

Although the change is not an easy one, tourism officials may have no other choice. For example, a major publication of the Meetings and Planners Association dedicated its entire March, 2004 edition to tourism security issues (March 2004 edition of *CMI (Corporate Meetings and Incentives) Magazine*). When corporate meeting planners choose a site at least partially by the level of tourism security that it offers, then it is clear that the paradigm shift is under way.

Concept Definitions

Age of Attention Deficit Disorder (AADD) The period of time in which the public demands instant answers to complex problems and seeks simple solutions to complicated issues.

Allocentric tourist One who is willing to accept a high level of risk as part of leisure travel.

Hegelian theory The assumption that social movements in history produce counter or opposite movements that in turn form a new historical worldview that then produces a new counter viewpoint.

Homo videns The idea that human beings are now defined by what they view rather than by what they know, thus replacing the idea of *Homo sapiens*.

Messianism An idealized state of mind, mixing religion and terrorism, so as to implement a form of restorative nostalgia.

Nostaglia A longing for a past that may be true, even if idealized. Nostalgia has two forms, restorative and reflexive.

Reflective nostalgia The more passive and less dangerous form of nostalgia; often used as a marketing tool within the tourism industry.

Restorative nostalgia A more militant form of nostalgia that often seeks to recreate past glories or perceived past glories through violent means or acts of terrorism.

Post-history The period of time after the end of the cold war when nostalgia for the past produces a series of low-grade or quasi-wars rather than nation vs nation wars.

Psychocentric The desire to avoid the greatest amount of risk possible in leisure travel.

Surety The probabilistic point in which security, economics, and safety meet.

TOPS Acronym for Tourism Oriented Policing (or Protective) Services. A philosophy that sees security as an integral part of the tourism industry.

Review Questions

1. How does nostalgia impact acts of terrorism against the tourism industry?
2. How do criminal acts differ from terrorist acts in regards to tourism?

3. What are the European roots of terrorism and how have these roots impacted the Middle East?
4. What are the differences among security, safety, and surety?
5. What are the differences between an apocalyptic terrorist and a practical terrorist?
6. What are four factors that may help to predict a major terrorism attack?
7. How did the tourism surety paradigm change after September 11, 2001?

References

Berman, P. (2003). *Terror and Liberalism*. New York: W.W. Norton & Company.

Boym, S. (2001). *Nostalgia*. New York: Basic Books.

Camus, A. (1942). *The Stranger (l'Etranger)*. New York: Republished by Random House.

Fukuyama, F. (1989). The end of history, *The National Interest,* 16(Summer), 3–18.

Hanson, V. (2004). *Between War and Peace*. New York: Random House.

Kramer, E. (2003). *Complicity: Terrorism in the News*. Montreal EMK Publishing.

Miller, J., Engelberg, S., and Broad, W. (2001). *Germs*. New York: Simon & Schuster.

Netanyahu, B. (2001). *Fighting Terrorism*. New York: Farrar, Straus, & Giroux.

Ottolenghi, E. (2004, May 10). http://www.jpost.com/servlet/Satellite?pagename=JPost/JPArticle/ShowFull&cid=1083998351689.

Peters, R. (2002). *Beyond Terror*. Mechanicsburg, PA: Stackpole Books.

Robertson, R. (1990). After nostalgia? Willful nostalgia and the phases of globalization, in B. Turner (ed.), *Theories of Modernity and Postmodernity*. London: Sage Publications.

Sartori, G. (1998). *Homo Videns*. Madrid: Taurus.

Sonmez , S. F., and Tarlow, P. (1999). Tourism in crisis: Managing the effects of terrorism. *Journal of Travel Research,* 38(1), 13–19.

Tarlow, P. (2003). Risk and crisis management in tourism. *Texas Town & City,* 80(4), 42–45.

Tarlow, P. (2003, March). Taking a realistic look at tourism in times of terrorism. *USA Today Magazine,* 131(2694), 52–54.

Tarlow, P. (2002, December). Los problemas para la seguridad para el turismo actual. *Policia y Criminalistica*. Buenos Aires, Argentina: Policia Federal Argentina, Vol. 344, pp. 50–56.

Tarlow, P. (2002). Tourism in the twenty-first century. *The Futurist,* 36(5), 48–51.

Tarlow, P. (2001). *Event Risk Management & Safety*. New York: John Wiley & Sons.

Tarlow, P., and Muehsam, M. (1995). Theoretical aspects of crime as they impact the tourism industry, in A. Pizam and Y. Mansfeld, *Tourism, Crime and International Security Issues*. Chichester, UK: John Wiley & Sons.

Veblen, T. (1963). *The Theories of the Leisure Class*. New York: Mentor Books.

3

War, Terror, and the Tourism Market in Israel

Aliza Fleischer and Steven Buccola

Learning Objectives

- To explain the nature of the tourism market as reflected in the market for hotel space.
- To investigate the dynamics of the tourism market.
- To delineate the difference between the supply and demand for hotel space and between the local and foreign demand for tourism.
- To demonstrate the interaction between local and foreign tourism demand.
- To analyze the impact of terror attacks on room prices and on the number of occupied rooms in hotels.
- To understand why hotels allow some capacity to go unused during war or terrorist intervals.
- To explain the importance of demand elasticity in hotels reactions to demand depressions during terror attacks.

Introduction[1]

By at least one reckoning, world tourism is larger than any other export industry. In 1999, tourism accounted for $532 billion in world sales, compared with $525 billion in auto exports and $399 billion in computer exports (World Tourism Organization, 2000). Tourism is important not only to developed nations but to many poorer ones, whose chief resources often include their visitor attractions.

Because of its impact on foreign exchange earnings and employment, tourism demand has been extensively investigated. Lim's (1997) review alone includes nearly a hundred demand papers, including O'Hagan and Harrison's (1984) analysis of US tourist expenditures in Europe and Syriopoulos and Sinclair's (1993) study of US and European tourism in the Mediterranean. As Sinclair (1998) and Lim note, most demand analysts use a single-equation format, focusing on relative prices, tourist income, and particular travel-generating events. The underlying assumption of these approaches is that capacity is infinite: demand at any location can be satisfied at constant cost. Tourism capacity may be regarded as depending on a measure of the "size" of the local attractions, on the associated infrastructure and transportation network, and on the number of beds at local hotels. Studies that include only demand components therefore assume that such capacity factors are large relative to the number of visitors. They also ignore pricing strategies that hotels employ to manage their fixed resources in the face of demand fluctuations.

Little work appears to have been done on the supply side of the tourism market, and even less on how supply and demand jointly react to exogenous forces. Among the few extant supply response studies is Borooah's (1999) on the Australian hotel sector, which characterizes the supply of hotel rooms as a function of earnings per room, occupancy rate, and the interest rate. To reflect the quasi-fixed nature of hotel capital, prices in Borooah's study lag and hence are predetermined, permitting the demand side of the market to be ignored. Wheaton and Rossoff (1998) have investigated whether cyclical behavior in the US hotel industry matches that in the overall economy. Presaging to some degree our own results, they find that bed night supply and demand often are in disequilibrium because hotels are reluctant to adjust prices in response to short-run occupancy rate changes.

The purpose of this chapter is to investigate a nation's tourism market, as reflected in the market for its hotel space. We examine Israel, where war and political climate appear to have had a significant impact on revenue and profits. We are interested in distinguishing between the local and foreign market for tourism, and in the interactions between these two sectors. We also seek reasons why hotels have allowed some capacity to go unused during war or terrorist intervals and, for that matter, during much of the year, despite an evident ability to price their services for full-capacity utilization. Our analysis owes much to Krakover (2000) who, along with Bar-On (1996), examined the influence of political climate on tourism, albeit in the absence of many demand or supply determinants.

Israeli Hotel Industry

Israel's hotels and guest houses are scattered throughout the country, concentrated, in descending order, in Eilat, Jerusalem, Tel Aviv, Tiberius, and the Dead Sea. During the past decade, hotel capacity in Israel has grown at 4.5% per year (Figure 1). The industry sells into two major markets: the local Israeli market and the international or foreign market. Bed nights to foreign tourists have trended slightly upward, fluctuating in correlation with regional political instability. They fell substantially in 1991, the year of the first Gulf War, and in the 1996–1998 period, when terrorism was frequent. On the other hand, foreign tourist nights rose rapidly when Israel signed peace treaties with Jordan and the Palestinians in 1995.

Israeli tourists are gaining an increasing role in Israel's hotel industry. Local bed nights have risen steadily in the past few years and now surpass the foreign market (Figure 1). While violent events temporarily depress international demand, they do not appear to affect Israelis' local tourism proclivities in any significant way. Instead, Israeli occupancy rates seem to rise somewhat during terror crises,

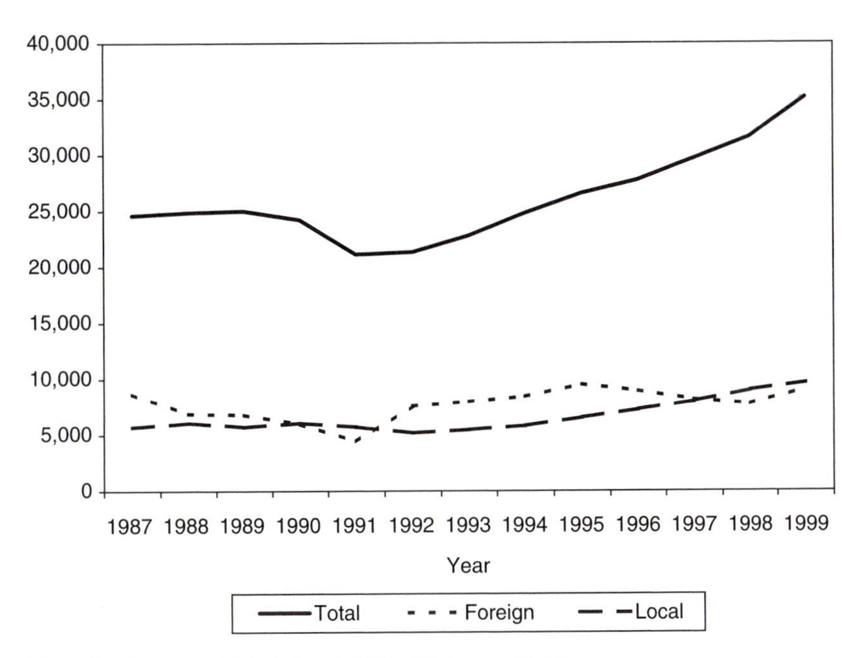

Figure 1 Hotel bed nights In Israel, 1987–1999, annual basis.

possibly in response to measures hotels take to offset shortfalls in the international market. The shape of the total occupancy rate curve (Figure 2) is similar to but smoother than that of the foreign rate curve, suggesting that the local market acts as a counterweight to swings in the international market but does not completely offset them.

Although international visitors' rationale for avoiding Israel soon after a terrorist incident may not be rational in the sense of minimizing the probability of their own injury, it is understandable in emotional terms. More problematic is the hotel industry's apparent failure to reduce local prices enough to compensate fully for the international shortfall. To explore this matter, we focus on Israeli hotel pricing behavior, and especially on how hotels adjust prices to both foreign and local guests in response to the demand shifts induced by exogenous events. We hypothesize that, because of inelastic local demand and the intangible costs of large price discounts, it generally is unprofitable for hotels to reduce local prices sufficiently for local demand to serve as a complete antidote to international market swings. Taken more generally, this study therefore provides insight into the common industry practice of permitting excess capacity during some seasons of the year.

Model

Our econometric model consists of supply and demand in the local market and supply and demand in the foreign market. The framework permits interactions

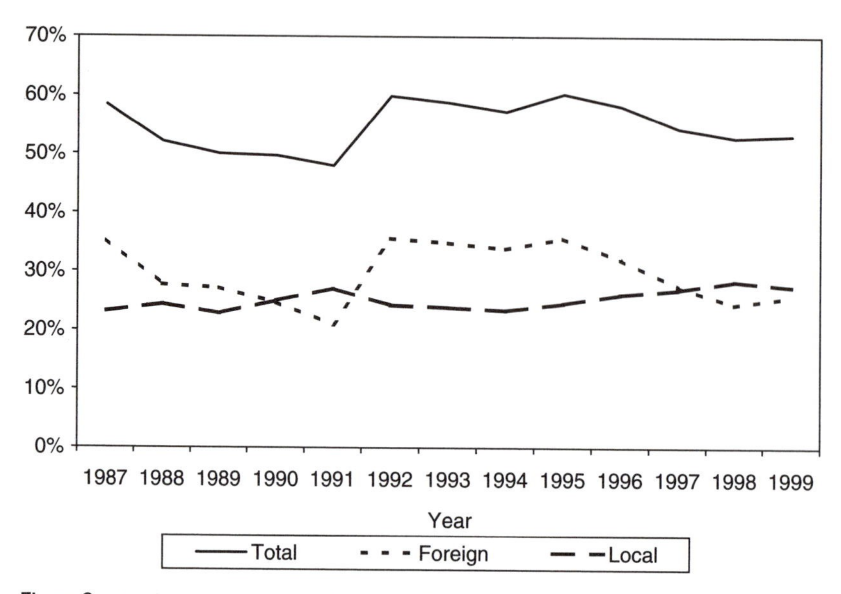

Figure 2 Hotel occupancy rates in Israel, 1987–1999.

between the two markets and between supply and demand in each market. To facilitate the discussion, the following variables are used, based on time period t:

PL_t	Mean monthly price charged to local (Israeli) tourists, in US dollars per bed night
PF_t	Mean monthly price charged to foreign (international) tourists, in US dollars per bed night
QL_t	Bed nights sold to Israeli visitors, thousands per month
QF_t	Bed nights sold to foreign visitors, thousands per month
$GNPL_t$	Israeli GNP, billions of US dollars per month
$GNPF_t$	Weighted mean GNP in foreign visitors' countries-of-origin, billions of US dollars per month
SCL_t	Average expenditure on Israelis' travel to non-Israeli destinations, in US dollars per person-trip
SCF_t	Average expenditure on foreigners' travel to non-Israeli destinations, in US dollars per person-trip
TER_{1-k}	Index of magnitude of terrorist or related political event, –9.0 to 9.0 (k is lag length)
S	A vector of seasonal (bi-monthly) dummy variables

In the local market, demand quantity QL_t^d should depend on Israeli hotel rates charged to Israelis (PL_t), on aggregate Israeli income ($GNPL_t$), on hotel and other travel costs charged to Israelis at sites outside Israel (SCL_t), and on seasonal factors **S.** Foreign demand quantity QF_t^d would depend on Israeli hotel rates charged to foreigners (PF_t), on hotel and other travel costs charged to foreign visitors at sites outside Israel (SCF_t), on seasonal factors **S,** and on the appropriately lagged terror index TER_{1-k}. We have then

$$QL_t^d = QL_t^d (PL_t, GNPL_t, SCL_t, \mathbf{S}) \tag{1}$$

$$QF_t^d = QF_t^d (PF_t, GNPF_1, SCF_t, TER_{1-k}, \mathbf{S}) \tag{2}$$

Prices to local tourists generally differ from those to foreigners, partly because many international visits are reserved 6 to 9 months ahead of time, while Israelis tend to reserve lodging much closer to the arrival date. Alternative or substitute costs are, in the foreign market, mean costs of foreigners' travel to destinations other than to Israel and, in the local market, mean costs of Israeli travel to destinations abroad.

Characterizing the supply side of the market requires distinguishing between long-run and short-run supply and between the supply of rooms to foreigners and the supply to Israelis. Because years are required to plan and construct an international-class hotel, capital stock can be altered only slowly in the lodging industry. In a given month, we accordingly assume hotels have a fixed total bed night stock. Capacity is not fixed, however, in the foreign or local market taken individually, since hotels have flexibility in allocating rooms across the two markets. In particular, because foreign visitors' early bookings permit hotels to plan ahead and hence reduce cost, managers book as many rooms as possible to international tourists, allocating the remainder to the local market. The short-run supply of bed nights in the local market is, then, a function of demand in the foreign market.

To reflect such stylized facts, it is useful to write the supply functions in price-dependent form. We further define the following variables:

E_t Number of empty (available but unsold) bed nights
QT_1 Total number of available bed nights ($QT_1 = QL_t + QF_1 + E_1$)
W_1 Weighted average wages at Israeli hotels, in US dollars per month

Hoteliers' minimum (supply) prices PF_t^s to foreign visitors should depend on the endogenous quantity of bed nights sold to them (QF_t), on current hotel industry costs such as wage rates W_t, and on any price-change sluggishness, represented here by lagged prices PF_{t-1}. Supply prices PL_t^s offered to Israeli visitors should depend similarly on the endogenous quantity of bed nights sold to Israelis (QL_t), on industry wage rates W_t, and on lagged prices PL_{t-1}. Because allocations to the local market usually are made only after most foreign demand is known, foreign bed nights QF_t are a further factor in the Israeli supply function. Declining sales QF_t to the foreign market should shift the bed supply to Israelis downward, reducing the minimum prices that hotels accept from Israeli visitors.

Finally, since Israeli hotels have excess capacity during much of the year ($E_t > 0$), total bed supply QF_t is independent of total demand $QF_t + QL$ and may also, therefore, affect the supply price to Israeli tourists. In a long-run decreasing (increasing) cost industry, rising total bed supply should put downward (upward) pressure on prices, and only if the industry is a constant-cost one will total bed supply have no price effect.

With these considerations in mind, we specify the supply functions as

$$PL_t^s = PL_t (QL_t, QF_t, QT_t, W_t, PL_{t-1}) \tag{3}$$

$$PF_t^s = PF_1 (QF_t, W_t, PF_{t-1}) \tag{4}$$

in which, as in demand specifications (1) and (2), the endogenous variables are QL, QF, PL, and PF.

Data Sources

Two principal data sources are used to estimate the system defined by equations (1) through (4). Total bed (room) stocks, quantities of bed nights sold to Israelis and foreigners, total revenues in each market, and total wage payments are published monthly by the Israeli Central Bureau of Statistics (1992–1999). All other variables, except for the terror index, were compiled from annual World Bank data (World Bank, 2000). In foreign country *i*, the substitute cost relevant to this analysis is an average of the expenses which *i*th-country tourists incur at sites other than in Israel, weighted by *i*th-country expenditures at those sites. Aggregate substitute cost SCF in foreign demand from equation (2) is then constructed as a weighted average, across the nations from which Israeli tourists are drawn, of those weighted expenditures, weighted by the shares of each tourist-generating country in total foreign tourist numbers in Israel in 1997. GNPF is similarly constructed as a weighted average of gross national products in the countries from which foreign tourists originate. Because GNP data are reported only annually, monthly figures were derived by linear interpolation, assuming constant GNP growth in a given year.

Our measure of the terrorist climate in Israel is taken primarily from Krakover (2000). In Krakover's index, zero represents a month with no terrorist event and nine a period of full-scale regional war. Intermediate values reflect the numbers of deaths or injuries in a given month or the magnitude of a limited war. To better reflect the euphoria surrounding especially positive events, we extended this index to include the signing of peace treaties, as shown in the Appendix. Between 1991 and 1998, such "peace" events occurred only twice, in each of which Israel signed a treaty with the Palestinians or with Jordan.

We divide the year into six parts and define a dummy variable for each part as follows: March and April (S_1); May and June (S_2); July and August (S_3); September and October (S_4); November (S_5); and December, January, and February (S_6).

Econometric Estimates

Demand Estimates

Demands and supplies from equations (1) through (4) were estimated in linear form using two-stage least squares and employing monthly data from March 1991 through December 1999 (EViews, 1994). Parameter estimates are given in Table 1 and the corresponding elasticities at sample means in Table 2. With the exception of a single seasonal dummy, all coefficient estimates in the Israeli demand equation are significant at the 10% level and have the expected sign. As Israeli hotels reduce prices, as overseas hotels that Israelis patronize raise prices, or as Israeli incomes rise, Israelis demand more local hotel space. The proportionate impact of Israeli hotel prices on Israeli demand is mild. The own-price elasticity averages only –0.63 (column 4, Table 2), ranging from –0.32 in midsummer to –0.94 in winter (Table 3). The cross-price elasticity in Table 2 is 1.00, implying a 10% increase in travel costs outside Israel boosts local demand in Israel by 10%. Income elasticity of demand is 1.3, suggesting Israelis regard hotel stays as a luxury. Local demand (column 4, Table 1) is lowest in the winter, highest in midsummer, and moderate in late spring and early autumn.

Table 1 Bed Night Supply and Demand Estimates, Israel, 1987–1999

	PL_t^s	PF_t^s	QL_t^d	QF_t^d
Constant	20.84 (4.20)	−2.43 (5.01)	−975 (448)	291 (506)
PL_{t-1}	0.042 (0.080)			
QL_t	−0.026 (0.003)			
QF_t	0.013 (0.004)	0.016 (0.004)		
QT_t	0.002 (0.002)			
W_t	0.045 (0.008)	0.014 (0.005)		
PF_{t-1}		0.611 (0.070)		
PL_t			−4.310 (1.570)	
$GNPL_t$			0.050 (0.007)	
SCL_t			0.390 (0.200)	
S_1			43.3 (22.5)	375.7 (54.0)
S_2			146.0 (26.8)	280.3 (59.0)
S_3			460.0 (35.9)	216.5 (52.0)
S_4			169.8 (25.6)	161.9 (45.9)
S_5			40.9 (28.2)	138.9 (51.2)
PF_t				−22.300 (4.490)
$GNPF_t$				0.001 (0.0002)
SCF_t				−2.130 (1.200)
TER_{t-2}				−12.400 (6.000)
R^2	0.78	0.57	0.89	0.46
N	94	94	94	94

Notes: $S_1 = 1$ if March or April, $S_2 = 1$ if May or June, $S_3 = 1$ if July or August, $S_4 = 1$ if September or October, $S_5 = 1$ if November, zero otherwise. The deleted (base) season $S_6 = 1$ if December, January, or February, zero otherwise. Numbers in parentheses are standard errors.

Coefficient estimates in the foreign demand equation (column 5, Table 1) likewise are significant at the 10% level. As one would anticipate, falling bed night prices in Israel, or rising incomes in nations from which Israel draws tourists, increase the foreign demand for Israeli hotel space. Own-price elasticities are substantially higher than in the local market, averaging −2.0 (Table 2) and remaining remarkably constant throughout the year (Table 3). On average, a 10% cut in the room price boosts foreign

Table 2 Supply and Demand Elasticities in Foreign and Local Markets, Israel, 1987–1999

	PL_t^s	PF_t^s	QL_t^d	QF_t^d
PL_{t-1}	0.04			
QL_t	−0.17			
QF_t	0.10	0.17		
QT_t	0.05			
W_t	0.57	0.23		
PF_{t-1}		0.61		
PL_t			−0.63	
$GNPL_t$			1.3	
SCL_t			1.0	
PF_t				−2.0
$GNPF_t$				0.035
SCF_t				−2.5

Note: Elasticities are calculated at sample mean.

Table 3 Own-Price Demand Elasticities in Local and Foreign Markets, by Season, 1987–1999

	Local Market			*Foreign Market*		
Season	*Price*	*No. of Bed Nights*	*Elasticity*	*Price*	*No. of Bed Nights*	*Elasticity*
S_1	88.1	421.6	−0.90	62.3	769.0	−1.80
S_2	77.8	565.5	−0.59	64.4	641.4	−2.24
S_3	70.0	932.0	−0.32	62.7	645.6	−2.16
S_4	79.9	584.2	−0.59	60.5	658.0	−2.05
S_5	87.6	416.5	−0.90	56.0	749.3	−1.66
S_6	84.9	386.8	−0.94	57.7	551.1	−2.33

tourism in Israel by 20%. That foreign demand is more elastic than local demand should come as no surprise. To an Israeli, the alternative to an Israeli hotel stay is to go overseas, requiring a passport, plane flight, and exposure to foreign language and culture. In contrast, the international tourist considering alternatives to Israel compares one overseas destination with another. For European tourists in particular, many of these alternative sites are as close to home as Israel is.

Income elasticity in the foreign demand equation (column 5, Table 2) is extremely low: at sample mean, a 10% rise in foreign incomes lifts foreign tourism demand in Israel by only 0.35%. Unexpectedly, foreign demand at Israeli hotels declines as expenses rise at the non-Israeli sites which foreigners frequent. The implication that, in the international tourist's mind, Israeli sites are Marshallian complements rather than substitutes for non-Israeli sites seems unaccountable and may only reflect errors in the construction of this rather complex variable. Seasonal dummies indicate that international tourism demand is lowest in the winter, peaks quickly in the spring (March through June), then declines gradually through the rest of the year.

Finally, and as anticipated, the effect of the terrorism index is negative: the more severe the war or terror climate, the lower the international visitor demand. A two-standard-deviation (four-unit) increase in the terror index reduces foreign demand quantity by 49,600 bed nights per month, a 7.5% drop from sample mean. The interval between terror event and this demand shortfall is important. We examined every lag length from 0 to 6 months; only the 2-month lag length was statistically significant. Interestingly, Krakover (2000) obtained the identical result in his examination of a much longer time series. It appears difficult for tourists to react on much shorter notice than 2 months, while over a longer period an event's psychological effect appears to subside.

Supply Estimates

The supply or minimum bed night price hotels charge to foreign visitors rises in Table 1 with the number of bed nights foreigners demand. At sample means, a 10% rise in foreign demand increases the foreign bed night price by 1.8%. Supply to the foreign market is the positively sloped function one normally expects of supply relationships: Israeli hotels' short-run marginal costs rise with the number of foreign visitors. Solving for QF and inverting equation (4) shows that hotels allocate more rooms to foreigners as the latter pay higher prices. Higher wage rates to hotel employees (W) lift the supply price to foreigners upward, shifting the foreign supply function to the left. Nevertheless, the 0.611 coefficient on the lagging price variable implies prices respond sluggishly to demand and wage rate changes.

Notably, the relationship between local demand quantity QL and local supply price PL^s in Table 1 is negative. That is, the short-run supply function in the local market has a negative slope in the neighborhood of the observed market equilibria, suggesting short-run marginal costs decline as local occupancy rates rise. Negatively sloped supplies are not uncommon in the lodging industry. Horwath Consulting (1994), for example, reports substantial evidence of high fixed costs and scale economies in British hotel chains. In the present short-run context, marginal costs consist of labor and material expenses plus any reputation costs implicit in the room rates charged. Marginal labor and material inputs may decline with increasing local guest volume if these inputs have an overhead component to them, as when a minimum housekeeping staff and power supply are required even when occupancy rates are low. The same size economies may not apply to foreign visitors, especially if services they demand are more labor-intensive than are services to Israelis. In any event, the local supply function is much flatter than the local demand function (in Table 1, $|-0.026| < |-1/4.31|$), so equilibria are stable in the Marshallian sense. That is, when local demand prices exceed local supply prices (to the left of an intersection point), hoteliers boost supply quantities, shifting the market in the direction of equilibrium.

Because, as discussed above, Israeli hotels normally allocate rooms to the local market only after most international demand has been satisfied, supply price PL^s to Israeli guests ought to depend not only on Israeli demand QL and on hotel labor costs, but on international demand QF and on the total number of beds QT as well. The positive and statistically significant coefficient of QF in the PL^s column of

Table 1 says that as foreign demand drops, the negatively sloped short-run supply to Israelis shifts leftward and prices to the local market fall. The shift, however, is not great: a drop in foreign demand of 100,000 bed nights per month reduces the price to Israelis by only $1.30. The situation is depicted graphically in Figure 7, which will be discussed later in this chapter. On account of the low short-run equilibrium occupancy rates, local demand intersects local supply where the latter is negatively sloped. Declines in international demand shift the local supply function downward but also raise the maximum number of rooms available to Israeli guests, in the vicinity of which local supply presumably has the normal positive slope.

The high standard error on the coefficient of QT in Table 1 suggests the long-run Israeli supply function is horizontal: the industry faces constant costs in the sense that changes in hotel capacity have no major effect on long-run marginal cost or thus price. Long-run marginal costs include capital as well as labor, material, and energy inputs.

Market Simulation

When the war or terror climate worsens, foreign demand for Israeli hotel space declines. The industry seeks to fill some of these vacancies by turning to the local market, and one way of doing so is to reduce prices to Israeli guests. Thus, terror incidents influence tourist activity and prices in both the foreign and local market sectors. It is useful to ask precisely how prices and sales volumes in each market are impacted by a terror incident, since this allows us to predict revenue changes in the two markets and to decompose such changes into price and quantity components. Such an analysis also permits us to see whether, or to what extent, the local market serves as a buffer to international demand swings. For example, because local demand is inelastic, it may not be desirable for hotels to reduce prices enough to compensate completely for the foreign demand shortfall.

To simulate a terror event's impact on equilibrium prices and quantities, we select a terror index level and set all other exogenous variables—including lagging prices—at sample means. Using trial values of lagged prices PL_{t-1}, PF_{t-1}, we compute trial values of the endogenous quantities QL_t and QF_t and prices PL_t and PF_t. We then employ the latter to update the lagging prices in the supply functions from equations (3) and (4) and recompute all four endogenous variables, proceeding in this fashion until cross-iteration changes in the endogenous variables are small. The simulation thus represents the asymptotic or equilibrium state of the endogenous prices and quantities. Our procedure for representing the impact of a terror-related event on predicted prices and quantities is to compare the price and quantity predicted in the presence of the terror event with those simulated in a terror-neutral environment (TER = 0). Differences between predicted and simulated levels are then linearly regressed against terror intensity and plotted in Figures 3–6.

As can be seen in these figures, both the foreign and local markets are affected by war and terrorist incidents. In the foreign market, the difference between the bed night demand quantity predicted at a given peace or terror intensity and that simulated in a terror-neutral state (Figure 3) falls as terror intensity rises, namely as one moves from left to right in the figure. A one-unit increase (along the –9 to +9 scale) in terror intensity reduces equilibrium foreign demand by 8,521 monthly bed nights compared to what it would be in a terror-neutral world. The predicted

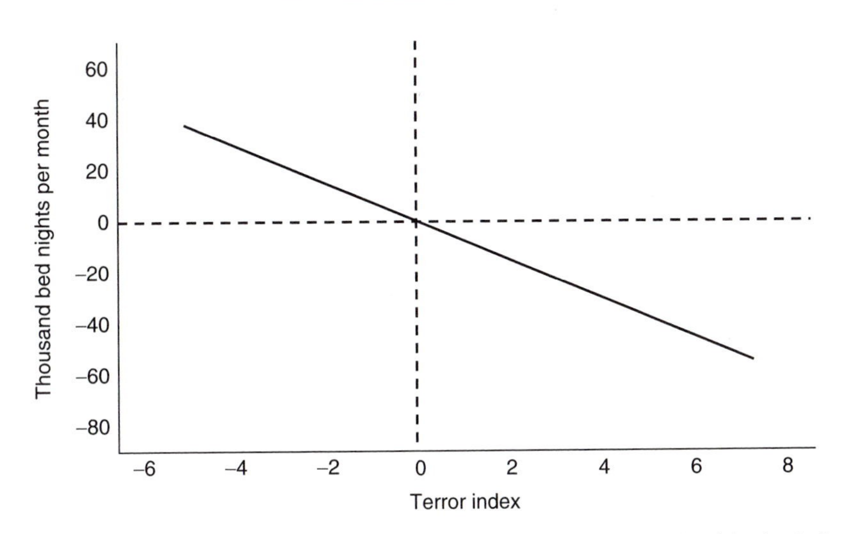

Figure 3 Difference between (i) predicted foreign bed nights and (ii) simulated foreign bed nights in a terror-neutral environment (TER = 0), as a function of terror index.

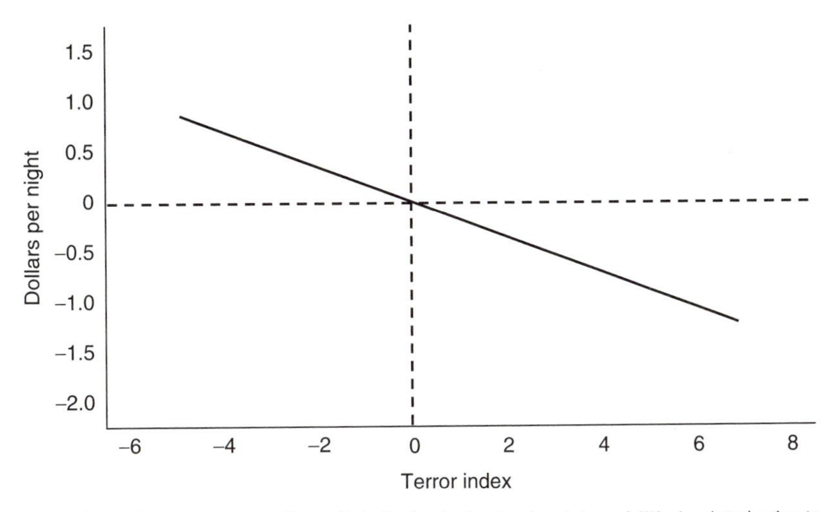

Figure 4 Difference between (i) predicted price to foreign tourists and (ii) simulated price to foreign tourists in a terror-neutral environment (TER = 0), as a function of the terror index.

bed night price similarly falls relative to the terror-neutral standard. Figure 4 shows that a two-standard-deviation (four-unit) boost in terror intensity cuts the equilibrium price to foreigners by an annual mean of \$0.68 per night. On both counts, hotels' foreign revenues decline.

Although terror attacks in our model have no direct impact on an Israeli's demand for a local hotel stay, they can reduce the hotel rates that Israelis pay, which in turn would influence Israeli hotel activity. In contrast to the demand-induced

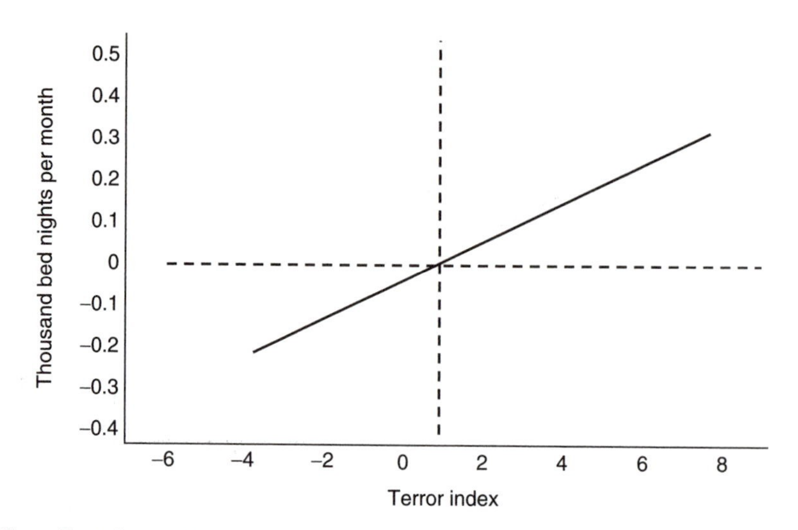

Figure 5 Difference between (i) predicted local bed nights and (ii) simulated Israeli bed nights in a terror-neutral environment (TER = 0), as a function of the terror index.

price declines in the foreign market, such effects would be supply induced, namely management's attempts to fill the rooms vacated by foreign cancellations. Figure 6 shows, however, that increases in terror severity reduce equilibrium Israeli prices by negligible amounts, reflecting the small size of parameter ($\partial PL^s/\partial QF$ = 0.013 in Table 1. Thus, although the difference between predicted and terror-neutral Israeli hotel stays does rise as terror grows (Figure 5; i.e., Israelis show a small tendency to patronize hotels more during high-terror periods), the buffer to foreign cancellations

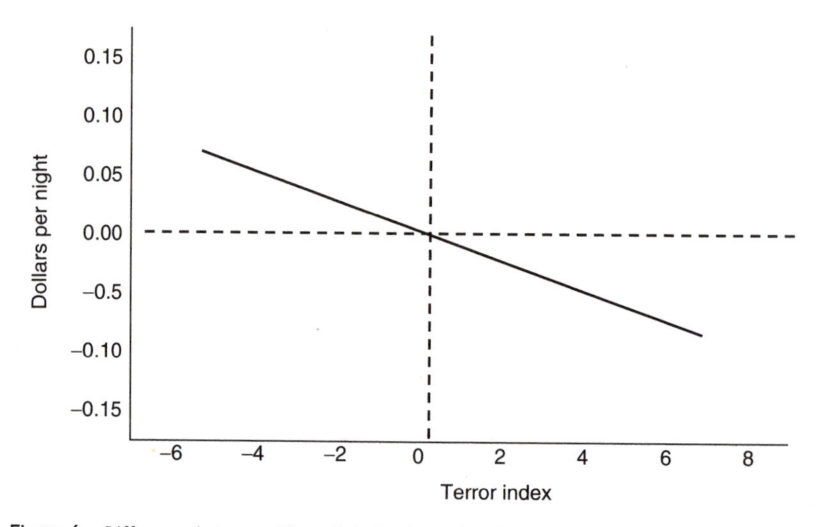

Figure 6 Difference between (i) predicted price to local tourists and (ii) simulated price to Israeli tourists in terror-neutral environment (TER = 0), as a function of the terror index.

Table 4 Annual Averages of Monthly Revenue Losses from Terror Events and Foreign and Local Markets, 1987–1999

	Foreign Market		*Local Market*		*Total*
Year	*($000)*	*%*	*($000)*	*%*	*($000)*
1992	−768.57	−0.21	−19.29	0	−787.86
1993	2,218.51	0.50	10.37	0	2,228.88
1994	−8,712.07	−1.87	−34.88	−0.01	−8,746.94
1995	−8,484.63	−1.64	−46.19	−0.01	−8,530.82
1996	−14,276.91	−2.55	−113.50	−0.02	−14,390.41
1997	−10,109.81	−1.75	−70.47	−0.01	−10,180.28
1998	−6,018.14	−1.03	−53.75	−0.01	−6,071.89
Total	−48,567.20		−342.72		−48,909.92
Average	−6,070.90		−42.84		−6,113.74

Note: Negative values refer to losses, positive values to gains.

is small. While foreign demand (Figure 3) falls 34,084 bed nights for every four-unit increase in terror, local demand (Figure 5) rises by a negligible 264 bed nights.

Why do local prices fail to decline sufficiently to fill many empty rooms in the wake of a terror incident? The same question applies in more general form to most countries' lodging industries, where excess capacity is at least seasonally evident. Table 4 shows the difference between predicted hotel revenue and that which would have been earned had the terror index been neutral (TER = 0), for each year between 1992 and 1998 and for both the foreign and local market. Only in 1993, the year a peace accord was signed with the Palestinians, was the terror index on average negative—indicating a better-than-neutral atmosphere—so only in that year did foreign revenues exceed the terror-neutral standard. Revenue shortfalls were greatest in 1996, a year of Middle East unrest.

Terror affects foreign revenue by first shifting the foreign demand function given in equation (2) leftward, reducing the number of bed nights demanded.

Table 5 Decomposition of Table 4 Losses by Percent Changes in Price and Quantity

	Foreign Market		*Local Market*	
Year	*% Change in Price*	*% Change in Quantity*	*% Change in Price*	*% Change in Quantity*
1992	−0.15	−0.04	0.00	0.00
1993	0.15	0.32	0.01	0.00
1994	−0.46	−1.50	−0.03	0.02
1995	−0.66	−0.74	−0.02	0.01
1996	−0.82	−1.44	−0.03	0.02
1997	−0.52	−1.16	−0.02	0.01
1998	−0.36	−0.69	−0.01	0.01

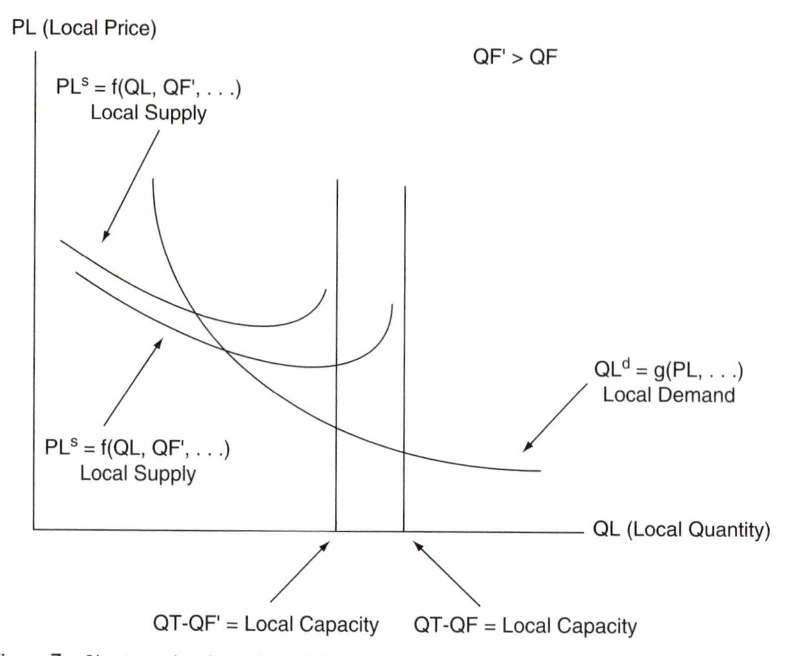

Figure 7 Short-run local supply and demand for Israeli tourism: a schematic view.

Declining demand then reduces hotels' marginal opportunity costs and thus their minimum or supply prices to foreign guests. In other words, foreign revenue falls because the foreign demand curve shifts down along the foreign supply function from equation (4). The respective contributions of quantity and price changes to foreign revenue shortfalls are shown for each year in Table 5. On average, changes in bookings (quantity changes) account for about twice as much of the revenue change as do price adjustments.

In contrast to those in the foreign market, revenue changes in the local market come about because the local supply function given in equation (3) shifts down along the local demand curve from equation (1). As the fifth column of Table 4 indicates, these changes have been rather minor; that is, price discounts to Israeli tourists in the wake of terror incidents have little effect on Israeli tourist revenues. Table 5, columns 4 and 5, make clear why. Because local market demand is inelastic (i.e., the absolute values of own-price elasticities in the local demand function are below unity in Tables 2 and 3), price discounts are proportionately no greater than the corresponding increases in bookings. Thus, hotels have little short-run incentive in the wake of a terrorist outbreak to reduce prices enough to fill many rooms.

Conclusions

To our knowledge, this has been the first attempt to formulate a supply and demand model of a nation's lodging industry. Focusing on supply as well as demand is fruitful inasmuch as it provides a view of management's price responses to demand shifts, of which war and terrorism are here given the greatest consideration. In our model, hotels face different costs in hosting foreign guests than they do in hosting local ones and can charge different rates in the two markets. Foreign demand is substantially price elastic, but unlike local demand is sensitive to regional terrorism. Thus, in the presence of violence, hotels seek to compensate for foreign demand shortfalls by reducing local prices enough to appeal to additional local tourists. Their success in doing so is constrained by the inelasticity of the local demand function. Indeed, management has had little success in countering terrorism's negative impact on hotel incomes, or on tourist revenue and foreign exchange.

Nevertheless, we find the terror effect itself to be modest. The empty hotel rooms and tourist shops popularly supposed to follow a terrorist outbreak do not well describe the Israeli experience, at least on an annual-average basis. Israeli political unrest normally is concentrated in urban centers such as Tel Aviv and Jerusalem, and foreign tourists appear able to some degree to dissociate events there from security expectations at Eilat, the Dead Sea, and other sites. We should, however, stress that the revenue impacts summarized in Table 4 and elsewhere in this chapter are annual means of monthly estimates, which may fluctuate widely in a given year. In addition, hotels are unable to eliminate many costs in the short run, so the revenue shortfalls indicated represent much larger proportionate changes in profit.

That hotels may charge different rates to foreign guests than to domestic guests suggests they may also price discriminate between the two markets, permitting them to boost revenues above the competitive norm. Strategic behavior of this sort requires inter-firm collusion. If hotels do price strategically, a price discrimination model might be a better vehicle for assessing the impacts of political unrest than is the present supply-and-demand approach, in which prices are cost-based. With that in mind, we tested a model in which foreign and local bed night prices are chosen to equate marginal revenues in the two markets. Statistical fits were poor and simulated prices and quantities diverged far from the actual values. Evidently, the collusion necessary for discriminatory pricing is not present in Israel, and a competitive model is a more accurate basis for the analysis of hotel behavior.

Note

[1] This chapter was previously published as: Fleischer, A., and Buccola, S. (2002). War, terror and tourism market in Israel. *Applied Economics,* 34(11), 1335–1343.

Review Questions

1. Why is it important to look at both demand and supply in the tourism market?
2. Why is it important to distinguish between local and foreign tourism markets when analyzing the impact of terrorist incidents?
3. How do hotels react when faced with a terrorist-based demand depression?
4. Why do hotels prefer, in a terrorist environment, to endure unused capacity rather than reduce prices?
5. Why is it important for hotels to estimate elasticities of demand?

Concept Definitions

Tourism demand and supply A tourism market exists when tourists and sellers of tourist services interact with one another. In order for supply and demand tools to be useful for analyzing such markets, each tourist and seller must be small relative to total market activity. Since tourists are a group distinct from sellers, we can analyze separately how the two groups behave. The interaction between the two determines market equilibrium price and quantity.

Price elasticity of tourism demand Demand elasticity is a measure of tourists' responsiveness to changes in the prices of tourist goods and services. It is reflected in the percentage change in a consumer's response in their buying decisions to a change in price.

References

Bar-On, R. R. (1996). Measuring the effects on tourism of violence and of promotion following violent acts, in A. Pizam and Y. Mansfield (eds.), *Tourism, Crime, and International Security Issues*. New York: John Wiley & Sons, pp. 159–171.

Borooah, V. K. (1999). The supply of hotel rooms in Queensland, Australia. *Annals of Tourism Research,* 26, 985–1003.

Central Bureau of Statistics. (1992). *Tourism and Hotel Services Statistics Quarterly*. Jerusalem: Government of Israel, quarterly issues 1992–1999.

Eviews. (1994). *User's Guide*. Irvine, CA: Quantitative Micro Software.

Horwath Consulting. (1994). *United Kingdom Hotel Industry 1994*. London: Horwath International.

Krakover, S. (2000). Estimating the effects of atrocious events on the flow of tourism in Israel, in G. Ashworth and R. Hartmann (eds.), *Tourism, War, and the Commemoration of Atrocity*. New York: Cognizant Communication Corp.

Lim, C. (1997). Review of international tourism demand models. *Annals of Tourism Research,* 24(4), 835–849.

O'Hagan, J. W., and Harrison, M. J. (1984). Market shares of US tourist expenditure in Europe: An econometric analysis. *Applied Economics,* 16, 919–931.

Sinclair, M. T. (1998). Tourism and economic development: A survey. *The Journal of Development Studies,* 34, 1–51.

Syriopoulos, T. C., and Sinclair, M. T. (1993). An econometric study of tourism demand: The AIDS model of US and European tourism in Mediterranean countries. *Applied Economics,* 25, 1541–1552.

Wheaton, W. C., and Rossoff, L. (1998). The cyclic behavior of the U.S. lodging industry. *Real Estate Economics,* 26, 67–82.

World Bank. (2000). *World Development Indicators 2000*. Washington, DC: International Bank for Reconstruction and Development.

World Tourism Organization. (2000). *Tourism Highlights 2000*. Madrid: World Tourism Organization.

Appendix

The terror magnitude index employed in this study, based largely on Krakover (2000), is as follows:

–5 Peace treaty signed with the Palestinians
–3 Peace treaty signed with Jordan
0 No terror event
1 Up to 10 persons injured outside main cities
2 From 10 to 19 injured and/or one killed
3 From 20 to 29 injured and/or 1 to 5 killed
4 From 30 to 59 injured and/or 6 to 15 killed
5 From 60 to 80 injured and/or 16 to 40 killed
6 Limited warfare across the border
7 Two or more terror events of security levels 3 to 5
8 A full-scale war between Israel and its neighbors
9 A war of regional or global concern

Data on the index were taken from the International Terrorism web site (http://www.ict.org.il). The use of –3 and –5 values for the two peace treaties is based on a judgment of the impact these treaties had on international press coverage of Israel. In a sensitivity analysis, equations (1) through (4) were re-estimated using –1 and –2 rather than –3 and –5, respectively. These substitutions had a negligible effect on regression estimates.

Index values 1 through 5 include injuries or deaths of private citizens only. Soldier casualties are excluded.

4

Fiji Islands: Rebuilding Tourism in an Insecure World

Brian King and Tracy Berno

Learning Objectives

- To understand both the short-term and long-term impacts of a particular type of civil disturbance (a military coup) on tourism in a small island destination.
- To understand the process of creating a tourism action group as a way of bringing relevant parties together to overcome adversity.
- To understand how military disturbances and coups d'état in particular impact tourism flows and the tourism sector as a whole.
- To understand the importance of crisis management strategies to destination management, and the need to adapt such strategies based on the experience of previous crises (e.g., the 1987 military coup).
- To have an improved understanding of the growing role of global and regional media in alerting overseas markets to problems experienced at a destination and the impacts of the problems.
- To understand the key element of a crisis recovery strategy (CRS).
- To be able to evaluate the relationship between destination image and positioning to changes (both negative and positive) in the reputation of a destination.

Introduction[1]

This chapter examines the similarities and differences between the military coups of 1987 and 2000 and aims to draw out some issues that are relevant to the handling of political and related security crises affecting tourism. It demonstrates the role of effective communication in overcoming problems that arise in small island nations as a result of natural or man-made crises. In light of the terrorist attacks in New York and Washington on September 11, 2001, understanding the impacts of security threats to tourism has gained added potency (Berno and King, 2002).

Tourism in Fiji

Situated in the southwest Pacific, Fiji is an archipelago of over 300 islands, one third of which are inhabited (TCSP, nd). The 1996 census recorded a total population of 772,655 with indigenous Fijians accounting for 51% of the population, Indo-Fijians 44%, and other ethnic backgrounds 5%. Most of the population (75%) lives on the main island of Viti Levu with the second largest island, Vanua Levu, accounting for a further 18%. The remaining 7% are spread across approximately 100 outer islands. According to the Ministry of Information, just under 40% are urban dwellers, concentrated mainly in the Suva-Nausori area, Labasa, Lautoka, Nadi, and Ba (Ministry of Information, 1999).

In terms of its tourism product, Fiji offers all of the attributes associated with a tropical island destination—sun, sea, sand, and surf—along with a reputation for friendly service and a welcoming resident population (Ministry of Tourism & Transport et al., 1998). Traditionally, Fiji has traded on this idyllic tourist image. Increased competition in main source markets (New Zealand and Australia in particular) from destinations with similar attributes (e.g., Queensland, Australia, beach resorts in Southeast Asia, and other South Pacific destinations) has forced Fiji to reassess its tourist product. This has been reflected in recent efforts to enhance the tourism image of the destination by placing more emphasis on Fiji's cultural and environmental richness and diversity. Two examples include a recent Fiji Visitors' Bureau (FVB) campaign "Discover the Fiji You Don't Know" followed later by "Discover the Fiji You'll Never Forget." These developments were further reinforced in 1998 with the release of *Ecotourism and Village-Based Tourism: A Policy and Strategy for Fiji* (Harrison, 1999), and by the inclusion of a section on ecotourism in the tourism sector development plan, which comprised a component of the National Development Plan (Berno and King, 2001).

Traditionally, Fiji's tourism industry has been predicated on the dominant short-haul markets of Australia and New Zealand. The visitor profile has, however, changed since 1987 with visitation from Australia declining as a proportion of the total. The positive element of this change has been that Fiji's range of source markets is now highly diversified. Short-haul travelers do, however, remain extremely important because of the longer average length of stay of visitors from Australia and New Zealand. This results in a disproportionate influence for Australia in responding to political crises in Fiji (e.g., the placing of sanctions). Fiji has no substantive domestic market and relies almost exclusively on visitors who arrive by air (McVey and King, 1999).

Sugar and tourism are the mainstays of the Fijian economy, though other industries such as manufacturing, forestry, fisheries, and mining also make significant contributions. Since 1989, tourism has consistently surpassed sugar as the primary source of foreign exchange. In 1998, tourism was directly and indirectly responsible for the employment of 45,000 people. The sector generated foreign exchange earnings of $568.2 million and contributed, directly and indirectly to 27% of GDP, up from a reported 20% in 1998. In 1999 the record number of visitor arrivals (409,955) generated over $600 million. The outlook for the year 2000 was promising. Until the events of May 19, 2000, the Fiji Visitors Bureau arrival target of 412,300 appeared quite realistic. These figures demonstrate the potential vulnerability of Fiji's tourism economy to political instability.

Figure 1 shows the long-term arrivals trend and highlights the severity of both coups on overall arrivals to Fiji. Table 1 indicates how arrivals for 2000 fell far short of the target in almost all markets—total arrivals were 28% down on the comparable figure for 1999. Double digit falls were recorded in all source markets with the exception of Korea, which surprisingly grew by 127%, albeit from a low base. It was not until June 2001, a year after the coup, that the monthly visitor arrivals exceeded those of the previous year.

The Events of May 19, 2000

Superficially, there were many similarities between the events of 1987 and 2000. As had been the case in 1987, the effect of the coup on the tourism industry was swift and far reaching. The circumstances were, however, quite different and tourism had changed considerably during the intervening period, prompting the need for different recovery tactics.

The swift conclusion to the so-called bloodless coups of 1987 allowed the tourism industry to focus on its recovery strategy. For example, the first of the two coups in 1987 was staged in May, but the annual Fiji Tourism Convention went ahead as scheduled during the following month. This gathering of local tourism industry leaders was used as a forum to develop plans for industry recovery. During the Tourism

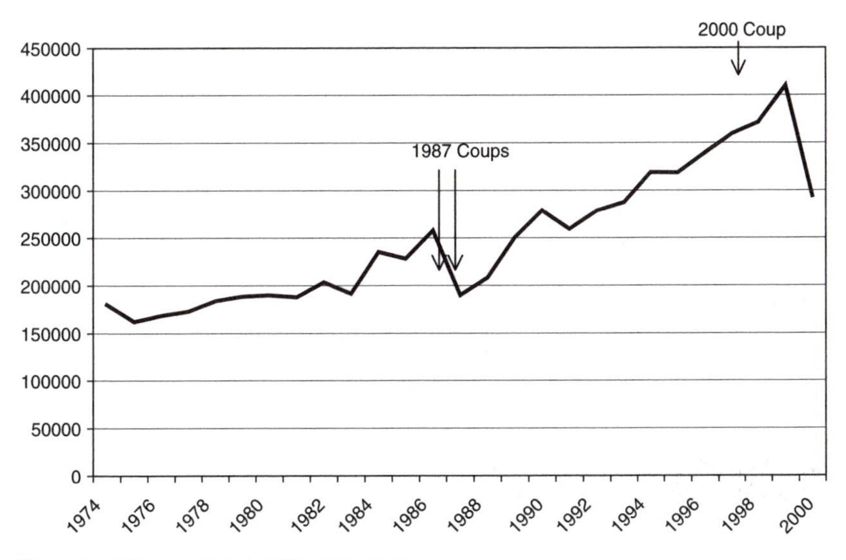

Figure 1 Visitor arrivals in Fiji, 1974–2000.

Table 1 Visitor Arrivals by Source, 1996–2000

Country	*1996*	*1997*	*1998*	*1999*	*2000*	*% change 1999/2000*
Australia	79,534	80,351	100,756	118,272	76,883	–33%
New Zealand	63,430	68,116	70,840	72,156	49,470	–31%
USA	38,707	43,376	48,390	62,131	52,534	–15%
Canada	11,431	13,359	12,837	13,552	10,532	–22%
UK	28,907	35,019	39,341	40,316	29,215	–28%
Europe	31,875	32,806	29,334	28,371	22,506	–21%
Japan	44,598	44,783	35,833	37,930	19,674	–48%
Korea	14,770	12,181	1,613	1,489	3,386	127%
Asia	6,334	6,345	7,708	7,797	6,750	–13%
Pacific	18,545	20,381	22,850	26,090	21,534	–17%
Others	1,429	1,724	1,840	1,851	1,586	–14%
Total	339,560	359,441	371,342	409,955	294,070	–28%

Source: Fiji visitors Bureau, 2001

Convention, an agreement was reached to set up a crisis management team within the National Tourism Office, to be called the Tourism Action Group (TAG). TAG had a single objective—to bring tourists back to Fiji as fast and effectively as possible (TAG, 2000a). To accomplish this, four immediate goals were established:

- Removal of travel advisories and union bans in Australia and New Zealand;
- A doubling of the National Tourism Organization marketing budget;
- Familiarization visits from the main markets for trade representatives; and,
- Marketing of special airfares and packages to the markets noted above.

Initially established for a six-month period, TAG was disbanded after 11 weeks, following evidence that the recovery was under way. A subsequent coup in September 1987 (after TAG had been disbanded) brought the recovery of the tourism industry to a sudden halt, but the TAG strategy was still relevant to a large extent and the industry experienced a significant recovery during 1988 (Lea, 1996).

Like many other tropical island nations, Fiji has long established procedures in place to deal with emergencies such as cyclones. This meant that the overall preparedness in 1987 was relatively good with provisions to contact hotels across the country to give advice on how to react. The experience of 1987 showed that a crisis management plan was a necessary contingency in the event of political unrest as well as for possible natural disasters. Consistent with this requirement, the crisis management group formed an integral part of both the marketing objectives and the marketing strategies of the Year 2000 Fiji Visitors Bureau Marketing Plan Summary (FVB, 1999).

The 2000 Recovery Plan

The 2000 coup had an immediate and significant impact on Fiji's economy and tourism industry. Many tourists to Fiji cut short their holidays and many prospective tourists made cancellations. The peak tourism months in Fiji are June through August. These three months include a school holiday period and many in the Southern Hemisphere take winter sun holidays at this time. The coup resulted in a decline of visitor arrivals in that period of almost 70%. The loss of tourism revenue had reached an estimated F$84.6 million by the beginning of August (excluding earnings by the national carrier Air Pacific). As a result, the hotel sector experienced a 44% reduction in employment. Major tourism operators such as Shotover Jet and the Sheraton Royal closed down as a result of low numbers. The tourism industry had gone into a free fall, and swift and effective action was essential.

The prior experiences of 1987 prompted the tourism industry to mobilize quickly. A meeting of key members of the tourism industry was convened jointly by FVB and Air Pacific during the week after the 2000 coup. At that meeting, several resolutions were adopted as part of a crisis management plan. The key issues were identified as:

- Communication with overseas source markets, overseas diplomatic missions, and the media;
- Operational issues such as how to deal with cancellations and travel insurance issues;
- Monitoring of key relationships for resorts and operations (e.g., landowners, unions, and staff); and
- Developing a long-term strategy to address tourism industry issues.

Consistent with the cooperative and cohesive response to the events of 1987, TAG 2000 was formed. Headed by the Managing Director of a tour operating company and including representatives from FVB, Air Pacific, Air New Zealand, the Fiji Hotel Association, and the Society of Fiji Travel Associates, TAG 2000 met for the first time in early June. The group implemented a series of initiatives intended to expedite a recovery. According to the TAG the objectives were to:

- Develop an International Plan that was fully coordinated;
- Establish that Fiji is safe and desirable;
- Finance the promotion of Fiji with cooperative support from the private sector and government;
- Seek the involvement of wholesalers and retailers in all source markets;
- Implement multiphase marketing strategies to achieve both short-term and long-term objectives; and
- Reinstate lost employment in the industry.

The following activities were scheduled:

- A public relations and media management consulting firm was contracted to assist in the management of the recovery program;
- Funding was solicited from private and public sector sources to support a short- and medium-term advertising and promotion campaign in all major visitor markets;
- Visits were made to consular representatives to encourage the lifting of travel advisories;
- Travel agent/wholesaler visits were arranged;
- A comprehensive media campaign was planned; and
- Special recovery fare packages were prepared.

TAG secured F$5 million in funding (FJ$3.65 million from public sector contributions and FJ$1.35 million from local industry) to support the campaign (FVB, 2000a; TAG, 2000b). With the funding and strategies in place, the recovery plan was ready to be operationalized as a two-phase campaign. Phase One was to be an intensive tactical promotional campaign in all key markets, followed by Phase Two, a follow-on coordinated, long-term advertising support program.

It was hoped that these strategies would be implemented immediately. The 2000 coup, however, proved to be very different from its predecessors because of the lengthy internment of the hostages (56 days) and the endemic political unrest and uncertainty. The chronology of events that followed the 2000 coup is outlined in Figure 2. Unlike the situation in 1987, the 2000 coup was not bloodless and an estimated 10 people died as a result of coup-related events. Graphic media images of unrest and violence were projected to millions around the globe. The situation was compounded by a pattern of political unrest that ensued in the Solomon Islands and Papua New Guinea. The incidents in these two Melanesian countries conveyed an impression of turmoil across the region, which further reinforced the negative images of Fiji and provided ongoing copy for the media. The perception of instability across the subregion is symptomatic of how one piece of bad news may be quickly picked up by the media and can become associated with often unrelated problems being experienced elsewhere. This was unusual because the media normally shows little interest in South Pacific issues. Despite the success of the recovery program in 1987 and lessons learned, it soon became clear that 2000 needed a different strategy.

In 2000, the timing of the implementation of the full recovery plan became a major preoccupation. Marketing could not bring about recovery in the tourism industry until the political situation was stable enough to warrant the lifting (or at least the downgrading) of the travel advisories issued by foreign governments. This scenario was unlikely until two key events occurred: (1) the release of the hostages, and (2) the return of illegally held arms to the Army. Some foreign governments warned that any preemptive launch of a full recovery campaign before the lifting of the relevant travel advisories would force them to react, and to actively discourage their citizens traveling to Fiji. In anticipation of such consequences, TAG determined that the recovery campaign would be delayed until the achievement of certain preconditions. This was defined as "the time that the hostages are released safely and that all the weapons are returned to the Army" (TAG, 2000a). Few anticipated how long it would take before the necessary preconditions were satisfied.

During the wait for preconditions to be satisfied TAG undertook a range of prelaunch activities. Regular contact was maintained with diplomatic representatives within Fiji regarding the status of travel advisories and a series of domestic public relations activities was undertaken. These public relations activities were seen as being essential in preparing Fiji for the recovery program and protecting Fiji's tourism product. One element of the campaign was the "Spirit of Fiji" (*Sa ka ga ni yalo*; FVB, 2000b). The Spirit of Fiji campaign was created in part as a response to the takeover by landowners of some key resorts in July. The campaign was aimed at the local indigenous community, and sought their help, assurance, and hard work to rebuild the idea of the "friendly Fiji islands." Advertisements appeared on television with employees and landowners featured prominently because they were considered to play a significant role in the hospitality industry. The advertisements highlighted the hardships being confronted by the industry, and communicated the human effects of the tourism downturn on employees who had been made redundant or were forced to reduce their working hours. The campaign was also taken out to villages through public meetings, school meetings, media releases, and visits to groups such as farmers, schools, landowners, elders, and tour operators. According to Silaitoga, another aim of the campaign was to encourage people to write to overseas diplomatic representatives to urge them to lift the travel advisories (Silaitoga, 2000). The final pre-pre-launch activity also involved a program of public relations overseas, however, and many events stood in the way of the recovery program.

The prolonged internment of the hostages was not the only issue with which the tourism industry had to contend. Land ownership issues were exacerbated by events associated with the coup. In early July, landowners from the Monasavu area seized control of the Wailowa power station and cut Fiji's electricity supply. It was to be nearly 8 weeks until power was fully restored and this impacted tourism operations outside the capital as well as in Suva.

The tourism industry was also affected by land ownership disputes. The coup leaders engaged in inflammatory rhetoric about indigenous land rights, which prompted disputes over resort-based tourism. In total, eight resorts were occupied by hostile landowners, the most notable being Turtle Island Resort which was taken over during "children's week." The resort had recently won the British Airways' *Tourism for Tomorrow* Award for the Pacific Region, thereby ensuring high-profile media coverage and a strong word-of-mouth effect. Initially the resort was occupied by Naisisili villagers, and then the owner and manager was taken hostage. The images broadcast internationally of families being evacuated from the island further damaged the tarnished image of tourism in Fiji and introduced the specter of personal threats to tourist security. The status of Turtle Island as a luxury property attracting the rich and famous meant that news of its difficulties caused the concern among travel distributors that no one anywhere in Fiji was entirely immune to the instability.

The Role of Media and Communications

Media coverage played an important role in both 1987 and 2000, but exerted a greater influence in the latter case. As has been widely observed, tourism thrives on positive publicity and the communication of negative and threatening images has been particularly vexing for a tourism industry that is more comfortable with images of happy, smiling faces. Shortly before the 2000 coup, the Fiji Visitors Bureau (FVB) had launched its new branding campaign, "Fiji—the world's one truly relaxing destination." Complementing this preferred positioning, the South Pacific Tourism Organization's branding of Fiji stated that "Hospitality is a way of life for the Fijian people and as a visitor you will be genuinely welcomed." This slogan was accompanied by photographs of smiling indigenous Fijians and such copy as, "[Fiji is] home to people who have been called 'the friendliest on earth.'" The images projected around the world of balaclava-clad gunmen and the looting and burning of Suva were in stark contrast to the touristic images of a peaceful, friendly, and above all relaxing South Pacific holiday destination. At the very least they undermined the credibility of the images being projected by the industry. The brand was very tarnished. Ironically the pre-1987 branding of Fiji was "Fiji—the way the world ought to be." The 2000 branding was in some ways a return to this theme, and had been prompted by the sense that Fiji had now moved beyond the political uncertainty and division of 1987. This renewed confidence has proved to be short-lived.

Negative media images had also been broadcast in 1987, but communication developments ensured a larger international audience and consequently greater market penetration and currency. In particular, the Internet provided up-to-the-minute information and commentary about the happenings in Fiji to a worldwide audience. Web sites such as *Fijilive!* and *Fijivillage* became much consulted sources of information within Fiji and overseas. *Fijivillage,* for example, received 4,373,893 hits in July 2000. Of those, 83% originated in the major source markets of visitors to Fiji, namely the United States (64%), Australia (13%), and New Zealand (6%) Fiji Village Statistics, 2000). In an on-line article, ABC News even referred to the political events as a Coup D'Net. "Last night, the world watched a coup attempt—on the Internet. *Fijilive.com* . . . posted the dramatic details of the coup attempt as they happened . . . even the major news services got much of their information from the site—including this one" (Mazza, 2000).

The Internet projected uncensored news in an uncontrolled fashion to a potential audience of millions. Supplementing this coverage, the various chat rooms and discussion sites ensured that the events in Fiji were deconstructed as well as reported. According to Narayan, attempts at gagging some Internet-based media during the coup were unsuccessful, as the blocked sites simply moved to a sympathetic mirror site (Narayan, 2000). Such coverage greatly impeded the ability of the government and the tourism authorities to control the message or to ensure selective coverage.

The recovery program has had to compete with a constant media barrage of negative image and copy. The easy accessibility of archival information on the Internet has also helped keep the situation in Fiji alive in the public's mind long after it might have faded. The ubiquity of the Internet also prevented tourism operators from identifying potential market sources that may have been unaware of the relevant political events. One notable resort on the island of Taveuni focused its post-1987 coup promotion on the US market on the basis that few agents and even fewer potential consumers in that market had any consciousness of the political events that had occurred. The assumption appeared to be correct since the resort returned rapidly to high occupancies. Such opportunities were not available in 2000.

In 2000, the experiences of the Korean and Japanese markets appeared to be contradictory. Korean Airlines recommenced services to Nadi on August 6, bringing in 200 passengers and 18 "key tourism related officers." These led to a steady recovery and to numbers that outstripped the previous year. In contrast, Air Pacific recommenced flights to Narita in Japan on August 10, but this action failed to prompt a recovery. The price sensitivity of the Korean market may have played a part in the different responses. It may have been that the discounted prices on offer were a major incentive to attract Korean operators. The responses may also indicate that the affected destination country (in this case Fiji through its national carrier Air Pacific) is less able to make a quick impact than an initiative launched in the source country (i.e., in Korea through the national carrier Korean Airlines).

Stalling of the Recovery

Fiji remained under a state of emergency and many regions were still under curfew. The lifting of travel advisories, which TAG had initially anticipated, did not occur. The benchmark for the launch of the recovery program was shifted to the time at which the travel advisories were downgraded. In a sense this made tourism a captive to external events and actions. The hostage-taking of two Air Fiji pilots (both New Zealand nationals) on Vanua Levu on July 27 increased the determination of governments in source markets to persist with their travel advisories. While the pilots were only interned for two days and were unharmed, the incident stiffened the resolve of the New Zealand government concerning its travel advisories. It was not until late August that the situation in Fiji was sufficiently settled for New Zealand and Australia to follow the actions of the United States, the United Kingdom, and Japan in downgrading their travel advisories. Although the organizers of the Spirit of Fiji campaign attributed their initiative as having played "an instrumental role in the lifting of travel advisories against Fiji," it is more likely that the more settled prevailing political situation resulted in the change.

The tourism industry looked forward to the start of the much-awaited recovery with the downgrading of the advisories. As originally planned, TAG immediately launched Phase One of the recovery with a range of price-driven recovery-marketing initiatives in Fiji's key markets. The implementation of Phase One, with its special recovery airfare and accommodation packaging and comprehensive media and public relations campaign, was followed by Phase Two in October. Phase Two focused on revised airfare and land content packages, continued public relations efforts, and market-specific cooperative campaigns.

The campaign achieved some success. Visitor arrivals for September and October exceeded the post coup forecasts, showing a slow but steady recovery, particularly in the New Zealand market. Australian numbers also increased marginally, but the Sydney Olympics, staged in September, had a diversionary effect on potential travel to Fiji. The TAG media visitation program was also deemed a success after positive international print and media exposure for Fiji was realized. Despite these optimistic signs of recovery, the challenges of the 2000 coup were not yet over.

Shortly after the launch of Phase One of the recovery program, a device was found outside a popular hotel frequented by foreign travelers in Suva. Although the device was safely disarmed, the wide reporting of the incident further reinforced the international image of Fiji as a volatile place. Then, after 2 months of relative calm in the country, an attempted rebel mutiny occurred at the headquarters of the Fiji military in Suva leading to a number of casualties. Twenty of the armed rebels escaped prompting the imposition of a 36-hour curfew in the greater Suva region. Most of the rebels were quickly apprehended. TAG immediately went into damage control, lobbying foreign embassies to downgrade travel advisories and disseminating the message that the upgraded advisories applied only to the greater Suva region. They made it clear that the rest of the country remained safe and that tourism establishments such as hotels, resorts, and attractions were unaffected by the uprising. Such arguments are difficult to sustain in tourism because the security consciousness of visitors makes them reluctant to accept arguments that attempt to confine security problems to particular localities, especially where normally tranquil resorts become targets (as was mentioned previously, Turtle Island was one of these).

The swift and decisive response by the military along with TAG's quick action resulted in the mineralization of the impact on the recovering tourism industry. Few cancellations were received from either New Zealand or Australia and resorts around the tourism centers reported good occupancies. All airlines in and out of Nadi and all tourism services continued to operate as normal.

Ongoing political unrest was not the only threat to TAG's recovery efforts. Phase One of the recovery campaign itself came under fire through an objection lodged with the Advertising Complaints Standards Board in New Zealand, which highlighted fundamental contradictions between the views of the tourism industry and those wishing to prolong the sense of political crisis. The recovery campaign, which showed an idyllic Fiji beach scene with the caption "Fiji Before and Fiji After . . . The only thing that's changed is the price," drew complaints from the Coalition for Democracy in Fiji (Pacific Media Watch, 2000). The latter claimed that the caption did not differentiate between Fiji's tourism product and the country itself. It was argued that this left room for the viewer to infer that the coup had inflicted no lasting effect on Fiji. The campaign was subsequently amended to read "Fiji holidays before and Fiji holidays after . . . The only thing that's changed is the price." The impact of the message was clearly reduced as a result of the qualification that had been added to the statement. Despite widespread media coverage for the incident in both New Zealand and Fiji, no detectable decrease in tourist numbers was evident. This confirms the view that tourists seeking relaxation take little notice of political matters at the destination, provided that personal security is assured.

The Recovery and Future Prospects

In spite of the setbacks, the tangible results of TAG's efforts suggest that their strategy for dealing with tourism in the face of political crises was successful. Average daily arrivals based on visitor permits at Nadi airport (the main gateway airport for Fiji, and after the coup, the only airport processing international arrivals) increased significantly after the launch of the campaign. As indicated in Tables 2 and 3, arrivals bottomed out at around 12,000 per month during the so-called peak season (June through August). From October to December numbers were approximately double this figure. Total airline seat sales out of New Zealand and Australia for Phase One ($499 and $599 fares sold between August 27 and September 30 for travel between August 27 and December 15) were New Zealand 14,900 seats and Australia 13,000 seats, a total of 27,900. The forecast foreign exchange revenue to government from tourism in 2000 is FJ$373.3 million. This represents a decrease of 45% compared to the 1999 earnings. It nevertheless represents a substantial return, given the circumstances.

Target Visitor Numbers as Presented in the FVB Year 2000 Marketing Plan Summary, (December 1999)

By October, many resorts were reporting occupancies in excess of 50%, with bookings improving rapidly. The public relations campaign saw over 100 media personnel, from over 50 media organizations worldwide, visiting Fiji through the TAG and FVB Media Visitation Program. Over 1,000 travel agents were projected will to have visited Fiji by March 2001. Industry employment was less resilient.

Table 2 Visitor Arrivals by Month, 1998–2000

Month	*1998*	*1999*	*2001 (Target)*	*2000 Actual*	*% change 1999/2000*
January	26,687	28,950	26,965	30,321	4.7%
February	22,556	25,263	24,253	30,058	19%
March	28,319	31,589	31,788	34,840	10.3%
April	30,570	29,082	29,839	38,069	30.9%
May	29,289	34,203	32,955	29,352	–14.2%
June	34,602	38,445	36,121	12,066	–68.6%
July	36,471	41,031	38,977	12,804	–68.8%
August	37,821	40,680	40,160	12,265	–69.9%
September	32,479	36,806	35,631	19,867	–46%
October	32,198	36,800	34,080	24,275	–34%
November	29,626	35,180	31,150	25,724	–17%
December	30,724	31,926	32,283	24,429	–24%
TOTAL	371,342	409,955	394,202	294,070	–29.1%

Source: Fiji Visitors Bureau, 2001.

Many hotel employees were still working reduced hours or were on leave until the situation improved. Although the situation was slowly improving, hotels lost more than FJ$100 million in the peirod since May 19, 2000.

Despite some grounds for optimism, several key concerns remained. As a result of the downturn in visitor numbers, airlines withdrew services or cut back on flight frequencies from market sources. Elections were finally held in mid-2001 and a Fijian-dominated government was installed, prompting the TAG to disband. The election of a legitimate government provided some stability and prompted the Australian government to remove sanctions. There have, however, been increasing reports of robberies leading to the need to reassure visitors about security.

At meetings in December 2000 and January 2001, TAG proposed that it should continue as a group for an additional 6 months under the auspices of FVB in a scaled-down format. After that, a full TAG committee would be reactivated if circumstances so dictated. What finally prompted the reconvening of the TAG was a fairly unrelated incident, albeit one that involved a security breakdown. In the face of declining tourism as a result of the New York and Washington terrorist attacks, TAG reemerged, an indication that Fiji's tourism industry was once again in crisis management mode.

In a presentation given to the Fiji Tourism Forum in December 2000, TAG summarized their vision of the way forward as follows:

- On May 19, 2000, the tourism tap was abruptly turned off;
- Tourism arrivals declined to a trickle;
- Fiji's economy suffered and so did Fiji's tourism industry;
- We are now in recovery mode;
- The flow of visitors is starting to increase;
- Fiji tourism needs to be united in our efforts to recover our lost status;
- We need to continue to deliver the promised service to our visitors at all levels (hotels, shops, tours, transportation, etc.); and
- Reinstatement of lost employment should remain a priority.

Table 3 Visitors by Source and Month of Arrival, 1999 and 2000

Country of Residence		Month of Arrival					
		Jul	*Aug*	*Sept*	*Oct*	*Nov*	*Dec*
Australia	1999	12,504	12,292	12,621	9,680	8,828	10,047
	2000	2,217	1,524	5,550	6,203	7,650	8,201
	% Change	−82.3	−87.6	−56.0	−35.9	−13.3	−18.4
New Zealand	1999	8,949	7,397	8,274	7,417	4,258	3,734
	2000	2,202	1,500	5,946	7,061	5,473	3,343
	% Change	−75.4	−79.7	−28.1	−4.8	28.5	−10.5
United States	1999	6,529	5,224	3,766	6,047	6,691	5,852
	2000	3,862	3,282	2,615	3,184	3,358	4,248
	% Change	−40.8	−37.2	−30.5	−47.3	−49.8	−27.4
Canada	1999	1,227	1,284	696	1,155	1,613	1,219
	2000	450	431	273	684	783	1,012
	% Change	−63.3	−66.4	−60.8	−40.8	−51.5	−16.7
Japan	1999	3,187	4,858	4,113	3,126	2,820	2,519
	2000	255	448	901	1,020	936	1,209
	% Change	−92.0	−90.8	−78.1	−67.4	−66.8	−52.0
Korea	1999	147	139	90	62	120	131
	2000	125	533	448	535	690	479
	% Change	−15.0	283.4	397.8	762.9	475.0	265.6
Continental Europe	1999	3,293	3,413	2,355	3,618	4,396	2,864
	2000	999	881	1,056	1,469	1,975	1,962
	% Change	−69.6	−74.1	−55.2	−59.4	− 55.0	−31.5
Pacific Islands	1999	2,244	2,967	1,793	2,651	3,088	2,643
	2000	1,033	1,127	825	1,534	2,218	1,543
	% Change	−54.0	−62.0	−54.0	−42.1	−28.2	−41.6
Others	1999	2,119	2,267	2,034	2,229	2,619	2,071
	2000	1,186	1,768	1,451	1,903	2,056	1,939
	% Change	−44.0	−22.0	−28.7	− 14.6	−21.5	−6.4
Total all visitors	1999	691	723	968	607	571	667
	2000	388	719	717	552	448	797
	% Change	−43.8	−0.6	−25.9	−9.1	− 21.5	19.4

It is critically important that all stakeholders in this industry from government, landowners, hoteliers, etc. understand that any serious destabilizing event attracting international attention will turn the tourist tap off again (TAG, 2000b).

Although the events of 2000 remain partially unresolved at the time of writing there are a number of lessons to be learned from the approach adopted by TAG, given the apparent effectiveness of its work in both 1987 and 2000.

Fiji Tourism in an Insecure World

It has been suggested by a Fiji academic that the country needs to consider what should be done to deal with the *next* coup, not make plans *in case* there would be another coup. One of the key lessons learned from the experience of the tourism industry is much the same: the need to be prepared. Most tourism plans, particularly those in tropical destinations, contain contingencies for natural disasters such as cyclones. They also need to consider other types of crises, such as political unrest. A key factor that facilitated the recovery program in 2000 was the speed at which the tourism industry was able to respond to the crisis. Within a week of May 19, key industry players had met, leading to the rapid formation of TAG and the formulation and operationalization of an initial strategy (TAG, 2000c). As a result, the industry was relatively well prepared to respond quickly to the mounting crisis. Beirman has highlighted this as an example of good practice, contrasting it with the cancellation of key trade events in Australia in the immediate aftermath of the terrorist attacks on September 11, 2001 (Beirman, 2001).

A coordinated and collective approach to the recovery process (i.e., one voice, one message) was essential. Coordinating the recovery effort through TAG, under the auspices of FVB, meant that a singular message was disseminated and that consistent responses were made toward situations. This was particularly important in dealing with the media and for downgrading the travel advisories. Although there was some internal debate about the size of the financial contributions demanded by the TAG and the promotion of a single product branding campaign, the strategy has ultimately been successful in attracting visitors back to Fiji. The ability of Fiji to reconvene the TAG at short notice following the terrorist attacks in the United States was a major competitive advantage.

The fact that funding was expected to involve a partnership between the public and private sectors is consistent with the "user pays" philosophy, which has become increasingly characteristic in developed countries in the 1990s. This difference between 1987 and 2000, then, is the result of the changing macroenvironment. Important lessons had been learned in 1987 about dealing with the media. As Scott (1988, p. 69) stated, "While the media clearly have a duty to inform, it was found that handling information with candor and integrity [could] in fact ensure that media coverage [could] be harnessed to aid the recovery process." Although TAG and FVB had solid strategies in place for media liaison and public relations, changes in information technology since 1987 meant that this was a much more difficult task in 2000. The uncontrollable nature of the Internet in particular posed significant challenges to Fiji's international image, and hence, to the recovery program.

As discussed above, land ownership also arose as a significant issue in 2000. Customary land ownership is particularly problematic in that land has traditionally been treated as a commodity not capable of alienation for individual ownership, but rather as an asset held in the communal realm. Private capital and freehold land ownership in tourism in Fiji, however, challenge this tradition. One of the manifestations of this juxtaposition was the takeover of various tourist resorts by indigenous landowners in 2000. Although outside the marketing-focused mandate of TAG and FVB, these events clearly impacted very negatively on the recovery program and needed to be addressed. The Spirit of Fiji campaign was one way of addressing the issues. However, grassroots initiatives were also affected. One of these was a Tourism Awareness Training Program initiated by the Savusavu Tourism Association and the Tui Wailevu after the hostage taking of the New Zealand pilots in Savusavu (Berno and Bricker, 2000; Carswell, 2000). Although not directly targeted at land ownership issues, this campaign addressed similar issues as the Spirit of Fiji, such as the economic importance of tourism to Fiji. Despite these efforts, land ownership will clearly be an ongoing concern that will need careful consideration.

Two periods of significant political instability within 13 years have done much to damage Fiji's image as a tourist destination. After the coups in 1987, a recovery to pre-coup tourist arrival numbers was realized by the end of 1988. Despite the success of TAG's campaign, a full recovery after the coup in 2000 is likely to take longer. It is going to take a continuing collaborative, concerted effort to recapture the image of "Fiji—the world's one truly relaxing destination." As long as the prospect of another coup cannot be ruled out, it may be difficult to offer tourists the prospect of absolute relaxation. In some respects, Fiji may have lost its innocence in 1987, leaving the tourism sector with no choice but to manage the new reality. Words such as *relaxation, paradise,* and *idyllic* can become tinged with irony. The events of September 11 have provided an extra layer of complexity. Prior to that date, Fiji may have appeared a very unsafe place. Post–September 11, Fiji appears like a haven of tranquility in a world where the most vulnerable targets for terrorist attacks appear to be the icons of Western capitalism, such as the former World Trade Center.

The Fiji experience is a timely reminder to other island destinations of the need to consider the full implications of terms such as *paradise, tranquility,* and *true relaxation*, which have been adopted by many island destinations. Fiji was better prepared than many to confront the crisis when it arose because of the familiarity of the country with cyclones and related natural disasters. Overcoming skepticism on the part of the overseas travel industry and the traveling public is a bigger challenge. Fiji's tourist sector recovered from the experience of 1987 and will do the same again, though the downturn in world tourism following the terrorist attack in the United States has been a further setback. It is clear that destinations are vulnerable to a confluence of local issues (e.g., the coups) and global concerns (the implications of the terrorist attacks). At least Fiji's experience of handling crises now makes it better prepared than many other tourist destinations that have never had to confront civil disturbances. Such destinations could learn from Fiji's preparedness for crisis, particularly the second time around.

From a research perspective, there is an urgent need for studies that investigate the experience of recovering from crisis in a variety of different settings. The development of widely applicable recovery models would be useful. Such research would be positively received in a world where travelers need increasing reassurance that industry and governments have the appropriate responses in place when crises strike.

Concept Definitions

Crisis management team Major crises require prompt remedial action. Typically the impact of a crisis extends beyond a specific sector of the economy and society. An integrated response will involve close collaboration between a range of key stakeholders. In Fiji such collaboration occurred through the formation of the Tourism Action Group (TAG).

Recovery plan Planning implies a longer-term perspective. Typically a crisis demands an immediate response. The recovery plan should relate to an established crisis management plan, but will involve a quick review of the particular circumstances of the crisis that has arisen. As soon as the conditions are right (in the case of the 2000 Fiji coup, "Ground Zero," the release of all the hostages, marked the first step), the plan will more immediately go into the implementation phase.

Coup d'état Literally a "blow" or "strike" against the state. While tourists are often untroubled by visiting destinations characterized by authoritarian regimes, the type of instability that is often prompted by a coup may be a major deterrent.

Review Questions

1. To what extent did the different durations and contexts of the two crises (e.g., hostage taking) impact the capacity of Fiji's destination authorities to rebuild international arrivals in the case of the post-1987 and post-2000 periods, respectively?
2. It is often observed that the process of globalization has increased in intensity. How did Fiji's relationship with the rest of the world influence the way in which the crisis was reported and communicated during 1987 and 2000, respectively?
3. To what extent was there evidence that the tourism sector (both public and private) learned from the experience of 1987 in handling the 2000 crisis?
4. After rebounding from the 2000 coup, Fiji has experienced a substantial growth of tourism at the expense of competitors in Southeast Asia. How should such experience inform future risk management strategies for remote destinations such as Fiji?
5. Summarize the role of the Internet in shaping the experience of the 2000 coup and its aftermath in terms of tourism development.

References

Beirman, D. (2001). *World Tourism under Siege*. PATA Occasional Papers Series No. 26. San Francisco: PATA.

Berno, T., and Bricker, K. (2000). *Tourism Awareness Training for Vanua Levu*. Suva, Fiji: Tourism Studies Programme, University of the South Pacific.

Berno, T., and King, B. E. M. (2001). Tourism in Fiji: After the coups. *Travel and Tourism Analyst,* 2, 75–92.

Carswell, C. (2000, December). Savusavu Tourism Association. The Waileve/Nasavusavu Village Tourism Awareness Workshops. Unpublished paper presented at the *Fiji Tourism Forum 2000*. Korolevu, Fiji.

Fiji Village Statistics. (2000). *Current Site Statistics—July 2000*. wysiwyg://69/http://www.fijivillage.com/stats/00/stats07.php3.

FVB (Fiji Visitors Bureau). (1999). *Year 2000 Marketing Plan Summary*. Suva, Fiji: Author.

FVB (Fiji Visitors Bureau). (2000a, August 24). Media release. *Special Bulletin,* 42. Unpublished manuscript.

FVB (Fiji Visitors Bureau) 2001. Fiji International Visitor Server. Ministry or of Tourism & Transport Suva., Fiji.

FVB. (2000b, September 5). Spirit of Fiji (*Sa ka ga ni yalo*) hosts ambassadors. *Fiji Visitors Bureau Bula News Bulletin*, 42. bulanews@brlafiji.com.au.

Harrison, D. (nd). *Ecotourism and Village-Based Tourism: A Policy and Strategy for Fiji*. Suva, Fiji: Ministry of Tourism and Transport.

King, B. E. M. & Berno, T. (2002) Tourism and Civil Disturbances. An Evaluation of Recovery Startegies in Fiji 1987-2000. *Journal of Hospitalits and Tourism Management.* 9(1), 37-45

Lea, J. (1996). Tourism, realpolitik and development in the South Pacific, in A. Pizam and Y. M. Mansfeld (eds.) *Tourism, Crime and International Security Issues.* Chichester, UK: John Wiley & Sons, pp. 123–142.

Mazza, E. (2000). *Coup d'web.* ABC news. http://abcnews.go.com/sections/world/DailyNews/fiji000519.html.

McVey, M. J., and King, B. E. M. (1999). Tourism in the Pacific Island Microstates. Regional experiences and prospects, in P. Kee and B. E. M. King (eds.). *Asia-Pacific Tourism: Regional Co-operation Planning and Development.* Melbourne, Australia: Hospitality Press, pp. 55–72.

Ministry of Information. (1999). *Fiji Today 2000.* Suva, Fiji: Author.

Ministry of Tourism and Transport, Tourism Council of the South Pacific, and Deloitte & Touche. (1998). *Fiji Tourism Development Plan.* Suva, Fiji: Ministry of Tourism and Transport.

Narayan, J. (2000, June 1). Wansol on-line barred. Wansolwara.

Pacific Media Watch. (2000). Fiji tourism television advertisement breaches code of ethics. http://www.pmw.c20.org.

Scott, R. (1988). Managing crisis in tourism: A case study of Fiji. *Travel and Tourism Analyst,* 2, 57–71.

Silaitoga, S. (2000, August 12). Efforts to revive tourism. *Fiji Times,* pp. 2–3.

TAG (Tourism Action Group). (2000a, October). *The Way Forward.* Paper presented to the Fiji Tourism industry, Nadi, Fiji.

TAG. (2000b, December). *Recovering Fiji Tourism.* Paper presented to the Fiji Tourism Forum 2000. Korolevu, Fiji.

TAG. (2000c). *News from TAG, July 9, 2000.* Unpublished manuscript.

TCSP (Tourism Council of the South Pacific). (nd). *The South Pacific Islands: Travel Planner.*

Note

[1] This chapter was previously published in an earlier version as: King, B., and Berno, T. (2002). Tourism and civil disturbances. An evaluation of recovery strategies in Fiji 1987–2000. *Journal of Hospitality and Tourism Management,* 9(1), 37–45.

II

Tourism and Crime Issues

Yoel Mansfeld and Abraham Pizam

This section discusses issues related to crimes committed against tourists or occurring at tourist destinations. As seen in the three chapters the impacts of crime on the destination are not uniform. In some cases, such as crimes related to drug use, the destination's image is tarnished and results in significant decline in tourist demand. On the other hand, thefts committed at tourist destinations did not necessarily reduce the tendency to visit or revisit the affected destination. The same was true for the circumstance of the criminal act. In some cases the tourists were innocent victims being in the wrong place at the wrong time, while in others the tourists contributed to their own victimization by soliciting some illicit activities.

In the first chapter of this section, Jerome McElroy discusses the spread of the narcoeconomy within the Caribbean islands and its short-term and long-term impacts on the tourism industry. Based on his analysis of primary and secondary sources the author concludes that the narcoeconomy with its violence, addiction, and corruption poses the greatest known threat to the viability of the tourism industry in the Caribbean islands. This occurs because:

- The perception of danger or irritation tarnishes a destination's reputation;
- Visitors have been shown to be more likely than residents to be victims of crime;
- Tourist attractions are more likely to become crime "hot spots";
- There is a known correlation between crime increase and visitor fallout;
- Drug peddlers constantly harass tourists;
- The presence of large numbers of uniformed police and sniffing dogs at major tourist attractions creates a negative security image and may affect word-of-mouth recommendations and preference for a return visit;
- The high cost of security, in both the private and public sectors, forces tourist operators to raise their prices and makes Caribbean tourism less competitive;
- Lack of tourism safety and security makes investment in the tourism industry unattractive; and
- Protecting tourists by placing them in gated all-inclusive resorts generates feelings of resentment and inequality from the local population.

Judy Holcomb and Abraham Pizam conducted an empirical study among US travelers, the purpose of which was to determine what, if any, effects past incidents of personal theft have on tourists' future decisions to travel to the affected destination. This was done by analyzing two study groups, those who have experienced personal thefts while on a trip, and those who had heard of such incidents through personal accounts of friends or family. The results of their study demonstrated that having been a victim of a theft or knowing of someone who has been such a victim while on a trip did not affect the likelihood to travel to the destination where the theft occurred.

The study found:

- No difference in the likelihood to travel to a crime-affected destination between those who have personally experienced a theft versus those who knew of someone who had;
- That the passing of time since a theft occurred had no effect on the decision to travel to the same destination;
- That the likelihood to travel to the affected destination was not significantly different between those who experienced a theft "on their person" and those who had experienced a theft "off their person";
- That the perceived severity of the incident did not have any significant effect on the decision to travel to the destination where the theft occurred;
- No significant correlation between previous "non-trip related crime victimization" and the likelihood to travel to the destination where a theft occurred;
- The passing of time, learning of someone's positive experience, and learning from the media about the destination's effort to make it safer, did not change the mind of tourists once they have decided not to travel to a destination where a theft occurred; and

- A positive correlation between the satisfaction with the manner in which the authorities handled the reporting of the theft and the likelihood to visit or revisit the destination where the theft occurred.

Based on the results of this study the authors suggest that destinations that are afflicted by thefts similar to the ones described in the chapter should not necessarily expect a reduction in tourist arrivals.

Dee Wood Harper analyzed street robberies against tourists in the Vieux Carré area of New Orleans. He concludes that many of the circumstances surrounding the robbery incidents—which he called contextual weaknesses—and especially the activities of the tourist themselves increased their vulnerability to attack. For example:

- In quite a few incidents the victims had developed some relationship with the offender before the robbery, such as going with the offender to procure a paid sex partner, to purchase drugs, or to have sex.
- In one out of every five cases the opportunity for the victimization was largely created by the choices made by the victim. By their actions, tourists placed themselves at a triple disadvantage of being a stranger, being isolated in an unfamiliar area, and being there in search of some illicit action.
- In numerous cases perpetrators used the desires of the victim for illicit action and suggested that the perpetrator can help the victim to find it. After luring the victim to an isolated location the robbery took place.
- Finally, because the robbed tourist is a transient person who resides elsewhere, the prosecution of the offenders is very difficult. By the time the case is brought to court the victim/witness has returned home and is not likely to return to press the prosecution. Thus the typical street robbery that involves out-of-town victims rarely leads to prosecution.

5

The Growth of the Caribbean Narcoeconomy: Implications for Tourism

Jerome L. McElroy

Learning Objectives

- To understand the factors that make the Caribbean an ideal trafficking corridor.
- To understand the macroeconomic problems that nourish drug trafficking.
- To become aware of past research results about island crime patterns.
- To appreciate the evidence on the spread of the narcoeconomy.
- To understand the long-run implications of drug trafficking on tourism.

Introduction: Three Waves of Globalization

All these people, these diversities of cultures, have been concentrated in that archipelago. You are very conscious of the ocean as a gateway out to other worlds. . . . Caribbean society was perhaps the first global experiment in human history. (George Lamming in E. J. Waters, 1999, p. 201)

The distinguishing feature of small island economies is their relative powerlessness in the global marketplace. They are so-called price takers and simulate the behavior of the atomistic firm in the classical economist's theoretical model of perfect competition. Their small size at the macro-social scale suggests that they are ipso facto vulnerable to policy initiatives originating in core economies. This is graphically illustrated by the profound impact which European imperialism, the first wave of globalization, had on the insular Caribbean in the post-Columbian era (Richardson, 1992). These external imprints include the stamp of dependent export monoculture on island economy, that is, a high volume, low value added production structure geared to foreign demand in the global network of center periphery trade. In addition, monoculture has had damaging effects on insular ecology and biodiversity, and its demand for cheap slave labor resulted in the creation of multiethnic societies of transplanted peoples. Unable to sustain population growth, the post-emancipation period saw widespread livelihood mobility/migration that has continued through the twentieth century. Since the 1950s the islands of the region have struggled to reduce dependence and achieve political and economic autonomy.

Although the Caribbean has experienced repeated rounds of externally imposed laissez-faire capitalism (Klak, 1998), the second epochal wave of globalization has been the sustained growth and worldwide spread of tourism in the post–World War II era. International visitor arrivals have increased at a robust rate above 5% per year since 1950 (Vellas and Becherel, 1995). Today tourism is the world's largest industry, accounting for over 10% of global GDP and trade and 8% and 9% of world employment and new capital formation, respectively (WTTC, 2004a). In part, this growth has been facilitated by the IMF–World Bank structural adjustment program favoring tourism as a lucrative export strategy for debt-burdened less-developed countries (LDCs) to improve loan repayment performance.

Nowhere has this process been more visible than in the island periphery. In addition to the pull of tourism opportunity, island decision makers have been pushed to alter reliance on traditional staples because of falling terms of trade, the dismantling of traditional colonial preferences, the worldwide expansion of staple supply, and the development of synthetic substitutes in the North. In the past generation, because of deliberate restructuring away from colonial crops like sugar and copra, island economies have diversified toward tourism, related construction, and offshore financial services. The intersection of this restructuring with multinational hotel/airline investment, the advent of low-cost jet travel, and the creation of aid-financed transport infrastructure has connected northern demand with southern supply and created the so-called pleasure periphery: the insular and coastal Mediterranean for Northern Europe, the Caribbean islands for North America and Europe, and North Pacific (Hawaii, Guam, Marianas) and the South Pacific for Japanese and other affluent Asian travelers.

This internationalization has especially intensified in the small island Caribbean where neoliberal free market policies favoring exports, deregulation, foreign direct investment, and political stability have long dominated the political economy (Gayle, 1998). As a result, the Caribbean has become the most tourism-intensive region in the world. According to 2004 estimates, tourism accounts for 15% of regional GDP and employment and roughly 20% of exports and investment (WTTC, 2004b). Landscapes from the Bahamas to Aruba are visibly altered by the ubiquitous hotel and condominium clusters and connecting road arteries, sun-hungry visitors, rental cars, and touring vans. Establishing the infrastructure to support the mass tourism visitor industry has altered insular coastlines and mountain faces, commercialized and quickened the once quiet pace of island life, and damaged the operation of natural terrestrial-marine weather and water buffering systems (McElroy and de Albuquerque, 1998). Such widespread socioenvironmental changes have raised questions whether tourism is a sustainable economic base, or just another short-run chapter in a long-period history of shifting export specializations and boom and bust economic and parallel demographic cycles.

Island tourism is also threatened by a third epochal wave encompassing the globalization of organized crime in general and the international restructuring of the production and distribution of illegal drugs (mainly cocaine and heroin) in particular. According to Pearson and Payaslian (1999, p. 427), "while prior to the 1960s the Italian and French 'connections,' extending from Palermo to New York, were the primary networks for smuggling drugs, during the past three decades Asian and Latin American drug production and smuggling have emerged as major competitors." This increasing cross-border traffic has led to the dollarization of some hinterland South American producers, the alleged decline of the Cosa Nostra in the United States, and, some argue, the increasing informalization of the traditional financial sector through the growing money laundering of drug profits. The spread of narco-trafficking across the Caribbean has created a cocaine corridor from Aruba to the Bahamas that threatens to overwhelm insular law enforcement and to destabilize the region's postwar economy. According to the West Indian Commission (1992, p. 343), "Nothing poses greater threats to civil society in CARICOM countries than the drug problem and nothing exemplifies the powerlessness of regional governments more."

Scope

This chapter focuses on the geographic and institutional infiltration of the global narcoeconomy across the insular Caribbean. It has six sections. The first reviews patterns of serious crime in the region that suggest the escalating presence of narco activity. The second presents some provisional new research. The third indirectly links these phenomena to the expansion of the archipelago as a corridor for South-to-North drug transit and money laundering and reviews the internal and external forces that have nourished the growth of narcoeconomy. The fourth describes the specific mechanics of the illicit trade routes and methods of conveyance through the region, while the fifth discusses implications for tourism and related activities. The conclusion broadly assesses the future.

Island Crime Patterns

It is clear that there is a different threat to island tourism other than natural amenity destruction and loss of cultural capital. Serious crime rates have risen in the region since 1970 in tandem with the globalization of Caribbean tourism and the rise of the region as a node in the global drug trade. Some trends illustrate this danger. Since 1970 there have been "significant increases in the rate of violent crimes in every Caribbean country for which data is available" (Harriott, 2002, p. 4). The most dramatic increases have involved theft/robbery, homicide, and serious assault, ". . . precisely the crimes associated with drugs" (Griffith, 2003, p. 394). Islands that have experienced either acceleration or high levels of such criminality tend to be both highly tourism penetrated and major destinations for narco traffic: Bahamas, Dominican Republic, Jamaica, Puerto Rico, and the US

Virgin Islands (USVI). In the special case of Jamaica, according to Harriott (2003), the growth of the narcoeconomy since 1980 has been associated with a rapid escalation in murder, the shift from dominantly property to dominantly violent crime, increasing gang warfare, multiple murders, and the killing of innocent bystanders.

Elsewhere, patterns have been similar. Since 1970 murder and rape rates have risen 50% in the USVI while robbery has tripled (de Albuquerque and McElroy, 1999b). Between 1985 and 1996, robbery rates doubled in Barbados and rape rates doubled in St. Kitts-Nevis. Similar sharp increases have been recorded for burglary and larceny over the same period for Antigua, Barbados, Dominica, and St. Kitts-Nevis (de Albuquerque and McElroy, 1999b). The presence of visitors also seems to be implicated in rising island crime. Similar to the experience of Hawaii (Chesney-Lind and Lind, 1986; Fuji and Mak, 1980), a case study of Barbados (de Albuquerque and McElroy, 1999a) revealed that visitors were much more vulnerable to property crime: triple the resident rate for robbery and for larceny. Residents were much more likely to be victims of violent crime: murder, rape, and major assault. These regional crime trends have also been associated with rising levels of drug addiction and a sharp increase in juvenile offenders. On the other hand, Harriott (2002) notes Caribbean rates are highly volatile and have shown some declines very recently, although this may be partly due to increased underreporting. In developing countries worldwide, only 50% of serious crimes are reported (Alvazzi del Frate, 2000).

The expanding literature on the determinants of crime in LDCs suggests that development is positively associated with property crime (notably theft). Various explanations are proffered, including Durkheim's *anomie* theory, which posits that rapid societal change produces new forms of wealth and the breakdown of traditional moral norms (Leggett, 2000). A variant, the subculture of violence, is based on blocked opportunity for marginalized urban youth who resort to criminality to survive. An increasingly important formulation is the opportunity cost or Chicago school model associated with the work of Becker (1968). It argues that development alters the benefit cost calculus of traditionally illicit activity. The combination of new wealth and the anonymity afforded by rural to urban migration produces rising benefits for crime and falling costs of detection/incarceration. This model seems especially suitable for the Caribbean, which has undergone a major postwar restructuring towards mass tourism. Visitors present lucrative targets because they tend to carry much portable wealth, ignore normal precautions, are unfamiliar with their surroundings, and are less likely to report crimes, correctly identify assailants, or return as witnesses at trial.

Caribbean research has suggested many of these same determinants. De Albuquerque and McElroy (1999b) have emphasized the link between the presence of visitors and property crime. In the same context, King (1997, p. 34) has touted relative deprivation, that is, the "sense of entitlement" among disenfranchised youth "because Americans have so much." Headley (1996) has stressed blocked opportunity and argued for the drug-and-guns subculture of violence. Bonnick (1994) believes a major conditioning variable is the bad guy image, so appealing to marginalized youth, portrayed by powerful American cinema and television in this most mass media penetrated subregion in the world. Robotham (2003) emphasizes the rising proportion of youth in the population and their falling labor force participation because of a failing macroeconomy. Finally, Rushton and Whitney (2002) have implicated rooted cultural values and family cultures as criminogenic in both African and Black Caribbean societies. However, rigorous empirical testing of these theories has been either lacking or inconclusive perhaps partly because of the imperfect quality/quantity of the data, the complexity of the crime phenomenon (Harriott, 2002), and/or the clandestine operation of an expanding narcoeconomy (Forst and Bennett, 1998).

This last conclusion is based on the following evidence: (1) police reports across the Caribbean that most serious crimes are drug-related (de Albuquerque, 1996a); (2) a noticeable change in the pattern of violent crime from domestic disputes to gang feuds and robbery (Harriott, 2003); (3) the increasing use of firearms (Harriott, 2002); (4) the increasing role of hard-core marginalized youthful offenders (Harriott, 1996); (5) the increasing number of apprehended perpetrators who are addicts (de Albuquerque and McElroy, 1999b); (6) the rising prominence of home-grown posses in organized crime; and (7) the increasing menace of returning felons deported from the United States and elsewhere (8,000 between 1993 and 2000; see Griffith, 2003) who are known to be engaged in narcotics and firearms importation.

Some Preliminary Evidence

As a preliminary test of some of the major determinants and to indirectly measure the presence of the narcoeconomy, a limited cross-sectional analysis was performed on 16 Caribbean countries for which complete data were available. The dependent variables were taken from uniform Interpol crime rates (per 100,000 population): murder and theft to represent violent and property crime, respectively, and the number of drug offenses to capture the influence of the narcoeconomy. The independent variables included population density to represent anonymity, average daily visitor density (per 1,000 population) to represent visitor presence, the unemployment rate to measure macroeconomic conditions, the percent of the population age 0–14 years as a surrogate indicator for the presence of juvenile offenders, and political status (independent = 1, dependent = 0) as a proxy for crime enforcement capacity. Independent countries were assumed to have more resources and stronger capabilities. All these data were taken from the CIA *World Factbook* except visitor density (see Padilla and McElroy, 2004). To avoid possible distortions introduced by the 2001 terrorist attacks, since US visitors dominate the tourist economy, years chosen ranged from 1995–2000 based on data available for each individual country.

Regression results showed none of the determinants influenced either murder or theft rates. Two factors had a statistically significant influence on drug offenses. According to Table 1, drug offenses were positively related to both visitor and population densities. According to equation (2), these two variables explain over 70% of drug offenses in the 16 islands. This may suggest that developed tourism destinations provide both anonymity and opportunity conducive to drug activity. Such a view corresponds to the increasing harassment and drug peddling observed across the region (WTTC, 2004b). On the other hand, neither local economic conditions (as measured by unemployment) nor varying levels of enforcement (as measured by political status) had any impact on drug offenses. Although quite modest, these limited results indirectly point to the presence of the narcoeconomy and the value of more extensive systematic empirical study.

$$\text{Drugs} = -103 + 0.632\frac{\text{pop.}}{\text{km}^2} + 6.910 \text{ visitor density}$$

Table 1 Determinants of Drug Offenses: Regression Results

Equation F-Value	*Independent Variable*	*Coefficient*	*t-Value*	*R^2 (adj.) (%)*
8.85	1 Population density	0.622	2.04*	67.7
	Visitor density	7.442	4.75*	
	Unemployment	5.020	0.28	
	Political status	64.80	0.28	
19.20	2 Population density	0.632	2.29	72.2
	Visitor density	6.910	5.72*	

Source: Interpol for crime data; *World Factbook* (CIA, 2002) for independent variables except visitor density (see Padilla & McElroy, 2004). Caribbean countries include: Antigua, Bahamas, Barbados, Bermuda, Caymans, Dominica, Dominican Republic, Guyana, Grenada, Jamaica, Montserrat, Puerto Rico, St. Kitts-Nevis, St. Vincent/Grenadines, Trinidad-Tobago, and Turks and Caicos.
Note: Asterisk (*) indicates statistical significance at the .05 level or better.

The Caribbean Narcoeconomy

The global drug trade is fostered by the general tendencies of globalization to reduce restraints on cross-border trade and capital flows and to uphold private property and bank secrecy. It thrives in an environment of secrecy, trade intensity, sophisticated finance and communications, anonymity, lax regulation, and external and internal economic vulnerability. The Caribbean closely fits this profile. The insular Caribbean lies between the Andean region of South America, where close to 90% of world cocaine is produced, and the major market of the United States where most cocaine is consumed. Proximity to South America plus a long history of British, Dutch, French, and Spanish commerce make the islands a natural corridor for supplying cocaine to Europe and for reexporting heroin produced in the Far East, and Ecstasy produced in Europe (UNODC, 2004). The geography of the region is conducive to the large-scale movement of drugs. The larger islands and mainland countries (Belize, Jamaica, Guyana, Dominican Republic, and Haiti) provide remote hinterlands to shelter trafficking. Many others are archipelagic states (Bahamas, Grenada, USVI, Turks and Caicos, St. Vincent, and Grenadines) with numerous unguarded points of entry and exit for transporting narcotics. The more remote outposts lack resources for either monitoring, much less intercepting contraband.

As a result, presently the value of narco traffic through the archipelago approaches $5 billion (UNODC, 2003). This exceeds the GDP levels of three fourths of the islands and the tourism impact of all but Cuba, the Dominican Republic, and Puerto Rico. It also exceeds the value of all merchandise exports (petroleum, aluminum ore, rum, etc.) of all Caribbean islands. By value cocaine is the most significant (85%) followed by ganja (cannabis 13%), and heroin and amphetamines (2%; Harriott, 2002). Two thirds of the cocaine flowing through the corridor is smuggled into the United States where it comprises nearly 50% of all cocaine imports, with the balance mainly from Mexico. The other third, accounting financially for a quarter of the market, is destined for Europe (UNODC, 2003). It is no wonder that island officials decry the drug menace as “an overwhelming phenomenon” (Sanders, 2003, p. 377).

The highly open character of the insular economies' structure is also a narco asset. Intense dependence on export-import trade provides traffickers with a wide array of air and sea transit opportunities. The burgeoning tourism industry based on on-site consumption of environmental amenities and imported goods significantly expands the menu for shipping contraband intercontinentally by freight as well as by courier because of the increasing anonymity created by busy docks and airports in the many popular tourist destinations across the region.

Money laundering—the conversion of drug profits into legal assets that cannot easily be traced to their illegal roots—is a third ingredient in the narco-triad along with drug transit and arms traffic. It likely exceeds $500 billion worldwide (Bryan, 2000) and is principally driven by the large 300–500% retail-over-wholesale price markups common for illicit drugs. It thrives under the same conditions that have shaped the Caribbean's successful offshore financial sector: proximity to markets, good communications systems, political stability, transactions secrecy, low taxation, exchange stability, and a competent cadre of local lawyers, bankers, and accountants. The islands (with Latin America) contain roughly 40% of the world's offshore banks and international business corporations (US DEA, 2003). In some small islands like Bermuda, offshore finance is the dominant sector and accounts for half of all economic activity. In the Caymans and British Virgin Islands, roughly one third of GDP is sourced in offshore services (Griffith, 1997). A number of islands where trafficking is on the rise are now considered top-tier money-laundering states. The US Drug Enforcement Administration (DEA) estimates that some $60 billion in organized crime proceeds are laundered through the islands every year (US DEA, 2003, p. 4).

The narcoeconomy is especially nourished by a climate of economic instability and uncertainty. During the past decade, the Caribbean has been buffeted by a series of external forces that have weakened traditional sectors, clouded the investment climate, and reduced the opportunity cost of trafficking. In recent years the region has become "increasingly irrelevant in economic and political terms" (Klak, 1998, p. 12). As a result of the diplomatic downgrading of the Caribbean (the Cuban threat in particular) with the demise of Communism, US aid to the region has declined over 80%. Export preferences for traditional staples like sugar and bananas have shrunk in the face of shifting US tastes toward artificial sweeteners and the consolidation of the European Union and World Trade Organization regulations.

Moreover, the islanders' attempts to diversify toward light industrial exports (textiles, assembly, manufacturing) have been thwarted on two broad fronts. On the external side, the heavy impact of NAFTA has resulted in the relocation of over 120 apparel plants to Mexico since the mid-1990s (Rohter, 1997). On the internal side, a host of domestic structural constraints continue to hinder viability. These include mineral scarcity, fragmented transport systems, and relatively expensive (by LDC standards) labor and utility costs. Even tourism has been damaged in some destinations in recent years because the strength of the US dollar, to which their currencies are tied, has discouraged long-staying, high-spending European visitors (Watson, 1997). In some islands, chronic emigration has spawned demographic imbalances (surpluses of old and young and deficits of technicians and innovators) that disfavor the skills and entrepreneurial talent required to prosper in a global economy (Conway, 1998). In others, problems have been further compounded by natural disasters.

In the larger heavily indebted countries like Jamaica, Trinidad, Guyana, and Antigua, some authors have argued that the IMF-designed structural adjustment programs (SAPs) to ensure loan repayment have weakened macroeconomic conditions particularly for the poor and indirectly nourished the narcoeconomy. Although no clear empirical links are established between debt burdens, SAPs, and the drug trade, there is a preponderance of coincident evidence. For example, Ghai (1991) associates substantial declines in Caribbean GDP with increases in debt repayment outflows. In Jamaica, Guyana, and Trinidad, Grant-Wisdom (1994) suggests that SAPs have coincided with infrastructure deterioration because of capital budget cutbacks, the contraction of public employment, and the proliferation of self-employment "indicative of the growth of the informal sector" (p. 171). Both Bernal (1992) and Phillips (1994) find that SAPs have negatively affected the delivery of basic services, and the latter concludes (pp. 147–148) that SAPs "are associated with the worsening of the health status of the people of the Caribbean."

These deteriorating conditions have combined with other internal problems, notably the growth of a subculture of violent youth, daily aware of the "good life" from the tourist hordes, but lacking the education, skills, and motivation to participate in the mainstream. The drug trade worldwide flourishes in such urban areas with an underclass of unemployed youth (UNDP, 1999). Their ranks are increasingly led by the organized posses that originated in the 1970s in Jamaica as local gangs in West Kingston trafficking in the domestic and regional marijuana trade. At roughly the same time, rival political parties began arming ghetto youth and pitting poor urban communities against one another (Stone, 1983). These "garrison" constituencies were largely responsible for the rise in violent crime in Jamaica culminating in 889 killings during the 1980 electoral campaign (de Albuquerque, 1996c). In the late 1970s, they came under the control of ganja gangs sustained by marijuana exports to the United States and were led by posse "Dons" who were responsible for internationalizing the ganja economy.

Fragments of these Jamaican-style posses have become embedded in other Eastern Caribbean societies like Antigua and St. Kitts and are responsible for escalating violence and setting up overseas branches of the narcoeconomy across the island chain. Dominican gangs in the north and Arubans in the east have also become involved (UNODC, 2003). They are heavily engaged in arms and drug running as well as "money laundering, fraud, kidnapping, robbery, burglary, prostitution, document forgery, and murder" (Griffith, 1997, p. 124). Their increasing responsibility for narco trafficking is a decentralization strategy employed by the Colombian-based crime organizations that dominate the regional drug trade (US DEA, 1998). The ranks of local gangs are swollen annually by drug felons from North America and elsewhere. These sophisticated criminals with extensive experience in the US cocaine trade are notorious for introducing American-style drive-by executions, the recruitment of juvenile runners, lookouts (Harriott, 1996), and a whole range of semiautomatic weapons in Jamaica and elsewhere. According to the Commonwealth Secretariat (1997, p. 105): "They leave our islands as high school criminals and return to us as postgraduates."

Mechanics

The mechanics of drug transit including both routes and modes of conveyance shift continually according to national, regional, and international counterenforcement measures (de Albuquerque, 1996b) as dealers display remarkable ingenuity keeping one step ahead of detection. During the 1970s and 1980s most cocaine was airlifted in small planes from Colombia either directly to the Bahamas (or indirectly through Jamaica) where air-dropped cargo was loaded onto high-speed "go-fast" boats for a final run to the US mainland. Because of increased air and sea anti-drug enforcement, now most cocaine is shipped in bulk cargo freighters along more circuitous routes: through Jamaica to Haiti, the Dominican Republic, and Puerto Rico, then on to the United States; or through South America to the Eastern Caribbean, and on to the United States and Europe (US DEA, 2003). In addition, small amounts of cocaine and heroin are transported by courier in commercial aircraft at busy tourist airports.

As an example of the ubiquity of trafficking activity, the Leeward Islands of the Eastern Caribbean have provided a new corridor to North America. The new routes include Puerto Rico, the "new Miami," and the USVI where shipments arrive by sea and air, are repackaged as domestic freight, and are transported north by cargo or courier, taking advantage of the anonymity created by large-scale tourist traffic and perfunctory Customs checks through these US territories. Today about a third of all cocaine in transit winds up in Puerto Rico (UNODC, 2003). Sophisticated satellite positioning systems are often employed by traffickers to coordinate drops in the least policed waters (US DEA, 1998). To elude US radar, traffickers often use stealth boats made entirely from wood and fiberglass as well as semi-submersible vessels (de Albuquerque, 1996b).

Further to the north, cocaine traffic has been deflected toward Haiti and the Dominican Republic—states ideally suited because of their location, poorly monitored coasts, mountainous and underpopulated interiors, significant poverty, and poorly paid and underequipped security forces. Cocaine shipped to Haiti quickly finds its way to the Dominican Republic, and then by go-fast boats or fishing boats to Puerto Rico or the Bahamas. The Dutch Antillean islands of Aruba, Curacao, and St. Maarten have also become prominent transit routes to the United States and Europe for heroin and cocaine smuggled by passenger and cargo flights as well as on cargo vessels and cruise ships. Their main attraction is that they function as free-trade zones that allow containerized contraband to escape inspection before reexport (US DEA, 2003). Grenada and St. Vincent and the Grenadines also function as bulk pipelines for drugs en route north via the American and the French territories (Guadeloupe and Martinique) while St. Lucia and St. Kitts-Nevis have become staging and stockpiling points for cocaine to be on-loaded into go-fast boats or fishing boats to Puerto Rico and the USVI.

These drug flows are facilitated in part by the lack of drug-sniffing dogs at Caribbean airports and by sophisticated concealment methods. In addition to hidden compartments and fuel tanks in maritime vessels, drugs have been stashed in every possible human orifice, especially the swallowing of cocaine sealed in condoms and heroin in latex-wrapped pellets. They have been hidden in every type of clothing, footwear, fruits, vegetables, furniture, appliances, vehicles, cigarette cartons, false amputee limbs, bibles, surfboards, live and dead animals, and even a "ganja guitar" fashioned completely of compressed marijuana (Griffith, 1997, p. 82). Lax customs inspection and local corruption facilitate the lucrative trade. The former is particularly true of cruise ship passengers and crew as they disembark in Puerto Rico or Florida ports. Couriers are often women recruited for up to $3,000 per trip, cruise ship crew members, or older couples on a cruise. Jamaica is the major source of air couriers. Over one third of the so-called mules arrested in the United States are from Jamaica (*Economist,* 1999) while British authorities claim that one in ten passengers flying Air Jamaica (so-called Cocaine Air) to Great Britain are transporting cocaine (UNODC, 2003, p. 15).

Short-Run Impacts

The success of transshipment across the region directed by experienced traffickers using sophisticated networks and methods suggest the Caribbean narcoeconomy—along with its associated evils of violence, drug abuse, corruption, and money laundering—is increasingly embedded in the insular political economy. This is evident in the rise in criminality across the region reviewed above, and the incidence of high-powered firearms and American-style tactics: gang warfare, drive-by shootings, and home invasions (de Albuquerque and McElroy, 1999b). The narcoeconomy is further suggested by the rise in citizen gun purchases, the spread of private security agencies and high-tech alarm systems, and the proliferation of guard dogs, high-wire fencing, and grilled windows. Griffith (1997, p. 123) reports the telling example of former Governor Roy Schneider of the USVI who carried "a Glock semiautomatic pistol when he did not have his official security detail."

Few segments of island life have gone untouched. The grim realities include: (1) the growing presence of drug-addicted youth in low-income urban areas and around resorts; (2) the active daily complicity of some local police, shipping, and Customs officials and airline personnel; (3) collusion at the highest levels implicating law enforcement and elected officials in what one regional writer calls "massive public corruption" (Reese, 1997) in nearly a dozen nations; (4) prison overcrowding primarily because of drug-related offenses; (5) the clogging of overtaxed court dockets and resources plus the associated problems of evidence tampering and witness and jury intimidation. In this last regard, one of the more celebrated examples is the case of Charles "Little Nut" Miller, a deportee, who is alleged to have successfully avoided extradition from St. Kitts to face narcotics violations in the United States by threatening American students at the island's offshore veterinary school (Larmer, 1998).

Certainly the most palpable evidence of narco traffic's role in island society is the daily economic impact. As the region's second largest export sector behind tourism, the narcoeconomy is "one of the few areas where global profits continue to flow from the developed to the developing world" (Leggett, 2000, p. 143). In the private sector, the infusion includes the substantial payroll to the drug workforce: growers (marijuana), pilots, boat captains, engineers, shippers, baggage handlers, couriers, lawyers, accountants, street pushers, and enforcers. Given the risks, these employees are paid above the going local wage, and their spending supports a diverse array of local businesses. In the public sector, clandestine payoffs to compliant police, Customs employees, and other officials are quite substantial. According to Harriott, (2002, p. 12), public corruption provides "Caribbean civil servants with some US $320 million in income annually."

These infusions circulate through the insular economy in a number of ways. They are fed by rising addiction, which is partly a product of the Colombian cartel's practice of paying traffickers in kind to avoid money laundering (DEA, 1998). They are also fed by protection and extortion rackets, which recently have become a major source of gang income (Robotham, 2003). On the other hand, drug profits are invested in legitimate enterprises serving residents and laundered into hotel facilities and casinos catering to visitors. As an illustration of the embeddedness of drug activity, marijuana's financial significance was exemplified by the December 1998 uproar in St. Vincent, the second largest Caribbean producer after Jamaica, when US troops helped destroy over one million plants; and growers began "a petition drive to ask President Clinton for damages" (Navarro, 1999, p. 4). The footprint of the narco dollar is deep.

Long-Run Implications

Evidence is mounting that the narcoeconomy will continue to grow and pose increasing threats to tourism and offshore finance. In the first case, Caribbean history has largely been the creation of global capitalism. In the same way the narco trade has been accelerated by the worldwide forces of globalization unleashed by the fall of Communism and nourished by the region's economic uncertainty, withering exports, and limited enforcement capability. The crime syndicates that control drug activity represent an almost intractable mega-multinational threat to local authorities because of their criminal sophistication, adaptability, financial power, and increasing political influence.

Second, the effectiveness of US-led counternarcotics activities in the region, which have not materially reduced drug availability over the past decade, has been hampered by fluctuating funding levels, conflicting objectives, and lack of operational coordination (US GAO, 1997). Specifically, it has been significantly blunted by the recent closure of Howard Air Force Base, the center of US anti-drug operations, as part of the Panama Canal Treaty, and will suffer further weakening with the proposed closure of the Roosevelt Roads Naval Base in Puerto Rico. Moreover, very recently the United States has diverted significant Coast Guard resources from drug interdiction efforts and patrolling Caribbean waters to shore up the Mexican border and bolster homeland security (UNODC, 2003, p. 22). As a result, today the United States has likely less than a third of the personnel and equipment it employed before 1990 (Abel, 1999). There has been a similar scaling back of European anti-narcotics efforts in the region (Sanders, 2003). The Caribbean has witnessed a concomitant rise in trafficking.

A third complication is coordination friction. Local autonomy is threatened by the perceived aggressive hegemony of the United States unilaterally implementing anti-drug countermeasures "in her backyard." These include sending in Marines to destroy marijuana plants at remote island outposts, as well as the recent Ship Rider Agreement that allows US incursions into Caribbean territory and waters "in hot pursuit" of traffickers (de Albuquerque, 1997). The sacrifices of local sovereignty caused by these US pressures to stanch narcotics trade have been difficult to swallow for Caribbean leaders who perceive the underlying problem to be primarily North American addiction. Moreover, the climate of cooperation has obviously been weakened since the 1990s by declining US aid to the region and by the deportation of known drug felons from overcrowded US prisons. For all these reasons, it is plausible to assume the narcoeconomy will continue to grow in a business-as-usual climate, without any sharp drop in North American drug consumption and/or dramatic success in enforcement and interdiction.

In the second case, the narcoeconomy with its violence, addiction, and corruption will undoubtedly pose the greatest threat to the sustainability of tourism. The paramount prerequisites of visitors spending their hard-earned income on holiday are safety and freedom from fear of victimization. Given the importance of visitor security as a condition for industry success, it is no wonder that crime, along with its drug-trafficking undercurrents, is considered the "most systemic internal threat to Caribbean tourism" (McElroy, 2004, p. 53).

This conclusion is evident from a variety of sources. First, even the perception of danger or irritation can tarnish a destination's reputation. In a survey of major US tour operators, King (2003) found the two most important factors deflecting visitors away from the Caribbean were fear of crime and harassment. Second, there is some reality to these fears since research has consistently shown that (1) visitors are more likely than residents to be victims of crime (larceny, theft, robbery), and (2) visitors are disproportionately targeted in "hot-spot" locations they are most likely to frequent (Harper, 2001). Third, there are numerous anecdotes documenting the links between major and widely publicized crimes and negative visitor fallout. One of the most notorious was the so-called Fountain Valley Massacre in 1972 in St. Croix, USVI. Eight tourists were gunned down by local thugs during a robbery. It took the island over 10 years to fully recover (de Albuquerque and McElroy, 1999a).

Fourth, academic research both on cross-national patterns and case studies covering long-term trends underline the negative impact of crime on visitor flows. In a multi-island study, Levantis and Gani (2000) found a 1% increase in crime rates was associated with a half percent decline in visitation in the Caribbean. In a longitudinal study of Jamaica, Alleyne and Boxill (2003) found a weak negative link between crime and US arrivals but a stronger relationship with Europeans, who tend to stay longer and spend more. The authors further argued that some of the negative fallout on tourism from the rising crime wave was likely offset by heavy promotional advertising as well as substantial hotel/airline discount packages, which are not sustainable practices in the long run.

Fifth, in a host of other less obvious ways, the narcoeconomy is impacting tourism throughout the region. It is no mystery that visitor harassment is more widespread in the Caribbean than in any other region across the globe, and that drug peddling is a major aspect of the problem (McElroy, 2003). Such behavior is fueled partly by local drug pushers seeking sales as well as by addicts needing income to satisfy their own habit. In destinations like Jamaica and St. Croix, USVI, harassment and theft against cruise passengers have become so dangerous that cruise lines have dropped such stops from their itinerary, costing the islands millions in lost revenue. Even attempts to control such problems may be somewhat unnerving for unsuspecting visitors. The presence of uniformed police patrolling beaches, docks, and resort properties and guard dogs sniffing for narcotics at air and seaports may seem incongruous with a typical vacationers' paradise image, and may negatively affect their word-of-mouth recommendations and preference for a return visit.

Sixth, the high cost of security, surveillance, and insurance in the private sector adds to the already relatively steep price of a Caribbean vacation, especially in the context of expanding global travel markets and intensifying competition from Asia and Pacific destinations. In the public sector, substantial resources devoted to resident and visitor protection are diverted from more productive developmental uses:

refurbishing the infrastructure, devising new attractions, human resource training, and tourism promotion. Harriott (2002, p. 5) estimates the overall cost of violent crime in Jamaica to be 6% of GDP per capita on an annual basis.

Seventh, a number of writers have argued that rising crime is particularly problematic for tourism since the industry is ultimately based on personal interaction and a welcoming experience (Dunn and Dunn, 2002). According to Harriott (2000, p. xv), a crimogenic environment breeds "insecurity among all segments of the population [including investors and developers] . . . declining public confidence in the criminal justice system, and growing cynicism among its functionaries," that is, a poor climate for business profitability and risk-taking. Sanders (2003, p. 384) cites anecdotal evidence that in the wake of a recent rash of kidnappings in Trinidad and Guyana "investment and tourism have been badly affected, and even social life curtailed." Sustained criminality also produces a loss of human capital through the emigration of middle-class professionals (accountants, bankers, realtors, planners, hotel managers, etc.) and entrepreneurs who provide the skill base for the tourism economy. They have the most to lose in a deteriorating climate and the wherewithal to succeed elsewhere.

Eighth, the hotel sector's principal response to shield visitors from potential harm has been the rapid expansion of all-inclusive resorts with their full complement of on-site gift shops, restaurants, night clubs, water sports, etc., for the convenience of a fixed daily fee. While these popular gated properties protect tourists and generate high foreign exchange and year-round employment, they have also "triggered a new wave of resentment" (Pattullo, 1996, p. 74) among taxi drivers, small shopkeepers, and other vendors who feel shut out. Adding to the charge of enclavism and low local income circulation, critics argue that all-inclusive resorts reinforce the worst perceptions of socioeconomic inequality attributed to Third World tourism, justify harassment among excluded street vendors and hagglers, and weaken community support for an industry that "heavily depends on the consent of the host population" (Dunn and Dunn, 2002, p. 31). Such a strategy undermines sustainability because it fails to address the "daunting task of creating a more egalitarian tourism in which all social strata share a viable stake and the motivation to sustain it" (McElroy, 2004, p. 53).

Ninth, all-inclusive resorts can also symbolize for unemployed youth that they lack access to legitimate channels of mobility and are left "condemned . . . to an economically marginal existence" (Headley, 1996, p. 39). For many of these disenfranchised outsiders, according to Block and Klausner (1987, p. 99), "prostitution and drug dealing have . . . become 'the only viable equal opportunity enterprises' for the lowest socioeconomic classes." Another ripple of debilitation, according to Griffith (2003)—fueled in part by in-kind cocaine payments for trafficking—is the associated sharp rise in the HIV/AIDS epidemic. Although it is concentrated in Haiti and the Dominican Republic, it is second only to sub-Saharan Africa in prevalence. In the short run this trend will place increasing pressure on island health budgets, and in the long run it will further deplete the insular labor pool.

Finally, the volume of drug-related cash crisscrossing the region has become so large that money laundering threatens to undermine legitimate offshore activity. Over the past decade, most islands with an offshore sector have either been implicated or placed under serious scrutiny. The three major problems have been excessive bank secrecy, poor supervision, and inhibiting international criminal investigations (Booth and Drummond, 1996). In response the OECD's Financial Action Task Force blacklisted a number of jurisdictions with the threat of sanctions until they agreed to adopt new anti–money laundering legislation created to tighten supervision and improve information exchange. All but one (Antigua) has complied. In the process, however, the international credibility of offshore banking was damaged (Bryan, 2000). According to Sanders (2003, p. 381), offshore banking in one jurisdiction has virtually collapsed and "there has been a significant reduction in the number of [offshore] businesses, revenue, and employment."

Conclusion

All three waves of Caribbean globalization were created by external capital and powered by foreign demand. As such they also share the same large intrusive scale that has left deep and lasting imprints on the fragile insular ecology and demography. Given this history, the highly lucrative nature of the drug transit trade, the durability of metropolitan consumer demand, the continuing problems of North–South cooperation and coordinated enforcement, and the ongoing American and European anti-trafficking retrenchment, the narcoeconomy should continue to flourish and weaken the region's political economy. According to Richardson's (1992, p. 131) "business as usual" comment, the narco trade resembles an extension of the colonial economy. "In producing and transporting narcotics for metropolitan consumption, Caribbean peoples simply are providing tropical staples for external sources, just as they have for the past five centuries."

This pessimistic prognosis is likely unless at least two critical tasks are accomplished. The first is a new comprehensive North–South anti-drug counteroffensive anchored by three principles. First, it must seriously address innovative ways to reduce burgeoning demand in the United States and Europe. Second, it must involve enhanced airport, border, and Customs security, the international distribution of narcotics intelligence, the relaxation of bank secrecy codes, and so on. Third, it must be a truly cooperative partnership combining the financial and technical resources of the North with the Caribbean's on-the-ground expertise. Only such a concentrated and concerted program can stanch the growth of a narcoeconomy that, in a sense, has been centuries in the making.

Second, serious efforts must be dedicated to strengthening the international competitiveness of tourism, as well as integrating all segments of society into the fabric of the industry. This will require considerable innovation along with a long-term commitment to training a new cadre of skilled young professionals. The goal is that in the future ordinary islanders would be at the forefront in creating, marketing, and delivering new products and services (Dunn and Dunn, 2002). Failing these two tasks, the alternative scenario may be a continuing syndrome of violence and instability that inhibits tourism's full potential (Sonmez, 2002). Far worse, over time, fragments of the region may slide unwittingly toward an accommodation with the disease, tacitly condoning a modicum of corruption that gradually erodes public integrity, the rule of law, and the culture of democracy (Bryan, 2000), and in the process spawning a lost generation of youth with "a decreased sense of the value of life, a lack of respect for property, and a lesser appreciation for honest work" (Griffith, 1997, p. 151).

Concept Definitions

Globalization Transborder economic activity embracing the exchange of goods and services between countries as well as the flow of labor (international migration) and capital (foreign investment).
Narcoeconomy The production, distribution, and sale of illegal narcotics. In the Caribbean, it principally refers to all the economic transactions that facilitate the trafficking of cocaine and heroin through the islands from South to North America.
Offshore finance Multinational banks, insurance, and business corporations dealing almost exclusively with metropolitan markets and clientele but located in islands for bookkeeping purposes attracted by their bank secrecy laws and low taxes.
Violent crime Includes crimes against persons: murder, rape, and aggravated assault (serious wounding).
Property crime Includes crimes against property: burglary, larceny, and robbery, which may also involve serious physical harm to the victim.

Review Questions

1. Why has the Caribbean become a major drug-trafficking route?
2. How large is the Caribbean narcoeconomy in dollar terms? Give a basis for comparison.
3. How is crime affected by development? What is the opportunity cost theory of crime?
4. What are the various evils associated with drug trafficking?
5. What assets do the islands have that facilitate money laundering? How big a problem is this?
6. In what ways will trafficking affect tourism in the long run?

References

Abel, D. (1999, July 9). Holes open in U.S. drug-fighting net. *Christian Science Monitor* (online), 1–3. http://www.csmonitor.com. Accessed July 9, 1999.

Alvazzi del Frate, A. (2000). The growth, extent and causes of crime: international crime overview, in M. Shaw (ed.), *Conference on Crime and Policing in Transitional Societies*. Johannesburg, South Africa: University of Witwatersrand (September), pp. 69–75.

Alleyne, D., and Boxill, I. (2003). The impact of crime on tourist arrivals in Jamaica, in A. Harriott (ed.), *Understanding Crime in Jamaica: New Challenges for Public Policy*. Kingston, Jamaica: University of West Indies Press, pp. 133–156.

Becker, G. (1968) Crime and punishment: An economic approach. *Journal of Political Economy,* 76, 169–217.

Bernal, R. (1992). Debt, drugs, and development in the Caribbean. *TransAfrica Forum,* 9, 83–92.

Block, A. A., and Klausner, P. (1987). Masters of Paradise Island: Organized crime, neo-colonialism, and the Bahamas. *Dialectical Anthropology,* 12, 85–102.

Booth, C., and Drummond, T. (1996). Caribbean blizzard. *Time Australia,* 2(26), 50–52.
Bonnick, B. (1994). Crime and violence: Its applications for economic expansion, in P. Lewis (ed.), *Jamaica: Preparing for the Twenty-first Century*. Kingston, Jamaica: Ian Randle, pp. 148–160.
Bryan, A. T. (2000). Transnational organized crime: The Caribbean context. Working Paper No. 1. Coral Gables, FL: University of Miami North–South Center. http://www.miami.edu/nsc/. Accessed September 15, 2004.
Chesney-Lind, M., and Lind, I. Y. (1986). Visitors as victims: Crimes against tourists in Hawaii, *Annals of Tourism Research,* 13, 167–191.
Central Intelligence Agency (CIA). (2002). *The World Factbook.* Washington, DC: USGPO. http://www/cia/gov/cia/publications/factbook/index.html. Accessed March 15, 2003.
Conway, D. (1998), Micro states in a macro world, in T. Klak (ed.), *Globalization and Neo-Liberalism: The Caribbean Context.* Lanham, MD: Rowman & Littlefield, pp. 51–63.
Commonwealth Secretariat. (1997). *A Future for Small States: Overcoming Vulnerability*. London: CS.
de Albuquerque, K., and McElroy, J. (1999a). Tourism and crime in the Caribbean. *Annals of Tourism Research,* 26, 968–984.
de Albuquerque, K., and McElroy, J. (1999b). A longitudinal study of serious crime in the Caribbean. *Caribbean Journal of Criminology and Social Psychology,* 4, 32–70.
de Albuquerque, K. (1997). New "big stick" policy—The Shiprider Agreement. *Caribbean Week,* February 14, 6–7.
de Albuquerque, K. (1996a). Give me a five dollar: The drug menace in the Eastern Caribbean. *Caribbean Week,* January 20–February 2, 1–2, 4, 5.
de Albuquerque, K. (1996b). How drugs get where they're going. *Caribbean Week,* February 3–16, 1–5.
de Albuquerque, K. (1996c). Looting and shooting and killing in a rampage. *Caribbean Week,* August 31–September 13, 1–2.
Dunn, H. S., and Dunn, L. L. (2002). Tourism and popular perceptions: Mapping Jamaican attitudes. *Social and Economic Studies*, 51, 25–45.
Economist. (1999). The Caribbean's tarnished jewel, 353(8139), 37–38.
Forst, B., and Bennett, R. R. (1998). Unemployment and crime: Implications for the Caribbean. *Caribbean Journal of Criminology and Social Psychology,* 3, 1–29.
Fuji, E., and Mak, J. (1980). Tourism and crime: Implications for regional development policy. *Regional Studies,* 14, 27–36.
Gayle, D. J. (1998). Trade policies and the hemispheric integration process, in T. Klak (ed.), *Globalization and Neoliberalism: The Caribbean Context.* Lanham, MD: Rowman and Littlefield, pp. 65–86.
Ghai, D. (ed.). (1991). *The IMF and the South: The Social Impact of Crises and Adjustment.* London: ZED Books.
Grant-Wisdom, D. (1994). Constraints on the Caribbean state: The global and policy contexts. *21st Century Policy Review,* 2, 153–179.
Griffith, I. L. (2003). Caribbean security in the age of terror: Challenges of intrusion and governance, in D. Benn and K. O. Hall (eds.), *Governance in an Age of Globalization: Caribbean Perspectives*. Kingston, Jamaica: Ian Randle, pp. 383–415.
Griffith, I. L. (1997), *Drugs and Security in the Caribbean: Sovereignty under Siege.* University Park, PA: Pennsylvania State University.

Harper, D. W. (2001). Comparing tourists' crime victimization. *Annals of Tourism Research,* 28, 1053–1056.
Harriott, A. (2003). Social identities and the escalation of homicidal violence in Jamaica, in A. Harriott (ed.), *Understanding Crime in Jamaica: New Challenges for Public Policy.* Kingston, Jamaica: University of West Indies Press, pp. 89–112.
Harriott, A. (2002). *Crime Trends in the Caribbean and Response.* Kingston, Jamaica: University of West Indies Press/UN Office of Drugs and Crime.
Harriott, A. (2000). *Police and Crime Control in Jamaica: Problems of Reforming Excolonial Constabularies.* Kingston, Jamaica: University of West Indies Press.
Harriott, A. (1996). The changing social organization of crime and criminals in Jamaica. *Caribbean Quarterly,* 42, 61–81.
Headley, B. (1996). *The Jamaican Crime Scene: A perspective.* Washington, DC: Howard University Press.
Interpol. (1995–2000). *International Crime Statistics.* Lyon, France: General Secretariat. http://www.interpol.com/Public/Statistics/ICS/Default.asp. Accessed March 14, 2003.
King, J. W. (2003). Perceptions of crime and safety among tourists visiting the Caribbean, in A. Harriott (ed.), *Understanding Crime in Jamaica: New Challenges for Public Policy.* Kingston, Jamaica: University of West Indies Press, pp. 157–175.
King, J. W. (1997). Paradise lost? Crime in the Caribbean: A comparison of Barbados and Jamaica. *Caribbean Journal of Criminology and Social Psychology,* 2, 30–44.
Klak, T. (ed.). (1998). *Globalization and Neoliberalism: The Caribbean Context.* Lanham, MD: Rowman and Littlefield.
Larmer, B. (1998, August 10). He's one tough nut. *Newsweek,* 41.
Leggett, T. (2000). Crime as a development issue, in M. Shaw (ed.), *Conference on Crime and Policing in Transitional Societies.* Johannesburg, South Africa: University of the Witwatersrand (September), pp. 141–150.
Levantis, T., and Gani, A. (2000). Tourism demand and the nuisance of crime. *International Journal of Social Economics,* 27, 959–967.
McElroy, J. (2004). Global perspectives of Caribbean tourism, in D. Duval (ed.), *Tourism in the Caribbean: Trends, Development, Prospects.* London: Routledge, pp. 39–56.
McElroy, J. (2003). Tourist harassment: Review and survey results, in A. Harriott (ed.), *Understanding Crime in Jamaica: New Challenges for Public Policy.* Kingston, Jamaica: University of West Indies Press, pp. 177–195.
McElroy, J., and de Albuquerque, K. (1998). Tourism penetration index in small Caribbean islands. *Annals of Tourism Research,* 25, 145–168.
Navarro, M. (1999). An outpost in the banana and marijuana wars. *New York Times* (April 4). http://www.nytimes.com. Accessed April 4, 1999.
Padilla, A., and McElroy, J. (2004). The TPI in large islands: Dominican Republic. *Journal of Sustainable Tourism* (forthcoming).
Pattullo, P. (1996). *Last Resorts: The Cost of Tourism in the Caribbean.* London: Cassell.
Pearson, F. S., and Payaslian, S. (1999). *International Political Economy: Conflict and Cooperation in the Global System.* Boston, MA: McGraw-Hill.
Phillips, D. (1994). The IMF, structural adjustment and health in the Caribbean. *21st Century Policy Review,* 2, 130–149.
Reese, C. (1997). War on drugs is a charade. *Caribbean Business,* (March 13), 34–35.
Richardson, B. (1992). *The Caribbean in the Wider World, 1492–1992.* Cambridge, UK: Cambridge University Press.

Robotham, D. (2003). Crime and public policy in Jamaica, in A. Harriott (ed.), *Understanding Crime in Jamaica: New Challenges for Public Policy*. Kingston, Jamaica: University of West Indies Press, pp. 197–238.

Rohter, L. (1997). Impact of NAFTA pounds economies of the Caribbean. *New York Times,* (January 30), 1.

Rushston, J. P., and Whitney, G. (2002). Cross-national variation in violent crime rates: Race, r-k theory, and income. *Population and Environment,* 23, 501–512.

Sanders, R. (2003). Crime in the Caribbean: An overwhelming phenomenon. *The Round Table,* 370, 377–390.

Sonmez, S. (2002). Sustaining tourism in islands under sociopolitical adversity, in Y. Apostolopoulos and D. J. Gayle (eds.), *Island Tourism and Sustainable Development: Caribbean, Pacific and Mediterranean Experiences,* Westport, CT: Praeger, pp. 161–180.

Stone, C. (1983). *Democracy and Clientilism in Jamaica*. New Brunswick, NJ: Transaction.

United Nations Office on Drugs and Crime (UNODC). (2004). *Global Illicit Drug Trends 2003*. Vienna, Austria: Office on Drugs and Crime.

UNODC. (2003). *Caribbean Drug Trends 2001–2002*. Bridgetown, Barbados: Caribbean Regional Office on Drugs and Crime.

United Nations Development Programme (UNDP). (1999). *Human Development Report 1999*. New York: Oxford University Press, UNDP.

US Department of Justice, Drug Enforcement Administration (US DEA). (2003). *The Drug Trade in the Caribbean: A Threat Assessment*. Washington, DC: DEA-03014. http://www.usdoj.gov/dea/pubs/intel/03014/03014.html. Accessed September 14, 2004.

US DEA. (1998). *The Supply of Illicit Drugs into the United States*. Washington, DC: Drug Enforcement Administration, National Narcotics Intelligence Consumers Committee (November).

US General Accounting Office (US GAO). (1997). *Drug Control: Long-Standing Problems Hinder U.S. International Efforts*. Washington, DC: GAO/NSIAD-97-75, Letter Report, February 27.

Vellas, F., and Becherel, L. (1995). *International Tourism: An Economic Perspective*. New York: St. Martin's Press.

Waters, E. J. (1999). "Music of language": An interview with George Lamming. *The Caribbean Writer,* 13, 190–201.

Watson, H. (1997). Global change: Restructuring the enterprise culture and power in contemporary Barbados. *Journal of Eastern Caribbean Studies,* 2, 1–47.

West Indian Commission (WIC). (1992). *Time for Action*. Bridgetown, Barbados: WIC.

World Travel and Tourism Council (WTTC). (2004a). *The 2004 Travel and Tourism Economic Research*. London: WTTC. http://www.wttc.org. Accessed August 30, 2004.

WTTC. (2004b). *The Caribbean: The Impact of Travel and Tourism on Jobs and the Economy*. London: WTTC. http://www.wttc.org/measure/PDF/Caribbean FULLTSA.pdf. Accessed September 2, 2004.

6

Do Incidents of Theft at Tourist Destinations Have a Negative Effect on Tourists' Decisions to Travel to Affected Destinations?

Judy Holcomb and Abraham Pizam

Learning Objectives

- To describe the effects of crime against tourists on tourism demand at affected destinations.
- To discuss the effects of crime against tourists on repeat visitation at affected destinations.
- To understand the reasons why crime occurs at tourist destinations.
- To know the specific effects that theft at tourist destinations have on visitation and revisitation.

Introduction[1]

Crimes against tourists occur throughout the world on a daily basis (Pizam and Mansfield, 1996). These incidences cause not only anguish and at times considerable financial loss, but can totally ruin the trip's experience. Tourists take trips to relax and get away from the stress of their lives or to conduct important business. But the feelings of vulnerability that can accompany victimization can alter perceptions about the destination, lead tourists not to return to the affected destination, or stop them from traveling altogether.

Media reports and anecdotal evidence suggest that theft is the most common crime experienced by tourists. But despite this, to date there have been few studies that specifically looked at theft against tourists and analyzed the effects of this victimization on future travel plans. In addition, it is not clear whether hearing that friends or relatives have been victims of theft incidences in a particular destination would affect tourists' intention to travel to that destination.

The purpose of this study was to determine the extent to which having been a victim of a theft while on a business or leisure trip, or having heard of such an account from a friend or relative, will affect a person's intention to visit or revisit the affected destination.

Crime and Tourism

Crime and Its Personal and Societal Effects

The impacts of crimes can be viewed from two perspectives, macro and micro. The macro impacts of crime pertain to how crime affects society in general. This relates to the effects of crime on a community or a tourist destination. The micro impacts relate to how crime affects a person or victim.

Crime affects society on a macro level in many ways. Economically, crime costs consumers billions of dollars each year; the economic effects are passed on to the consumer in the form of higher prices. "Economic costs of crime arise when crime causes society to divert time, energy, and resources from more productive resources" (Entorf and Spengler, 2002). Because public authorities are given the charge of preventing crimes and prosecuting criminals, the citizens are the ones who end up paying for these services through increases in local, state, and national taxes. Another economic cost to society is the expense of incarcerating convicted criminals.

The impacts of crime on societies as a whole can be devastating. Societies can cease to function normally if crime becomes the overriding concern. For example, the sniper shootings of 2002 in the Washington, DC, area interrupted the function of normal life for the majority of citizens during that period. Fear spread throughout the area causing citizens to avoid gas stations, craft stores, and other places where previous shootings had occurred. Parents stopped taking children to parks, school recess was cancelled, and virtually all outdoor activities were halted (Sernike, 2002). Hence, crime in many forms can interrupt normal life.

The micro impacts of crime involve its consequences on individuals. Fear of crime can have a major effect on an individual's life whether or not the fear is actually rational. Fear of crime can cause people to remain in their home, curb activities, and even avoid travel altogether. Victimization of individuals can have various physical as well as psychological effects. Medical costs, counseling, and lost wages can cause great burdens for victims. In 1997, it was estimated that in the United States alone firearm injuries related to crimes cost $802 million in hospital charges (Srikameswaran, 2003). In 1998, according to the FBI, the cost of criminal victimization—excluding medical treatments—in the United States was said to exceed $15.8 billion (USDOJ and BJS, 1999).

The financial costs of crimes to victims may be minor compared to the pain and suffering experienced by victims and their families. These effects can be serious and long lasting. Even victims of burglary, which is a crime conducted against a dwelling, can cause an individual to become fearful of being victimized again. Those fears may lead to short- or even long-term interruption in normal life functions.

Crime at Tourist Destinations

Unfortunately, tourists and travelers have always been victims of both violent and nonviolent crimes. Violent crimes such as robbery, rape, assault, and murder have a more negative effect on tourists than do personal property crimes. Violent crimes are, at times, life altering and can be more devastating to a victim than nonviolent crimes such as thefts (Pizam, 1999). Incidents such as the killing of a disabled American tourist on the *Achille Lauro* cruise ship bring massive media attention. This incident caused tourists to avoid Mediterranean cruise travel and opt for the Caribbean and Alaskan lines instead (Caribbean and Alaska Cruise, 1986). Other incidents of violent crime against tourists were the killing of a pregnant German mother in Miami and a male English tourist near Tallahassee, Florida. These murders caused a decrease in tourist arrivals to the state of Florida and tarnished the image of the state as a safe tourist destination (Pizam, 1996, pp. 1–2).

Fujii and Mak (1980) studied crime in Hawaii and found a statistically significant relationship between tourism and violent crimes such as murder/homicide and rape. Chesney-Lind, Lind, and Schasfsma (1983) found a similar significant relationship between tourism and the rate of assault in Honolulu. Another study performed in Hawaii by Chesney-Lind and Lind (1986) looked at violent crimes in two different locales, Honolulu and the island of Kauai. This study showed that tourists were significantly more likely to become victims of the violent crimes of robbery and rape than local residents. In Honolulu, violent crimes against tourists were 27% higher than against residents. More specifically tourists were 11% more likely to be a victim of rape than were residents. The island of Kauai showed similar results. The rape rate of tourists was nearly three times that of residents and twice that of the national average.

Several theories of criminology have been used to understand how tourist destinations expose visitors to the risk of criminal victimization (Crotts, 1996). Two predominant theories, the Hot Spot theory and the Routine Activities Theory (Cohen and Felson, 1979) were used to attempt to place the location and incidence of crimes against tourists into theoretical contexts. The Routine Activity Theory suggests three elements for a predatory crime to occur: a suitable target, a motivated offender, and the absence of capable guardians (Cohen and Felson, 1979). Tourist destinations seem to fit into this framework. Usually, tourists make a suitable target because of their tendency to carry large amounts of cash and unknowingly roam in areas that residents would consider dangerous. The offenders are motivated to victimize tourists because of their view of them as the "haves" and view of themselves as the "have-nots," the so-called Robin Hood theory. In addition, because many tourist destinations are not willing to admit that tourist crime is a problem, they may not have the capable guardians or law enforcement agents in place in order to deter tourist crime.

The Hot Spot theory can also be used to explain the reason why certain locations in tourist destinations seem to expose tourists to incidences of victimization. As suggested by Crotts (1996), "Places where tourists are at the greatest risk of becoming victimized have been shown to cluster in a few specific types of places" or hot spots (Crotts, 1996). These hot spots provide a place of opportunity at which predatory crimes can occur. From the Uniform Crime Reports, Crotts provided statistical information for the state of Florida. For example, Dade County (the greater Miami area) was the location of 30% of all reported property crimes against tourists and 37% of all reported violent crimes against tourists in the state of Florida during 1993. To put these statistics in perspective, in the same year Dade County hosted 16% of the state's tourists, but was the location of 30% of the total crimes committed against tourists.

Cochran, Bromley, and Branch (2002) also used the Hot Spot theory in their study analyzing victimization and fear of crime in the entertainment district of Ybor City in Tampa, Florida. The study tested Meithe and Meier's (1990) "structural choice" model. This model examines structural features such as proximity to offenders and exposure to risk and choice features such as target attractiveness and guardianship. The results of the study indicated that victimization and fear of crime were significantly associated with two constructs of the structural choice theory, proximity and guardianship. However, victimization and fear of crime were not significantly associated with exposure or target attractiveness. The proximity construct was demonstrated by findings showing that living or working in Ybor City carried a higher risk of victimization. Ybor City patrons who report having victimized friends were 4.6 times more likely to be crime victims themselves than those who did not live or work in Ybor city. The guardianship construct of the theory was demonstrated by findings that patrons who perceived higher levels of guardianship were less likely to be crime victims. This is the case, even though the Ybor City entertainment district tends to host a large number of local residents.

In a comparison of tourist crime and tourist seasons in Miami, seven major crimes, namely murder, rape, robbery, assault, larceny, burglary, and auto theft, were studied from 1963 to 1966. The study investigated whether seasonal tourism generated increased crime rates against persons and property. The results indicated that the tourist season (which is December through April) and the crime season seem to show a similar pattern. More specifically, major economic crimes, such as robbery, larceny, and burglary, had a similar season to tourism, while auto theft and crimes of passion did not (McPheters and Stronge, 1974).

Harper (2001) suggests that crime against tourists is a rational, rather than a spontaneous, process. To prove this, Harper interviewed police informants who provided valuable information regarding how they perceive tourists as victims and their thought process of preying on those victims. The author concluded that criminals seem to take advantage, in a rational and calculated manner, of tourists in vulnerable situations (Harper, 2001).

Tourists can be considered vulnerable to criminal victimization because they:

1. Are obvious in their dress;
2. Carry items of portable wealth;
3. Are relaxed and off guard;

4. Are less likely to press charges should the criminal be caught;
5. Are not familiar with the surroundings;
6. Engage in risky activities; and
7. Have no social support at the destination

The Impact of Crime on Tourism Demand

The impact of crime on tourist arrivals is a major concern to tourist destinations. Most tourist destinations try to paint a beautiful picture of their areas in order to entice travelers to visit. But governments as well as the media frequently warn the public about the dangers of traveling to destinations that are frequently plagued by criminal activities against tourists. In the United States, the State Department issues travel warnings for certain countries and areas to warn potential tourists of safety/security concerns, sometimes suggesting that Americans avoid traveling to certain areas altogether. The news media also plays a role in the dissemination of this information. As an example, an Annual Personal Safety Survey by Mercer Human Resource Consulting, cited in the British *Daily Telegraph* (Starmer-Smith, 2003) listed the world's most crime-prone tourist destinations. In order of severity, they were Kingston, Jamaica; Rio de Janeiro, Brazil; Cape Town, South Africa; Mexico City, Mexico; St. Petersburg, Russia; Buenos Aires, Argentina; Bangkok, Thailand; Washington, DC, USA; Rome, Italy; and Athens, Greece. Information obtained by tourists from media accounts or articles such as these could greatly affect a tourist's travel decisions.

Hong Kong, another tourist destination known for crime prevalence, had experienced in 2003 an escalation of street crime, which may have been the cause of a downturn in demand. Hong Kong had three times as many pickpocketing cases in 2003 as for the same period in 1998 (Fraser, 2003).

Negative media attention on a tourist destination in relationship to crime incidents can sometimes be blown out of proportion (Crystal, 1993). In the case of the murder of a pregnant German mother in Miami and a male British tourist near Tallahassee, Florida, tourism to Florida declined significantly. Media accounts suggested that crime was rampant against Florida tourists while official statistics told another story. These statistics showed that crime reported by nonresidents had in fact been on the decline in Florida since 1990; however, the perception of crime became reality following the much publicized shootings and tourist arrivals declined (Schiebler, Crotts, and Hillinger, 1995).

Although often ranked by travel intermediaries as one of the top three US urban destinations, New Orleans, another destination known for its high crime rates, has been threatened by the negative perceptions of potential tourists and meeting planners. During the early 90s New Orleans had a murder rate eight times that of the national average and five times that of New York City. Residents as well as visitors experienced theft and muggings, even in the streets of the French Quarter. Negative media coverage as well as anonymous letters sent to meeting planners regarding high crime rates contributed to this negative perception. In a study of 350 meeting planners and tour operators conducted by the University of New Orleans, New Orleans scored the lowest of eight cities regarding perception of visitor safety (Anonymous, 1996). These safety concerns seem to be more prevalent among travel intermediaries than among tourists themselves. At the time of the article, a New Orleans visitor profile report indicated that 82.4% of the respondents perceived New Orleans to be a somewhat safe or safe destination (Dimanche and Moody, 1997). The authors note that the results could be explained by the fact that visitors who have actually made the decision to travel to New Orleans could be considered risk takers and they have chosen the destination for its risqué image (Dimanche and Lepetic, 1999).

Alleyne and Boxill (2003) examined the relationship between tourist arrivals and crime rates in Jamaica. The results showed that crime rates had an overall negative effect on most markets. Though the impact of crime on the overall market was relatively small, the impact on European arrivals was large and significant. The authors provided several reasons for the above results, the most important of which was that with the exception of the Europeans, most tourists in Jamaica stayed in all-inclusive resorts and had very little contact with the outside world. The Europeans, however, did not frequent all-inclusive resorts, making them more susceptible to crimes (Alleyne and Boxill, 2003).

The Impact of Crimes on Tourists' Decision to Return to a Destination

Little information is available in the area of the effects of crime on revisitation. Mawby (2000) looked at revisitation as a very small aspect of his study. In his study Mawby looked at victims and nonvictims in relation to revisitation. The results indicated that 56% of the subjects who were victims of crimes at a tourist destination indicated that they would definitely return to the destination affected. Only 14% of victims said they would probably or definitely not return. These results were somewhat surprising, as one would have expected that most, if not all, of those who were victimized by a crime at a tourist destination would choose to not revisit it in the future. But in this study, victimization does not seem to have had a significant influence on the decision to return to a tourist destination.

In a study of international travelers, Sönmez and Graefe (1998) examined the influences of international travel regarding past experiences, types of risk, and the overall degree of safety felt during international travel. These factors were examined in conjunction with the likelihood of individuals to travel to or avoid various regions on their next international trip. The findings showed that past travel experience and risk perceptions influenced future travel behavior. Regarding safety, interest in future international travel was influenced by the degree of safety individuals felt during past international travel trips. As well, past travel experiences, in which perceptions of risk and feelings of safety were experienced, influenced the avoidance of those areas rather than the likelihood to return. Results also showed that travelers who have more experience of a particular region may become more confident and thus more likely to return.

Last but not least, in a study conducted by George (2003) tourists' decisions to return to a destination were analyzed. Overall, a little more than 50% said that they would very likely return to Cape Town, a South African city that is currently experiencing a high rate of crimes against tourists and residents alike. As expected, a negative association was found between having been a victim of crime or feeling unsafe and the likelihood of return. More than half of those who claimed that they were very likely to return to Cape Town had not encountered an incident of crime, or had not felt that their life was in danger during their stay. The purpose of the respondent's visit played a factor in their perception of safety as well. Respondents who were traveling on holiday and visiting friends and family were less likely to return than were business travelers. The author felt that this response was possibly because business travelers most likely do not have a choice of whether they return.

Theft at Tourist Destinations

Theft is the most prominent crime against tourists as shown in many studies regarding crime and tourism (Brayshaw, 1995; Chesney-Lind and Lind, 1986; de Albuquerque and McElroy, 1999; Harper, 2001; Mawby, 2000). The history of theft against tourists or travelers can be traced back to medieval times when highway robbers were the fashion. Their victims were well-to-do travelers in carriages, stagecoaches, or on horseback. One of the most famous highway robbers was said to be Robin Hood, who was known for robbing the rich (travelers) and giving to the poor (Brandon, 2001).

Theft encompasses several types of crime, the broadest being larceny. Larceny involves the taking of property from a person without that person's consent and with the intent to deprive the person permanently of the use of the property (Brown et al., 1996). The two types of larceny are *grand larceny* and *petty (petit) larceny,* depending upon the value of the property stolen. Another form of theft is robbery. Brown et al. (1996) define robbery as "a form of theft in which goods or money are taken from a person against that person's will *through* the use of *violence or fear*. The key words in this definition are *violence or fear*. These words differentiate this type of theft from larceny. Burglary in general is defined as a crime against a dwelling. In the case involving tourists away from home, a dwelling can be considered as a hotel room or a place where a tourist resides (except a private residence) while traveling.

Several studies tried to analyze the prevalence of thefts experienced by tourists while traveling away from home. In a study by Mawby (2000), burglary was found to be the most prominent crime experienced by tourists with theft from person and attempted theft from person being the next most common. The study showed that 4.7% of tourists were victims of burglary, 3.5% experienced an attempted theft from their person, and 3.3% were actual victims of theft from their person. To compare the responses from their survey against victim survey data from the national British Crime Survey (BCS), the responses were multiplied by 26 (52 weeks per year/2) in order to achieve approximate annual incidence rates using an average length of vacation of 2 weeks. When the authors compared these results against the BCS, they found the rates estimated in their survey to be much higher than the BCS rates and concluded that people generally experience considerably higher rates of victimization while they travel than they do at home (Mawby, 2000).

In a study that compared crimes against tourists with crimes against residents in the Caribbean, the results indicated that burglary and larceny against tourists in Barbados between 1989 and 1993 were far more prevalent than murder, wounding, rape, or robbery. The results showed that residents were much more likely to be victims of violent crime (over six times more likely to be murdered or to be a victim of aggravated assault) than were tourists. On the other hand, tourists were disproportionately victims of property crimes and robbery (four to six times more) than residents were. Tourists were also much more likely to have valuables stolen from their persons, rooms, or vehicles than residents were. The study also indicated that police in several Caribbean destinations sometimes expressed a rather offhand attitude toward tourist property crimes and tended to respond in an apathetic way to this type of crime. Police often looked upon tourists as being incredibly naïve, at times even blaming the victims (de Albuquerque and McElroy, 1999).

The results of the above study were confirmed in another study that was conducted by Chesney-Lind and Lind (1986), and that was mentioned earlier in this chapter. Using police data from the locations of Honolulu and Kauai in Hawaii, the authors found that tourists were significantly more likely to be victims of crime than local residents. In both cities, data showed that tourists experienced a higher rate of larceny than residents did. In Honolulu, rates for burglary, larceny, and robbery were substantially higher (62%) for tourists than for residents. Furthermore, the results also indicated that tourists in Honolulu were robbed at a rate significantly higher than the national average. As to the island of Kauai, the results illustrated even a more astonishing picture. As it turns out, tourists were victims of robbery at a rate six times higher than that of residents. Similar to Honolulu, tourists had higher rates of robbery, larceny, and rape on the island of Kauai than residents.

Barker, Page, and Meyer (2002) conducted a study of crimes against tourists attending the 2000 America's Cup in Auckland, New Zealand. The main purpose of the America's Cup study was to compare the rates of crimes committed against international tourists to those against domestic tourists. Data was collected by interviewing nonresident tourists attending the America's Cup as well as the use of tourist victim information reports (TVIRs), which are police reports used for reporting crime against tourists. The results showed that property crime was predominant over violent crimes, 98.5% versus 1.5%. Fifty-five percent of crimes reported were theft from vehicles; other theft comprised 39.1%. The study also compared victimization of overseas versus domestic tourists. It was found that overseas tourists were more likely to experience theft from accommodations or their person and the value of items stolen was higher than that of domestic tourists. On the other hand, domestic tourists were more likely to experience theft from vehicles. With regard to location, 55% of thefts occurred in public places, 15.8% in accommodations, and 10.4% in camper vans. Of the total crime reported in the tourist victim information reports, 50% involved theft or burglary from accommodations and 29.4% theft from vehicles.

In an attempt to analyze the results of several studies, Harper (2001) compared tourist and resident populations' crime experience in five international locations. The study confirmed the findings of previous studies, stating that tourist victimization is higher than nontourist victimization. The author also confirmed that tourists are more likely to experience larceny, theft, and robbery than residents (Harper, 2001).

As demonstrated by the studies cited above, theft is undeniably the most prevalent crime against tourists. Studies have shown time after time that (a) tourists are more likely to be victims of thefts than are local residents, and

(b) tourists are more likely to be victims of theft of personal property than victims of violent crimes.

Study Objectives

By now it should be evident that incidents of crimes and especially various forms of thefts are most prevalent at tourist destinations. It is also evident that a high incidence of violent crimes such as murder, rape, assault, or robbery affects tourist arrivals through negative publicity that is caused by government warnings, media reports, and word of mouth. What is not so clear is whether incidents of nonviolent crimes such as theft at tourist destinations have a negative effect on tourist future decision to travel to an affected destination.

To answer this question, we undertook a study, the purpose of which was to determine what, if any, effects past incidents of personal theft on tourists have on future decisions to travel to the affected destination. This was done by analyzing two study groups, those who have experienced personal thefts while on a trip and those who heard of such incidents through personal accounts of friends or family.

More specifically we were interested to find whether:

- Tourists who experienced a personal theft while traveling will be less likely to travel to the affected destination than those who did not experience such theft, but knew someone who did.
- The more time passes between the incidence of a personal theft that occurred to tourists or their friends or relatives at a tourist destination and the decision to travel to the same destination, the higher is their likelihood of revisiting that destination.
- Tourists who have experienced a theft on their person while traveling will have a lower tendency to return to the same destination than tourists who have experienced a theft of their belongings in a property they were occupying (i.e., hotel room, car, etc.).
- The more severe is the incidence of theft that occurred to tourists or their friends or relatives while traveling, the lower the likelihood of visiting the affected destination.
- The more satisfied tourists are with the way the destination's authorities handled a crime report at a tourist destination, the higher is the likelihood that they will revisit the same tourist destination.
- Tourists who have previously been victims of crimes in their own community will be less likely to travel in the future to a destination where either they or their friends or family members experienced a crime incident, than those who have not been such victims in the past.
- The passage of time will positively affect the decision to revisit or visit a tourist destination where a theft crime occurred to tourists or their friends or relatives.
- Learning of someone's positive experience at a destination where tourists or friends or relatives experienced a theft in the past will positively affect the decision to visit the affected destination.
- Learning from the media about the efforts to improve security in a destination where tourists or their friends or relatives experienced theft in the past will positively affect the decision to visit the affected destination.

For the purpose of this study, personal theft is defined as "theft of personal property, either from one's person, a car, or lodging establishment, while a tourist is traveling."

Methodology

Sample

The sample consisted of a net of 1,017 respondents who were interviewed on the telephone during the third week of May 2004 by a professional polling firm that was engaged to add 13 questions on the subject of "incidents of theft while on a trip" to their weekly telephone omnibus survey. The sample is based upon a random digit dialing probability sample of all telephone households in the continental United States and represents households with both listed and unlisted numbers. All sampled numbers were subject to an original and at least four follow-up attempts to complete the interview. All completed interviews were weighted to ensure accurate and reliable representation of the total population 18 years of age and older according to the latest US Census figures.

Research Instrument

The survey consisted of 13 questions that were added on to the weekly omnibus household survey. With the exception of one open-ended question that asked for the location of the incidence of theft, all other questions were closed-ended. To be included in the survey, the respondents had to pass one of two qualifying questions as follows: (1) Having traveled over 50 miles outside of their home for business or pleasure—excluding commuting to and from work—within the past 10 years; and/or (2) knowing someone who has experienced a theft while on a business or leisure trip. Respondents were eliminated from the survey if they neither traveled over 50 miles nor knew anyone who had experienced a theft while on a business or pleasure trip. Those who traveled over 50 miles were further asked whether they experienced a theft while on a trip or not. Those who did not experience a theft were further asked if they knew anyone who did. Thus, the actual survey consisted of two groups, those who experienced a personal theft while traveling and those who knew of someone who had such an experience.

The theft incidence questions related to:

1. Geographical location of the theft.
2. Date of theft.
3. Whether the items were stolen off the respondent's person or from a location occupied by the person.
4. Perceived severity of the theft.
5. Physical location of the theft (car, hotel room, bus, etc.).
6. What other crimes (not while on a trip) the respondent has been a victim of in the past.
7. Whether the theft has been reported to the local law enforcement agency and how was it handled.
8. Size of the group in which the respondent (or person that they knew) traveled.
9. Future intention to travel to the destination where the theft occurred. And for those who did not intend to travel to those destinations:

10. What factors might change the respondent's decision to travel to the affected destination.

In addition, the following general demographic questions related to the theft survey participants were obtained from the omnibus survey:

- Family size and composition;
- Marital status;
- Gender;
- Race;
- Employment status; and
- Education.

The survey questions were tested in a pilot study conducted over the telephone with respondents who were past victims of thefts while traveling.

Results

Respondent's Profile

Of a total sample of 1,017 respondents, 215 were found to have either experienced personal theft or knew of someone who had. Of these 215 respondents, 69 or 32% had experienced personal theft while traveling and 146 or 68% had known of someone who had experienced personal theft while traveling. Fifty-three percent of the respondents were male and 47% were female.

A large majority of the respondents (68%) were under the age of 49. Of those who personally experienced theft while traveling, 73% of the respondents were under the age of 49 and only 4.5% over the age of 65. Of those who knew of a friend or relative who had experienced theft, 65% were under the age of 49, and 15% were over the age of 65.

As far as household income is concerned, 51% of the respondents had a household income larger than $50,000 per year.

The majority of the respondents or their friends/family who incurred theft (69%) were traveling with friends, family, or significant others when the theft occurred, while 23% were traveling alone.

Length of Time Since Thefts Occurred

More than three quarters of the respondents (78%) indicated that the crimes addressed in the survey occurred less than 5 years ago. Only 5% indicated that the thefts occurred between 5 and 10 years ago.

Location of Thefts

As can be seen from Table 1, the most prevalent locations indicated by the study respondents were "in a vehicle" and "in a hotel" with 25.1% and 25.7% of the responses, respectively. Of the specific answers to this question (N = 181) 76.4%

Table 1 Location of Theft

Location of Theft	*Percent*
At an airport	9.3
In a vehicle (rented or self-owned)	26.3
In a hotel	24.9
While sightseeing (off your person/off the person of whom you learned of the theft)	3.9
While on public transportation (off your person/off the person of whom you learned of the theft)	9.8
While visiting a tourist attraction, park, or museum (off your person/off the person of whom you learned of the theft)	5.9
Bar/restaurant	3.9
Home	4.4
Other	11.7
Total	100.0

of the responses indicated that the theft did not occur off the victim's person (i.e., the items were not stolen from somewhere on their body).

Severity of Thefts

Respondents were asked to rate the severity of the theft that they or their friends/relatives experienced on a scale of 1 through 7, where 1 indicates "not severe at all" and 7 "very severe." The mean rate of severity reported by the respondents was 4.7, with a standard deviation of 1.87 and a median of 5.0. Fifteen percent of the respondents indicated that the theft was not severe and 35% of the respondents indicated that the theft was either somewhat or very severe.

Reporting Thefts to Authorities

Seventy-nine percent of the respondents indicated that the thefts were reported to the authorities. When the respondents were asked to rate the manner in which the theft report was handled by the authorities on a scale of 1 through 7, where 1 indicates that the theft was handled extremely poorly and 7 indicating that the theft was handled extremely well, the mean and median were both 4.0, and the standard deviation was 2.1. Of the 138 respondents who indicated that the theft was reported, 32% claimed that it was handled "poorly" by the authorities and 31% indicated that the theft was handled "very well" or "extremely well."

Likelihood to Consider Traveling to the Affected Destination

When asked to rate the likelihood that they would consider traveling to the affected destination now or in the future on a scale of 1 through 7, where 1 indicates "definitely not" and 7 indicates "definitely yes," the mean response was 5.5 ("most likely yes") with a standard deviation of 2.2 and a median of 7.0. A very large portion of the respondents (55%) said that they would definitely travel now or in the near future to the destination where the theft occurred.

Factors That Might Change Respondents' Minds about Traveling

Table 2 lists the factors that might change respondents' minds about traveling to a destination where a theft occurred. This table represents only the responses of those respondents who indicated that they would "definitely not travel to the destination" where the theft occurred. Of these respondents, an absolute majority (67.3%) said, "nothing would change my mind" followed by "passing of time" (12.2%), and "learning of someone's positive experience" (12.2%).

Previous Crime Victimization

An absolute majority of the respondents (57%) indicated that the person (the respondent or the person the respondent knew who experienced the personal theft on a trip) had been a victim of personal and non-trip-related theft in the past. Twenty-three percent of the respondents had been a victim of thefts as well as more serious crimes.

Geographic Location of Theft

Tables 3 and 4 list the geographical location of the thefts discussed in the survey. Of the 204 responses to this question, 157 or 77% of the theft incidents took place in a domestic location and 47 or 23% took place outside of the United States. As can be seen from Table 4, the top US cities/districts where the incidents took place were Washington, DC (6.3%), followed by Las Vegas (5.1%), Atlanta (3.8%), and Chicago (3.8%).

Table 2 Factors That Might Change Respondents' Mind about Traveling*

Factor	*Percent*
Passing of time	12.2
Someone's positive experience	12.2
Learning from media	4.1
Other	4.1
Nothing will change my mind	67.3
Total	100.0

*Restricted to those respondents who indicated that they would "definitely not travel to the destination where the theft occurred."

Table 3 Top Five US Cities (District) Where Thefts Occurred

City	*Percent*
Washington, DC	6.3
Las Vegas	5.1
Atlanta	3.8
Chicago	3.8
New York City	2.6
Other cities*	78.3
Total	100.0

*Each individual city in the "other cities" category had a frequency of less than 2%.

Table 4 Top Five Countries Outside of the US Where Thefts Occurred

Country	*Percent*
Italy	12.7
Mexico	12.8
Canada	10.6
France	6.4
Spain	6.4
Other countries*	51.1
Total	100.0

*Each individual country in the "other countries" category had a frequency of less than 5%.

Internationally, Italy, Mexico, Canada, France, and Spain respectively were the most prominent with 49% of the incidents occurring in these countries. Both Italy and Mexico consisted of 13% of the responses each.

The Impact of Thefts on Future Intentions to Travel to Affected Destinations

Differences between respondents who have personally experienced thefts and those who knew someone who has experienced thefts on intentions to travel to affected destinations

To test for the statistical differences between those who have experienced theft themselves and those who have heard of such experiences, an independent t-test was conducted. The results show that there was no statistically significant difference between the two groups. The mean likelihood to revisit the affected destinations for those who have experienced a personal theft was 5.47 (1–7 scale where 1 indicates "definitely not likely to return" and 7 indicates "definitely likely to return") and the mean likelihood for those who knew someone who has experienced such thefts was 5.52. Both groups had expressed a relatively high likelihood to visit or revisit destinations in which they or their friends/relatives experienced an incidence of theft.

Relationship between length of time since theft occurred and likelihood to travel to the affected destinations

To test for this relationship, a Pearson product moment correlation was conducted between the length of time since the theft occurred and the likelihood to travel to affected destinations. The results indicate that there was no statistically significant correlation ($r = .07$) between the two variables ($p = .34$). Hence, the length of time since a theft occurred does not seem to influence the likelihood of visiting or revisiting the affected destination.

Differences between respondents who have experienced or knew others who have experienced "thefts off one's person" and those who have experienced or knew others who have experienced "thefts not off one's person," and likelihood to travel to the affected destinations

To test for the above difference, an independent t-test was performed comparing the type of theft—whether "off one's person" or "not off one's person"—and the likelihood to travel to the destination where the theft occurred. Those who experienced or knew of someone who experienced a theft "not off their person" (i.e., hotel room or tourist attraction) had a mean likelihood to visit or revisit the affected destinations of 5.52 as compared to a mean of 6.00 for those who experienced or knew someone who experienced a theft "off their person" ($t = -1.972$; $p = .06$).

Relationship between the perceived theft severity and likelihood to travel to the affected destinations

One would normally expect that the more severe the incidence of theft that occurred to a person or his/her friends or relatives while traveling, the lower the likelihood would be of revisiting or visiting the affected destination. But the results of the Pearson product moment correlation indicate no significant correlation ($r = -.09$; $p = .17$) between the two variables. In an effort to determine if both groups (those who experience a theft and those who knew someone else who experienced a theft) would produce the same results, a Pearson product moment correlation was separately run for only the group of respondents who experienced personal theft themselves. The results were similar to the one with both groups and showed no significant correlation ($r = -.12$; $p = .35$) between the perceived severity of the theft and the likelihood to travel to the affected destination.

Relationship between satisfaction with the manner in which the authorities handled the reporting of the theft and the likelihood to travel to the affected destinations

To test for above the relationship a Pearson product moment correlation was conducted. The results showed that the relationship is positive and statistically significant ($r = .46$; $p = .003$). Therefore it is possible to infer that the better the crime was perceived to be handled by the authorities, the higher the expressed likelihood to travel to the destination where the theft occurred. In other words satisfactory handling of theft reports tends to increase the likelihood of visiting or revisiting a destination where a theft occurred.

Relationship between previous victimization and likelihood to visit or revisit a tourist destination where a theft occurred to tourists or their friends or relatives

The results of the t-test comparing those who have been previous victims of crime in their own communities ("not while traveling") and those who have not been victims of crime in their own communities indicate that there was no significant difference between those who have been a previous victim of crime (mean likelihood to visit or revisit = 5.42) as compared to those who have not been a victim of crime (mean likelihood to visit or revisit = 5.52; $t = -.12$; $p = .91$). Thus it is possible to conclude that previous crime victimization has no effect on the likelihood to visit or revisit a destination where a theft occurred.

The relationships among (1) passage of time since a theft occurred; (2) learning of someone's positive experience at a destination where tourists or their friends or relatives experienced an incidence of theft in the past; and (3) learning from the media about the efforts to improve security at the affected destination and the likelihood to visit or revisist this destination

As shown elsewhere, of the 215 respondents in this study only 49 (22.8%) indicated that they would choose not to return to the destination. With this subset of 49 respondents, a one-way ANOVA was performed to compare the above three variables (passing of time; learning of someone's positive experience; and learning from the media of efforts to make the destination safer) as well as a fourth variable (nothing would change their minds). A post hoc SNK (Student-Neumann-Keuls) test found no statistically significant difference between each of the four groups on their likelihood to visit or revisit the affected destination. All four groups fell into one subset. As can be seen from Table 5, it is possible to conclude that for those who had decided not to return to the destination, the passing of time, learning of someone's positive experience, or learning from the media of efforts to make the destination safer would have no differential effects on changing the respondents' minds about not returning to the destination where a theft occurred.

Conclusions

This study was set up to determine what, if any, the effects are of having experienced or knowing of someone who has experienced an incidence of theft on a trip, on the likelihood to travel to the destination where the theft occurred.

The results of the study clearly demonstrated that having been a victim of a theft, or knowing of someone who has been such a victim while on a trip, did not affect the likelihood to travel to the destination where the theft occurred. Furthermore, with the exception of one factor, namely the manner in which the crime report was

Table 5 One-Way ANOVA for Factors That Would Change One's Mind to Travel to the Destination by Likelihood to Travel to the Destination Where the Theft Occurred

Factors	*N*	*Mean**	*SD*
Passing of time	6	2.66	1.21
Learning of someone's positive experience	4	2.75	1.50
Learning from the media of efforts to make the destination safer	0		
Other	2	1.0	.00
Nothing would change their mind	33	1.94	1.27
Total	45	2.07	1.29

	Sum of Squares	*df*	*Mean Square*	*F*	*p•*
Between groups	6.84	3	2.28	1.42	.25

*Based on a 1–7 scale where 1 = definitely not likely to return and 7 = definitely likely to return.

handled by the authorities, no other factor had any positive or negative effect on the likelihood to travel to the destination where the theft occurred.

More specifically the study did not find any difference in the likelihood to travel between those who have personally experienced a theft versus those who knew of someone who had. Second, the passing of time since a theft occurred had no effect on the decision to travel to the same destination. Third, the likelihood to travel to the affected destination was not significantly different between those who experienced a theft "not off their persons" and those who had experienced a theft "off their person." Fourth, the perceived severity of the incident did not have any significant effect on one's decision to travel to the destination where the theft occurred. Fifth, the study found no significant correlation between previous "non-trip related crime victimization" and the likelihood to travel to the destination where a theft occurred. Sixth, it was found that the passing of time, learning of someone's positive experience, and learning from the media about the destination's effort to make it safer did not change the mind of tourists once they have decided not to travel to a destination where a theft occurred.

Last, but not least, as mentioned above, the study found a positive correlation between the satisfaction with the manner in which the authorities handled the reporting of the theft and the likelihood to visit or revisit the destination where the theft occurred. The results of this study indicated that over 41% of the respondents gave a negative evaluation to the manner in which the authorities handled the crime report. But for the remaining 59%, proper handling of the crime incident lessened the negative feelings experienced by the theft victims and led to increased likelihood of future visitation. Destinations should take lessons from the city of Amsterdam that has developed a very effective victims' assistance program developed to reduce crime victims' negative feelings towards the city (Hauber and Zandbergen, 1999).

In summary, this study provided some valuable information for tourist destinations regarding one's decision to travel to a destination where a theft occurred. A majority of the respondents (55%) indicated that, even though they experienced theft or knew of someone who did, they would definitely travel to that destination.

These results support the findings of Mawby (2000) who in a survey of British tourists, found that 56% of victims and 55% of nonvictims would definitely return to the destination where the theft occurred. These findings are also similar to results obtained by George (2003) who, in his Cape Town study, found that 54% of his respondents stated that they would very likely return even though they had encountered an incident of crime or felt that their life was in danger during their stay. Hence, the results of this study confirm that experiencing a personal theft or knowing of someone who had such an experience was not a deterrent for visiting a destination where a theft occurred. Therefore, destinations that are afflicted by such thefts should not necessarily expect a reduction in tourist arrivals.

Study Limitations

Although this study investigated numerous issues relating to theft at tourist destinations, there are nevertheless some limitations to it. First, regarding crime severity, one must take into account that one's evaluation of crime severity is subjective and can be interpreted in many different ways.

Second, the study did not ascertain how often the respondents traveled and whether they actually traveled to the destination before they or their friends/relatives experienced the theft. Since this information was not ascertained from the respondents in this survey, it is difficult to deduce if it would have had any effect on the results.

Third, another aspect that might have affected the likelihood to travel to an affected destination could have been the purpose of travel and more specifically whether it was for business or leisure. For example, if a respondent was traveling on a business trip he or she may not have had any choice as to whether or not to travel to the affected destination.

Fourth, the survey did not ascertain the commercial and sentimental value of the loss from the theft. This could have had a significant effect on how the theft was perceived. If the theft equated to a small percentage of the total trip expenditure, respondents might have seen the theft as having a nominal effect on their pocketbook and thus not hesitate to visit the destination in the future. However, if the stolen items had a sentimental value or had a high commercial value, this might have a caused a significant distress and affected the likelihood of future visitation.

Concept Definitions

Larceny The taking of property from a person, without that person's consent, and with the intent to deprive the person permanently of the use of the property. The two types of larceny are *grand larceny* and *petty* (*petit*) *larceny,* depending upon the value of the property stolen.

Robbery A form of theft in which goods or money are taken from a person against that person's will through the use of *violence or fear*. The key words in this definition are *violence or fear*. These words differentiate this type of theft from larceny.

Burglary A theft against a dwelling.

Tourist dwellings A hotel room or a place where a tourist resides (excluding a private residence) while traveling.

Review Questions

1. What is the "Routine Activities Theory" and how is it related to crime against tourists?
2. What is the "Hot Spots Theory" and how is it related to crime against tourists?
3. What is the "Structural Choice Model"?
4. Describe the effects of crime on society and the individual victim.
5. What is the relationship between the perceived theft severity and likelihood to travel to the affected destinations?
6. What is the relationship between satisfaction with the manner in which the authorities handled the reporting of the theft and the likelihood to travel to the affected destinations?
7. What were the factors that were found to change the respondent's decision to travel to the affected destination?
8. Describe the main findings of the study reported in this chapter.

Note

[1] This chapter is based on the MS thesis of the first author.

References

Alleyne, D., and Boxill, I. (2003). The impact of crime on tourist arrivals in Jamaica. *International Journal of Tourism Research,* 5, 381–391.

Anonymous. (1996). *The Impact of Gambling on the City of New Orleans.* New Orleans, LA: University of New Orleans, Division of Business and Economic Research.

Barker, M., Page, S. J., and Meyer, D. (2002). Modeling tourism crime: The 2000 America's Cup. *Annals of Tourism Research,* 29(3), 762–782.

Brayshaw, D. (1995). Occasional studies: Negative publicity about tourism destination: A Florida case study. *Travel and Tourism Analyst,* 5, 61–71.

Brandon, D. (2001). *Stand and Deliver! A History of Highway Robbery.* Gloucestershire, UK: Sutton Publishing.

Brown, E. S., Esbensen, F.-A., and Geis, G. (1996). *Criminology: Explaining Crime and Its Context* (2nd ed.). Cinncinnati, OH: Anderson Publishing.

Caribbean and Alaska Cruise Report Record Demand. (1986, January). *Advertising Age,* 1.

Chesney-Lind, M., and Lind, I. Y. (1986). Visitors as victims: Crimes against tourists in Hawaii. *Annals of Tourism Research,* 13(2), 167–191.

Chesney-Lind, M., Lind, I. Y., and Schaafsma, H. (1983). *Salient Factors in Hawaii's Crime Rate.* Manoa, HI: University of Hawaii.

Cohen, L., and Felson, M. (1979). Social changes and crime rate trends: A routine activity approach. *Sociological Review,* 44(4), 588–608.

Crotts, J. C. (1996). Theoretical perspectives on tourist criminal victimization. *Journal of Tourism Studies,* 7(1), 2–9.

Crystal, S. (1993). Welcome to Downtown USA. *Meetings and Conventions,* 28(3), 42–59.

de Albuquerque, K. D., and McElroy, J. (1999). Tourism and crime in the Caribbean. *Annals of Tourism Research,* 26(4), 968–984.

Dimanche, F., and Lepetic, ??. (1999). New Orleans tourism and crime: A case study. *Journal Of Travel Research,* 38, 19–23.

Dimanche, F., and Moody, ?? (1997). *New Orleans Area Visitors Profile: January–June Results*. New Orleans, LA: University of New Orleans, Division of Business and Economic Research.

Entorf, H., and Spengler, H. (2002). *Crime in Europe: Causes and Consequences*. Berlin: Springer-Verlag.

Fraser, N. (2003, May 25). Street crimes tarnish HK image. *South China Morning Post,* p. 5.

Fujii, E. T., and Mak, J. (1980). Tourism and crime: Implications for regional development policy. *Regional Studies,* 14, 27–36.

George, R. (2003). Tourists' perceptions of safety and security while visiting Cape Town. *Tourism Management,* 24(5), 575–585.

Harper, D. W. (2001). Comparing tourists crime victimization; Research note. *Annals of Tourism Research,* 28(4), 1053–1056.

Hauber, A. R., and Zandbergen, A. (1999), Victimized in Amsterdam: The organized reaction. *Crime, Law and Social Change,* 31(2), 127.

Mawby, R. I. (2000). Tourists' perceptions of security: The risk–fear paradox. *Tourism Economics,* 6(2), 109–121.

McPheters, L., and Stronge, W. (1974). Crimes as an environment externality of tourism: Miami, Florida. *Land Economics,* 50(3), 288–292.

Meithe, T. D., and Meier, R. F. (1990). Opportunity, choice and criminal victimization: A test of a theoretical model. *Journal of Research in Crime and Delinquency,* 27, 243–266.

Pizam, A. (1999). A comprehensive approach to classifying acts of crime and violence at tourism destinations. *Journal of Travel Research,* 38(1), 5–12.

Pizam, A., and Mansfeld, Y. (eds.). (1996). *Tourism, Crime and International Security Issues*. Chichester, UK: John Wiley & Sons.

Schiebler, S. A., Crotts, J. C., and Hillinger, R. C. (1995). Florida tourists' vulnerability to crime, in A. Pizam and Y. Mansfield. (eds.), *Tourism Crime and International Security Issues*. Chichester, UK: John Wiley & Sons, pp. 37–50.

Sernike, K. (2002, October 24). The hunt for a sniper: The children; What Would be Best? *New York Times,* p. 33.

Sönmez, S. F., and Graefe, A. R. (1998). Determining future travel behavior from past travel experience and perceptions of risk and safety. *Journal of Travel Research,* 37(2), 171.

Sönmez, S. F., and Graefe, A. R. (1998). Influence of terrorism risk on foreign tourism decisions. *Annals of Tourism Research,* 25(1), 112–144.

Srikameswaran, A. (2003, January 1). Gun violence carries a high tag: $800 million plus. *Pittsburgh Post-Gazette,* p. 3.

Starmer-Smith, C. (2003, June 21). 10 cities to beware. *The Daily Telegraph*. (London).

US Department of Justice (USDOJ), and Board of Justice Statistics (BJS). (1999). *The 1998 National Crime Victimization Survey*. http://www.ojp.usdoj.gov/bjs/cvict.htm.

7

The Tourist and His Criminal: Patterns in Street Robbery

Dee Wood Harper

Learning Objectives

- To learn what criminologists know about robbery in general.
- To learn what criminologists know about offender motivation.
- To understand the situational elements of tourist robbery victimization.
- To understand the behavioral context of tourism as it relates to crime victimization.
- To understand why tourists may be more vulnerable to street crime than non-tourists.

Introduction

This chapter describes three interrelated situational elements in the tourist victim pattern in street robbery: (1) tourist/criminal convergence in space and time; (2) the physical features of locations where victimization tends to occur; and (3) the behavioral context of tourism and tourist victimization. Data on street robbery events involving tourists in and around the Vieux Carré (French Quarter) area of New Orleans in the years 2000 and 2001 provides the empirical basis for this discussion. The chapter includes a review of relevant criminological and tourism literature as it relates to crime victimization. Research evidence suggests that being a tourist creates a contextual weakness that places the tourist at greater vulnerability of victimization. Looking at the tourist as a special victim type highlights the contribution made by the victims to their victimization, perhaps more so than would be the case in other victimization contexts. The status and role of tourists and their behavior place them outside the bounds of the mundane and create a special relationship with the host location and host population. While their relationship with the actors in the institutionalized tourist settings is standardized and for the most part legitimate, their search for the authentic and backstage experiences brings them into contact with an altogether different set of actors who may be in a position to exploit them.

What Do We Know about Robbery?

Research on robbery has explored a variety of issues including the notions that victimization may be partially explained by exposure to certain situations (Gottfredson, 1976), personal characteristics of the victim (Hindelang, Gottfredson, and Garafalo, 1978; Blose, 1978), and the victim's gender, age, and race (Hindelang and McDermott, 1981). Other issues explored in the study of robbery have included trends, patterns, and types of robbers and robbery (Roebuck and Cadwallader, 1961; Normandeau, 1968). Moreover, street robbers have been characterized in the criminological literature as "semi-professional," "unskilled," "novice," "addict," "amateur," "alcoholic," and "garden variety." Conklin's (1972) influential typology of robbery focuses on the robber's commitment to robbery as a major source of livelihood. The most common type of robber for Conklin is the *opportunist robber*. This type of robber appears to act randomly; however, considerations such as victim vulnerability, potential for getting money, location suitability, and escape opportunities indicate some planning and thought occurs before the commission of the crime. The fact that planning does occur has been consistently identified in other research (Petersilia, Greenwood, and Lavin, 1978; Harper, 2000a). Planning in street robbery appears to be rudimentary and most likely occurs just before the robbery event (Lejeune, 1977).

The classification of a robber as an opportunist is also understandable from the *routine activities perspective* (Cohen and Felson, 1979) that emphasizes the structural features of opportunity largely apart from the motivation of either offender or victim, even though the victim and a motivated offender is part of the model. In other words, robberies occur when victim and offender occupy the same physical space at the same time without the presence of capable guardianship. The opportunist robber is also understandable from the *rational choice perspective* (Heineke, 1978; Cornish and Clarke, 1986; Eide, Aasness, and Skjerpen, 1999), which suggests that opportunities are created by the choices offenders make.

Previous research has found tourists to be overrepresented as victims of street robbery and larceny/theft (Harper, 1983; Chesney-Lind and Lind, 1986; Stangeland, 1998; de Albuquerque and McElroy, 1999). This suggests that tourists may be singled out and targeted for victimization by the criminal element. Other researchers have noted that the vulnerability of tourists is greater in areas already experiencing high levels of conventional crime (Schiebler, Crotts, and Hollinger, 1996). Put another way, increasing the number of tourists in a high crime area will likely increase the number of tourists' victimizations.

Researchers have also noted that tourist locations tend to be *hot spots* for crimes against tourists (Miethe, Stafford, and Long, 1987; Roncek and Maier, 1991; Ryan, 1993; Harper, 2000b). Crimes against tourists will cluster in locations near hotels, motels, bars, restaurants, and other tourist attractions. New Orleans' Vieux Carré (French Quarter) is a prime example of a hot spot. The hot spot perspective focuses on a physical location that places offender and potential victim in contiguity, thus increasing the probability and the opportunity for the crime to occur.

To understand the tourist–victim pattern requires a thorough examination of the situationally grounded experience of both the tourist and the criminal predator in the tourism context. To grasp the meaning of this pattern requires an examination of the behavioral and cultural context of the actors and the action as it unfolds (Becker, 1963).

The Tourist

Some sociologists have begun to develop a theoretical frame of reference for viewing tourism within the broader context of basic structural and cultural themes of modern society (MacCannell, 1973, 1976; Cohen, 1979a, 1979b; Apostolopoulos, Leivadi, and Yiannakis, 1996). By attempting to come to grips with what it means to be a tourist, MacCannell, for example, has conceptualized tourism as the modern equivalent of the religious pilgrimage or a quest for *authentic experience* (1973) or the *second gaze* (2001). In his view this modern quest for authentic experience parallels a more primitive concern for finding the sacred. Modern society is for MacCannell shallow, inauthentic, and alienating. Therefore, authenticity is thought to be elsewhere and beyond the immediate and mundane experience of most people—it is something that has to be sought after. From this perspective people in modern society are induced by their alienated circumstances to become a tourist, a seeker of the authentic. The second gaze simply emphasizes the desire on the part of the tourist to get beyond touristic representations.

MacCannell couples this conceptualization of the modern condition with the notion that sightseeing or tourism is the activity of seeking out authentic attractions but finding instead what may be described as staged authenticity. Realizing this, the tourist often continues the quest by seeking out the *backstage* or to put it another way, trying to find out what the real back regions are like, or, how the natives really live. In this quest the tourists may be presented with a contrived or staged backstage, thus making the search for authenticity even more futile. Nevertheless, this search for the authentic is what motivates the tourist to continue to visit touristic locations and, in large measure, what makes their behavior understandable once they are in the touristic location.

New Orleans provides staged authenticity (horse-drawn carriages, voodoo museums, and Cajun everything, with Mardi Gras parades often staged for large conventions) with standard commercial attractions such as the Aquarium of the Americas, Harrah's Casino, and Jazzland and unique authentic attractions such as the French Markets, the architecture of the French Quarter, St. Louis Cathedral, Jackson Square, and the nearby St. Louis Cemeteries. The New Orleans tourist experience is perhaps not complete without encountering some of the characters that populate the French Quarter area.

The Predator

Victimization of tourists seems to reflect a calculated decision-making process on the part of the predator (Harper, 2000b). Based on interviews conducted with erstwhile offenders, the tourist robbery pattern is a reflection of a body of knowledge possessed by the robber, that is, how to target and rob tourists, cues to look for in selecting a potential victim, developing a relationship with the victim, in some instances, placing himself and the potential victim in a location favorable to committing the robbery, and finally, committing the robbery and making good his escape. The robber also seems to be well versed in the ways of the tourist, the sights they want to see, their desire to party, and, in some instances, their desire for illicit action such as sex and drugs.

Robbery is an offense chosen easily by the offender when faced with the need for fast cash. Jacobs and Wright (1999) have noted that offenders find themselves in a cycle of expensive habits (gambling, drugs, and heavy drinking) that requires a steady flow of cash. In the context of street culture, robbery is understandable because it produces cash that is immediately translated into illicit action. Being involved in this type of action produces a self-identity of "coolness," "hipness," and "badness" (Katz, 1988). Therefore, being a street robber goes beyond the actual act of committing a robbery to include the creation of a distinctive lifestyle.

The Setting

The Vieux Carré is the centerpiece of New Orleans's tourism. The Quarter, or Quarters as it is locally referred to, is a collection of mostly eighteenth and nineteenth century West Indian Creole, Spanish, and Victorian influenced architecture. It combines a residential area, which has undergone many ethnic and cultural transitions, with a high concentration of historic sites (notably, Jackson Square with the St. Louis Cathedral and the Pontabla Apartments), art galleries (Royal St.), restaurants (everywhere), and bars and adult entertainment (Bourbon St.).

On the one hand, locals seem to view tourists with a tolerant ambivalence. On the other hand, tourists are recognized as a crucial asset to the economy of the city and are accorded gracious and hospitable treatment. At the level of street culture, however, preying on and ripping off tourists in a variety of ways including robbery is seen as a way of "getting over" (outsmarting the naïve out-of-towner). These activities can include everything from tap dancing and miming to various hustles and con games (e.g., "I betcha I can tell you where you got dem shoes"), theft of uncontrolled property, "breaking" cars, grabbing chains and purses, pickpocketing, and armed robbery. Tourists are viewed as legitimate targets and are vulnerable. They behave with great naïveté in their search for authentic out of the way, backstage experiences, people, and things.

Findings

For purposes of the study I combined armed robbery (taking anything of value from a person by use of force or intimidation while armed with a dangerous weapon) and simple robbery (not armed with a dangerous weapon) and refer to the combined crime as street robbery. Street robberies of tourists or other visitors to the city from outside of the state of Louisiana for the years 2001 (N = 377) and 2002 (N = 175) gathered from police reports form the empirical basis of the study. (Following the events of September 11, 2001, the number of tourism and convention visitors to the city dropped dramatically during the remainder of that year and did not begin to recover until the winter of 2003–2004.)

For the 552 cases there were 979 perpetrators; meaning it was not uncommon for the victim to be outnumbered in the robbery event. The most common robbery event (84%) involved a single male as a victim. In the small number of robberies where females were listed as the victim (first named in the incident report) only 2% were women alone; all the others were accompanied by a male companion. Racial differences in victimization essentially modeled the racial composition of tourist and convention visitors to the city with 86% being identified by the police as white. Four hundred thirty (86%) of the incidents involved white victims. All victims were over 19 years of age and under 60, with 47% in the 25–35 age group. Table 1 provides data on the origin of tourist victims.

Offenders, as described by victims, were 100% male (with 27 incidents involving female accomplices). In 80% of the incidents the perpetrators were described as black with the most common pattern being two black males (46%) followed by 28% of the incidents with a single black male perpetrator (this percentage is consistent with the racial composition of the city). All offenders were described as being young (16 to 30 years of age) with 2% described as maybe in their late 50s or early 60s. In 94% of the cases, some type of weapon was present, with 58% of the incidents involving handguns. The second most common category of weapons was a knife, box cutter, razor, or sharpened screwdriver (22%).

The map shown in Figure 1 identifies the location of the robberies in this study. Seventy percent of the robberies occurred in perimeter locations such as North Rampart Street and beyond. Police have focused on this area with what they informally refer to as border patrol. As one police informant described it, "I look for people who don't 'fit'; I stop them, the would-be perpetrator often runs away and I warn the out-of-towner that they are in a dangerous area and should return to the French Quarter" (Nolan, 2004).

The robberies that occurred within the bounds of the French Quarter also tended to be on the perimeter and in areas with dim lighting, low pedestrian density, and no apparent police presence. These areas are perhaps attractive as robbery venues because of easy proximity to the crowds of Canal Street and the anonymity it affords as well as easy egress to the nearby public housing projects, the Treme, or

Table 1 Origin of Tourist Victims

Origin	*N*	*Percent*
California	70	13
Foreign nationals	65	12
Texas	58	11
Florida	40	7
New York	32	6
Georgia	24	4
Illinois	21	4
Mississippi	19	3
Alabama	19	3
Other states	204	37
Total	552	100

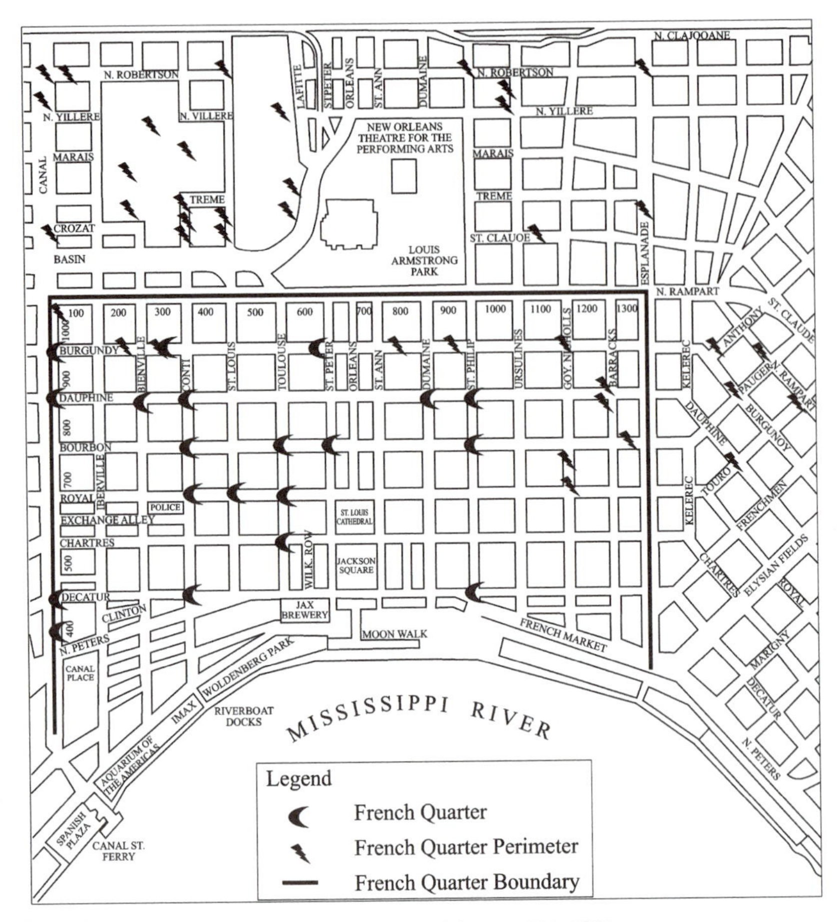

Figure 1 Typical 90-day robbery pattern, New Orleans, 2000–2001.

the Fabourg Marigny residential areas bounding the French Quarter on the East and North flanks. Other robbery sites with similar characteristics included a parking garage (2%), parking lots (2%), a cemetery (8%), and a park (4%). By contrast, Bourbon and Royal Streets are well lighted, have heavy pedestrian traffic and numerous police foot and mounted patrols, and, with the exception of pickpockets, are virtually crime free.

Table 2 provides data on the time of day of occurrence of the robberies. Street robbery in the French Quarter is, not surprisingly, a nighttime activity with 3 out of 4 nighttime robberies occurring between midnight and six AM. The French Quarter bars and clubs are, for the most part, open 24 hours a day and 7 days a week. Midnight to 6 AM finds mostly tourists (a few tourism workers) and the potential predators on the street.

The physical features of locations where robberies occur seem to support a nonrandom hot spot pattern influenced by certain population and physical characteristics of the French Quarter and surrounding areas. The French Quarter is a

Table 2 Offense Time of Day

Time	*Number of Cases*	*Percent*
Midnight–1:59 AM	144	26
2:00 AM–3:59 AM	77	14
4:00 AM–5:59 AM	110	20
6:00 AM–7:59 AM	11	2
8:00 AM–9:59 AM	0	0
10:00 AM–11:59 AM	22	4
Noon–1:59 PM	11	2
2:00 PM–3:59 PM	0	0
4:00 PM–5:59 PM	33	6
6:00 PM–7:59 PM	22	4
8:00 PM–9:59 PM	66	12
10:00 PM–11:59 PM	55	10
Total	551	100

grid of narrow, dimly lit streets. The architecture throughout the area is ancient and similar and for many visitors somewhat disorienting. It is an adult tourist attraction where people overeat, overdrink, and generally have a great time. Bourbon Street is the entertainment heart of the French Quarter and as mentioned earlier, is relatively crime free with the exception of pickpocketing.

Two thirds (368) of the robberies reported in this study occurred on side streets away from the main areas of pedestrian traffic. Seventy-six percent of the victims reported that they were walking back to their hotels or to another entertainment venue. In 37% of these robberies the location of the robbery event would indicate that the victim was either lost or not being completely truthful about where he was going (the robbery occurred outside of, and away from, areas where hotels, restaurants, and other entertainment venues are located). With only one major exception, hotels are located outside the Bourbon Street entertainment corridor and thus the tourist must walk to and from his or her hotel. While there are some smaller hotels within the French Quarter, the larger hotels are located on the middle to lower Canal Street perimeter of the area. Getting back to one's hotel can present a problem even for a cold-sober visitor to the city. Tourists often remark how easy it is to get turned around in the Vieux Carré.

The behavioral context of all the robberies in this study suggests what Katz (1988, p. 170) refers to as *contextual weaknesses*. This notion may be applied to the present cases in a number of ways. First, as was mentioned earlier, 60% of the incidents occurred between midnight and 5:59 AM. The potential victim is returning to his hotel in a strange city, most likely fatigued and under the influence of alcohol, winding through dark and unfamiliar streets.

Many of the circumstances surrounding the incidents clearly increased the victim's vulnerability to attack. Police reports indicate that in at least 20% of all robbery events the victims had developed some relationship with the offender before the robbery. In 40% of these instances the victim went with the offender to get a woman. In another 40% the victim went with the offender to purchase drugs. In another 10% the victim reported going with the offender to the housing projects (no reason specified). And in one case the victim, after having sex with a man she picked up in a bar, passed out drunk in her hotel room and awoke later to discover that she had been robbed.

In one out of every five cases in our study, victims "collaborated" in their own victimization, in a sense. That is to say, the opportunity for the victimization was largely created by the choices made by the victim. These victims most likely would not find themselves in the circumstances that led to their subsequent victimization on their home turf. By their actions, within the situational context of tourism, they placed themselves at a triple disadvantage of being a stranger, being isolated in an unfamiliar area, and being there in search of some illicit action. Because of the nature of this context, it is likely that a larger number of incidents of this type go unreported to the police. In a recent case a tourist picked up someone who he thought was a prostitute. The "prostitute" turned out to be male in drag who lured him to a darkened side street and robbed him at gunpoint. In 18% of all tourist robbery incidents, perpetrators, working alone, were found to be responsible. They took advantage of the victims' desires for illicit action, and insinuated that they could help them achieve this objective. After luring the victim to an isolated location the robbery took place. All of these robberies were accomplished with a handgun and no one was injured. In most of the remaining cases the crimes were committed by two or more offenders with some injuries occurring, with the victim being struck or shoved to the ground and in some cases kicked. In all of the robbery incidents in the study the robbers achieved dominance either through weapons or dominant numbers (or both) or through the action of the victims. The victims placed themselves in a contextually weak position by searching for authentic, albeit illicit, action.

Finally, the fact of having the status of tourist, being a non-permanent, transient person, adds another dimension to contextual weakness. Assuming an arrest is made in the incident, the likelihood of prosecution being pursued is low. The victim/witness has returned home, and, unless he or she was seriously injured or experienced a large but recoverable loss, he or she is not likely to return to press the prosecution. The local prosecutor has only limited resources to bring witnesses back to the city. These resources are typically reserved for high-profile cases where, fearing adverse publicity, it is politically (and economically) important for the city to appear to be vigilant. Unfortunately, most garden-variety street robberies that involve out-of-town victims rarely reach the prosecutorial stage.

Case Studies

Contextual weakness also involves being isolated either through misadventure or lack of knowledge. The tourists find themselves in a dangerous part of the city and vulnerable to attack. The following examples illustrate types of contextual weaknesses. Two Swiss tourists wandered into an area behind a public housing project near the French Quarter. They had visited one of the old cemeteries with the above-ground tombs and taken some photographs; upon leaving they soon became victims. Dusk was beginning to fall on this clear warm Saturday evening in June when a young black male approached the couple. Feeling lost, they asked the young man for directions to the French Quarter. After he points in the general direction of the Vieux Carré, he produces a blue steel revolver and demands their property. The victims, "fearing for their lives," hand over $2,000 in US and Swiss currency, a 14k gold necklace, earrings, a 35mm camera, and a video camera.

In the following example the victims are rendered contextually weak on three points: the lateness of the hour, the isolated location, and a tactic employed by their robber. A couple from Houston, Texas, after having a late dinner and drinks, decide to go for a stroll along the Moonwalk on the levee between Jackson Square near the Mississippi River. It is 3:00 AM on Sunday in an area that is poorly lit. The couple is returning from their stroll by the stairs over the floodwall by the fountain next to the closed Café Banquette. Going up the stairs as they descend is a male, 6 feet tall and 180 pounds, who appears to be about 30 years of age, neatly dressed in a black leather jacket and blue jeans. After he passes them on the stairs he says, "Hey, what time is it?" The couple stops and turns around to face a .38 caliber revolver. "Give me your wallet," the man demands. The male victim complies. "Take off all your jewelry." They each remove all of their jewelry and hand it over to the robber, who then departs in the direction of the river making good his escape. The bandit's earnings for less than one minute of work: $24,090 in cash and jewelry.

The innocent enough sounding request, "Hey, what time is it?" serves an important function in this robbery event. Researchers have often pointed to the use of civil requests as a stock-in-trade tactic in street armed robbery (Katz, 1988, p. 174). The request is used to connect with the victim with what seems to be a reasonable request but one that also allows the would-be robber to momentarily "hold" his target in place and scan his would-be victim without causing undue alarm.

Roger provides an example of "collaborative" victimization. Roger, a 21-year-old white male from a small Texas town, is on his first visit to New Orleans. His New Orleans connection, a student at a local university, has returned to his uptown apartment and has left Roger to continue drinking with a stranger, a friendly male who has engaged him in conversation and has even bought a round of beers. It is 4:00 AM. In the course of the conversation the friendly stranger casually asks Roger if he would like to score some marijuana. Roger answers in the affirmative, and after some further conversation focusing on the high quality of the marijuana the stranger has in his car that is parked just a few blocks away, they finish their beers and leave the bar. In the 1100 block of Burgundy Street, where the friendly stranger's car is supposed to be parked, he stops, looks up and down the street, produces a small caliber revolver and tells Roger he will blow his head off if he doesn't give him his money. Roger complies and gives the friendly stranger $476.00 in cash and a brand new "K-Bar" pocketknife worth another $50.00.

The following case, while technically not a street robbery, does offer another illustration of the victim as an active collaborator in their victimization. Judy and Marie are also in New Orleans in search of the authentic; in this case they are here to "do some serious partying." Judy and Marie have been drinking nonstop since their late afternoon arrival from a neighboring state. It is 12:30 AM and they have met two male dancers in a Bourbon Street saloon. They pay the bartender $40.00 to get the men off from their jobs early. The foursome goes for drinks at several bars in the Vieux Carré. Around 2:30 AM Marie returns to their hotel with one of the men while Judy and her newfound friend continue their pub crawl. Arriving at her hotel room Marie and her friend have consensual sex; then Marie promptly falls asleep (passes out). Upon awakening the next morning her male dancer is missing along with $180.00 in cash, and her girlfriend Judy has not returned to their hotel room. At this point Marie reports the robbery and her concern for her missing friend to the police. While the police are taking the report, Judy returns to the hotel room with her male dancer friend. They have spent the night at his apartment on Bourbon Street. It seems that Judy's dancer had only known Marie's dancer by his first name and for only two days. Despite some obvious weaknesses in the case the police say they will seek a warrant for the arrest of Marie's dancer.

Discussion

Looking at tourists as a special victim type highlights the contribution made by the victims to their victimization, perhaps more so than would be the case in other victimization contexts. The status and role of tourists and their behavior places them outside the bounds of the mundane and creates a special relationship with the host location and host population. While their relationship with the actors in the institutionalized tourist settings is standardized and for the most part legitimate, their search for the authentic and backstage experiences brings them into contact with an altogether different set of actors who may be in a position to exploit them. As Ryan and Kinder (1996, p. 24) have suggested, being a tourist includes a range of behaviors that, perhaps, would not be tolerated in a non-tourist context, but in the tourist context which includes tourist space and place, carousing late at night, drunkenness, and seeking sexual liaisons and drugs is tolerated. The tourist, because of the nature of the role, is allowed greater latitude and, in the present context, this latitude increases vulnerability to criminal victimization. Their behavior is expected to be different because they are tourists and are to some degree expected, at least in the context described here, to be acting at the margins of legality. Being tourists, in terms of the role expectations and anonymity coupled with the nature of touristic places, increases the opportunities to look for and find illicit action and decreases the chances of being caught and exposed to significant others. Concomitantly, the presence of tourists in large numbers creates many vulnerable targets and opportunities for motivated offenders to commit robberies.

In Cohen's theoretical work on the phenomenology of tourism experiences, he develops what he refers to as a recreational mode of tourism experience (Cohen, 1979a). The recreational form views the touristic experience as a type of entertainment akin to other types of entertainment; an activity to be enjoyed for its restorative and recreative powers, in a sense, not unlike the religious pilgrimages of old. The present research expands the recreational mode of tourism experience by adding the dimension of tourism as an escapist activity that allows tourists to play out certain hedonistic desires away from the eyes of significant others (e.g., family, friends, and workmates). Furthermore, tourists, in this instance, may or may not be motivated by feelings of alienation from their mundane everyday lives or feel they need to be recreated for that matter, they simply use the tourist role as an opportunity to seek and find experiences not normally available to them in their everyday lives.

Finally, the experience of crime victimization as a tourist perhaps gives some concrete meaning to the notion of the tourist moment (Cary, 2003). Victimization is a serendipitous moment in the sense that the outcome of the search for authenticity is unanticipated. No one ever really intends to be a crime victim. Yet the risk taken by tourists in seeking out the authentic has led to their victimization. The victimized tourists have now gone beyond the subjective experience of being simply tourists and have discovered themselves in an experience they never anticipated.

Concept Definitions

Armed robbery The taking of anything of value from a person by use of force or intimidation, while armed with a dangerous weapon.
Authentic experience Modern society is conceptualized as shallow, inauthentic, and alienating. Therefore, authenticity is thought to be elsewhere and beyond the immediate and mundane experience of most people—it is something that has to be sought after. From this perspective people are induced by their alienated circumstances to become tourists, seekers of the authentic.
Contextual weaknesses The circumstances surrounding the crime incident that clearly increases the victim's vulnerability to attack.
Hot spot perspective The focus on a physical location that places offender and potential victim in contiguity, thus increasing the probability and the opportunity for the crime to occur.
Opportunist robber This is the most common type of robber. They appear to act randomly but with consideration for victim vulnerability, potential for getting money, location suitability, and escape opportunities, which indicates some planning and thought occurs before the commission of a crime.
Rational choice perspective This perspective suggests that opportunities are created by the choices offenders make.
Routine activities perspective This perspective emphasizes the structural features of opportunity largely apart from the motivations of either offender or victim, even though the victim and a motivated offender are part of the model. Robberies occur when victim(s) and offender(s) occupy the same physical space at the same time without the presence of capable guardianship.
Second gaze The second gaze simply emphasizes the desire on the part of the tourist to get beyond touristic representations.
Simple robbery The taking of anything of value from a person by use of force or intimidation without being armed with a dangerous weapon.
Tourist robbery pattern A pattern that reflects a body of knowledge possessed by the robber, that is, how to target and rob tourists, cues to look for in selecting a potential victim, developing a relationship with the victim, and in some instances actively placing the robber and the potential victim in a location favorable to committing the robbery and finally, committing the robbery and making good an escape.

Review Questions

1. Assume you are a tourist for a moment. What precautions would you take to avoid being a victim of a street robbery?
2. Assume you are a robber for a moment. What would you look for in selecting a victim?
3. What are some of the contextual elements of being a tourist that can increase vulnerability to crime?

4. What does it mean to be a collaborative victim of crime?
5. What are some of the elements of the tourist role that allow or encourage behavior that people would not engage in on their home turf?

References

Apostolopoulos, Y., Leivadi, S., and Yiannakis, A. (1996). *The Sociology of Tourism*. London: Routledge.

Becker, H. (1963). *Outsiders: Studies in the Sociology of Deviance*. New York: Free Press.

Blose, J. (1978). *Criminal Victimization in Massachusetts*. Boston: Committee on Criminal Justice Analysis.

Cary, S. (2004). The tourist moment. *Annals of Tourism Research,* 31(1), 61–77.

Chesney-Lind, M., and Lind, I. (1986). Visitors as victims: Crimes against tourists in Hawaii. *Annals of Tourism Research*, 13, 167–191.

Cohen, E. (1979a). A phenomenology of tourist experiences. *Sociology,* 13, 179–201.

Cohen, E. (1979b). Rethinking the sociology of tourism. *Annals of Tourism Research,* 6(1), 18–35.

Cohen, L., and Felson, M. (1979). Social change and crime rate trends: A routine activity approach. *American Sociological Review,* 44, 588–608.

Conklin, J. (1972). *Robbery and the Criminal Justice System*. Philadelphia: Lippincott.

Cornish, D., and Clarke, R. (1986). *The Reasoning Criminal*. New York: Springer-Verlag.

de Albuquerque, K., and McElroy, J. (1999). Tourism and Crime in the Caribbean. *Annals of Tourism Research,* 26(1), 968–984.

Eide, E., Aasness, J., and Skjerpen, T. (1994). *Economics of Crime: Deterrence and the Rational Offender*. Amsterdam: North Holland.

Gottfredson, M. (1976). *Classification of Crimes and Victims*. Doctoral dissertation, State University of New York at Albany.

Harper, D. (1983). *The Tourist as Crime Victim*, unpublished paper, Academy of Criminal Justice Sciences, San Antonio, Texas.

Harper, D. (2000a). Robbery, armed. *Encyclopedia of Criminology and Deviant Behavior* London: Taylor & Francis.

Harper, D. (2000b). Planning in tourist robbery. *Annals of Tourism Research,* 27(2), 517–520.

Heineke, J. (1978). *Economic Models of Criminal Behavior*. Amsterdam: North Holland.

Hindelang, M., Gottfredson, M., and Garafalo, J. (1978). *Victims of Personal Crime: An Empirical Foundation for a Theory of Personal Victimization*. Cambridge, MA: Ballinger.

Hindelang, M., and McDermott, M. (1981). *Juvenile Criminal Behavior: An Analysis of Rates and Victim Characteristics*. Washington, DC: U.S. Government Printing Office.

Jacobs, B., and Wright, R. (1999). Stick-up, street culture and offender motivation. *Criminology,* 37(1), 149–173.

Katz, J. (1988). *Seductions of Crime*. New York: Basic Books.

Lejeune, R. (1977). The management of a mugging. *Urban Life,* 6, 123–148.

MacCannell, D. (1973). Staged authenticity: Arrangements of social space in tourist settings. *American Journal of Sociology,* 79(3), 589–603.

MacCannell, D. (1976). *The Tourist, a New Theory of the Leisure Class*. New York: Shocken, McKean.

MacCannell, D. (2001). Tourist agency. *Tourist Studies,* 1(1), 23–37.

Miethe, T., Stafford, M., and Long, J. (1987). Social differentiation in criminal victimization: A test of routine activities/lifestyle theories. *American Sociological Review,* 52, 184–194.

Nolan, G. S. (2004, April). Personal communication.

Normandeau, A. (1968). Patterns of robbery. *Criminologica,* 1, 2–13.

Petersilia, J., Greenwood, P., and Lavin, M. (1978). *Criminal Careers of Habitual Felons*. Washington, DC: U.S. Department of Justice.

Roebuck, J., and Cadwallader, M. (1961). The negro armed robber as a criminal type: The construction and application of a typology. *Pacific Sociological Review,* 4, 21–26.

Roncek, D., and Maier, P. (1991). Bars, blocks, and crimes revisited: Linking the theory of routine activities to the empiricism of "Hot Spots." *Criminology,* 29, 725–760.

Ryan, C. (1993). Crime, violence, terrorism and tourism: An accidental or intrinsic relationship? *Tourism Management,* 14, 173–183.

Ryan, C., and Kinder, R. (1996). The deviant tourist and the criminogenic place—The case of the tourist and the New Zealand prostitute, in A. Pizam and Y. Mansfeld (eds.), *Tourism, Crime and International Security Issues*. New York: John Wiley & Sons, pp. 23–36.

Schiebler, S., Crotts, J., and Hollinger, R. (1996). Florida tourists' vulnerability to crime, in A. Pizam and Y. Mansfeld (eds.), *Tourism, Crime and International Security Issues*. New York: John Wiley & Sons, pp. 37–50.

Stangeland, P. (1998). Other targets or other locations? *British Journal of Criminology,* 38(1), 61–77.

III

Tourism and Safety Issues

Yoel Mansfeld and Abraham Pizam

This section is composed of five chapters that discuss the issue of the physical safety of tourists while visiting certain destinations or participating in a variety of dangerous tourist activities. The exposed dangers contained in this section range from participation in commercial adventure tourism activities, encounters with wildlife, infection from various diseases, and/or being injured by natural disasters and unsafe travel conditions.

Geoffrey Wall, in a chapter describing the unfolding of the crisis that occurred in Toronto, Canada, as a result of the SARS epidemic, analyzes the influence that the media had on public perception of the gravity of the event. He suggests that the Canadian media has been somewhat responsible for arousing fears where minimal risks, in fact, existed. In addition, he raises some serious questions about the role that the various governments should have in crisis management and disaster reduction and mitigation plans and he advocates the creation of such plans for the private sector and especially the tourism industry. Last, but not least, he describes the various recovery methods that were used in the SARS case and analyzes their short-term and long-term effectiveness.

In their chapter on Australia's commercial adventure tourism, Damian Morgan and Martin Fluker review the research findings and common practices relevant to risk management for operators in this sector. This review highlights two salient points: first, a model of the risk management operating environment for commercial adventure tourism operations, and second, suggestions for research to enhance understanding of this dynamic tourism sector. The authors conclude by suggesting that managers in this industry have the responsibility of: (a) providing their clients with experiences that are adventurous but within an acceptable margin of safety; and (b) informing their clients about the physical risks involved in these activities before they commit to the adventure. Because future clients will be seeking new and more challenging adventure experiences, this industry, in the authors' opinion, will become more regulated through legislative and accreditation requirements, in order to improve safety standards and meet client needs.

Wilson Irvine and Alistair Anderson's chapter explores the impacts of foot and mouth disease on peripheral tourist destinations in the Grampian region of Scotland, which was indirectly affected by the disease; and the Cumbria region of England, which was directly affected by the presence of the disease. The empirical data, collected by two surveys, showed that Cumbria, which was closed to visitors, was very badly affected, but the Grampian area experienced no cases of foot and mouth. Despite this, the impact on the tourism industry of both destinations was similar. In the authors' opinion this seems to confirm that perceptions, rather than facts or real circumstances, created the disastrous effects of this catastrophe on the tourism industry. The authors conclude with the claim that tourism decisions seem to be made in the heart, not in the head, and therefore it is the responsibility of public agencies to mitigate against these unrealistic perceptions through tourist education.

Based on the information provided by travel advisories of the main English-speaking generating markets to China, Zelia Breda and Carlos Costa identified the following travel risks/hazards to China:

Crimes

- Frequent pickpocketing and minor thefts at train and bus stations of large cities.
- Infrequent violent crimes such as robberies and murder of tourists in some rural locations.
- Frequent attacks and robberies of foreigners in popular expatriate bar and night-club areas in Beijing and Shanghai and in the shopping district of Shenzhen.
- Minor thefts and sexual harassment on overnight trains and buses.
- Robberies by armed bandits in the most remote areas of China.
- An increase in drug smuggling and related crimes in the Yunnan province.

Health Hazards

- Possible infection by the following diseases:
  - Cholera;
  - Hepatitis A;
  - Malaria;
  - Tuberculosis;
  - Typhoid fever;
  - Mosquito-borne diseases such as dengue fever and encephalitis B;
  - Rabies;
  - Avian influenza;
  - HIV/AIDS; and
  - SARS.

Local Travel Safety

- Road and traffic accidents;
- Pedestrian accidents;
- Air accidents on internal flights;
- Incidents of overcrowded ferries, which sank resulting in the loss of lives; and
- Attacks of piracy (against private yachts) in the South China Sea.

Natural Disasters

- Earthquakes; and
- Severe rainstorms that can cause flooding and landslides.

Political Situation

- Restrictions on undertaking certain religious activities, including preaching and distributing religious materials.
- Regulations against any public demonstrations that do not have prior approval from the authorities.

The authors conclude that the tourism industry in China was in the past negatively affected by health hazards (i.e., SARS), changes in politics (i.e., Tiananmen Square incident and the cross-strait relations), and economics, but only for a short duration, after which it recovered quickly and continued to grow at an accelerated pace.

Gianna Moscardo, Matthew Taverner, and Barbara Woods' chapter discusses the potential impacts of wildlife on tourist safety. It is based on the results of two studies, one of which was a critical incidents study of the worst wildlife tourist experiences, while the second was a survey of tourists' perceptions of the risks associated with different wildlife in the North Queensland region of Australia and the impact these risk perceptions had on their travel behaviors. The study found that different tourists react differently to the same species of wildlife. Some see them as dangerous but attractive, others see them as dangerous and unattractive, while for still others they can be perceived as attractive but not dangerous. The study also discovered that although tourists overestimated the level of risk associated with contact with wildlife, a large proportion of them did not intend to change their behaviors in response to potential risks from wildlife encounters.

8

Recovering from SARS: The Case of Toronto Tourism

Geoffrey Wall

Learning Objectives

- To suggest that the tourism industry is operating in a turbulent environment and must expect to have to respond to a diversity of extreme events.
- To illustrate the importance of the media in conveying information about extreme events.
- To demonstrate that perceptions of risk, even where little risk may occur, have far-reaching consequences for tourism.
- To indicate the range of responses made by one city to address the negative consequences of an adverse situation on its tourism industry.
- To stimulate thought about the roles of the public and private sectors in risk management for the tourism industry.

Introduction

The tourism industry is operating in a turbulent environment in which both global and local events have major consequences for economic success. Whether they be wars, political crises, terrorist activities (real or imagined), natural disasters, health emergencies, or other extreme events, they all have swift and usually negative repercussions for those involved in tourism. Of course, this is nothing new. What is novel is that, fueled by the speed and reach of communications, information, in both words and pictures, is disseminated almost instantaneously throughout the world. As a result, the rapidity of the onset of the consequences of even distant events is virtually immediate, leaving little time to put effective damage-control initiatives in place, and the magnitude of the repercussions of the events is exaggerated. Fortunately, there is evidence that tourism can rebound fairly quickly once the threatening situation has passed. However, in the meantime, a great deal of damage occurs, economic dislocation of destinations and businesses takes place, costs of responding to the crisis increase at a time when income is curtailed, the market is soured, misinformation and incomplete information abound, uncertainty prevails, and recriminations occur at a time when calm and measured responses are called for and a cohesive set of responses is required.

It is also novel that the events that individually appeared to be unusual and extreme are being seen as examples of a class of "unexpected" damaging events that are less surprising in the increasingly turbulent environment in which we find ourselves. In fact the grouping of natural disasters, wars, terrorist events, chemical spills, epidemics, and other such undesirable phenomena into a class of catastrophic events requiring forethought, planning, and management, while not new, is now widely recognized as being appropriate. While differing in detail, such phenomena have in common the destruction of lives and property, to different degrees and at different speeds, but often on a massive scale. Recognition of the similarities and differences between such damaging occurrences is at the root of an important new research and policy direction for it changes the focus from discrete, unique events that are unmanageable, to a recognition that these are rare but recurring phenomena that may be planned for and whose adverse consequences, if not avoided, may be mitigated.

In a turbulent environment, one strategy that merits consideration is risk management. Risks cannot be eliminated but they can be reduced. However, it is important to consider who should bear the risks—and the answer will vary from place to place and with the nature of the risks. In the Western world and increasingly elsewhere, the tourism industry is dominated by the private sector but it operates in an environment that is also managed by the public sector. What risks should be assumed by the private sector and what is the responsibility of governments for risk mitigation in an increasingly turbulent environment?

The availability of accurate and timely information is necessary to inform decision makers, but such information is not easily acquired in times of stress. It has been said before that "perception is reality"—tourists respond to what they perceive the situation to be and this may or may not mirror reality depending in large part upon the accuracy of the information to which they are exposed. The very ability of potential tourists to exercise choice—to decide whether or not to go, when or where to go, and what activities to participate in—creates problems for most tourism operators with a fixed plant that makes it difficult to respond to short-term vicissitudes in the market.

Thus, given the above discussion of an increasingly turbulent world, this chapter examines the decisions made particularly by governments and, to a lesser extent, by the private sector, in the aftermath of an extreme event with far-reaching consequences for the tourism industry. It is also concerned with the management of information. The event of concern is the SARS (Severe Acute Respiratory Syndrome) outbreak and, though global in its repercussions, the perspective is primarily from Toronto, Canada. Tourism is an important activity in Toronto. The Greater Toronto Hotel Association alone represents establishments employing more than 30,000 people and its hotels bring in approximately $Can 2 billion annually in revenues (all $ figures in this chapter are in Canadian dollars). Of course, this does not include tourism activity in many attractions, transportation, shops, and other tourism-related businesses. Altogether, Toronto receives about 16 million visitors annually who generate about $7.2 billion in revenue (other sources indicate half this expenditure). The volume of tourism economic activity that is at risk is substantial.

Context and Methods

In the context of extreme events, SARS has more in common with a drought than a major earthquake. In the case of SARS, globally, not many people died compared with many other disasters, the recognition of the challenge posed by SARS was fairly slow, the disaster event lasted a relatively long time (allowing for management strategies to be implemented during the event), but millions of people were affected. Nevertheless, the accumulation of many small dislocations and the costs of response measures were massive. SARS was a new disease but, aided by advanced communication systems, it was not the first to have a global reach, and, from that perspective, it was not without precedent. At the time that this chapter was being prepared, the media was engrossed with bird flu. Not long before that it was mad cow disease and shortly before that it was foot and mouth disease.

Although SARS originated in China, at the time that SARS was first reported in the Western press, the author was in Beijing and first learned about it in an e-mail from Canada. Like many others, he subsequently had his Asian travel curtailed and his Asian projects disrupted—it was difficult to undertake tourism research or training in China in the summer of 2003! Thus, the author has personal experiences of the consequences (but not the disease) that he can draw upon. However, the main sources of information for this chapter come from elsewhere. This contribution is based primarily upon a systematic analysis of the contents of two Canadian newspapers from April 2003 to the present (September 2004). The two newspapers are the *Globe and Mail* (GM) that is published daily in Toronto but has broad national and international coverage; many regard it as Canada's national newspaper. The second paper, *The Record,* is published daily in Kitchener, Ontario, which is located approximately 100 km west of Toronto. It contains international stories but primarily serves a regional market.

For many years the author had been clipping and saving all of the articles on tourism in the two papers. Thus, data acquisition was in progress prior to the onset of the event. It was therefore only necessary to sort the clippings to take out those pertaining to SARS. The clippings were then arranged chronologically and by newspaper. The articles were then reread to identify the stakeholders involved, critical events, actions proposed, and actions taken. The empirical part of the chapter reports the results of these activities.

SARS in Toronto

Forty-four people died in Toronto as a consequence of SARS. The first case of SARS in Toronto was identified in a hospital on March 7, 2003, and the first death occurred when the son of a woman who attended a wedding in Hong Kong died in hospital on March 13, 2003. However, when the author returned from Beijing on March 23, 2004, SARS was not a concern in Toronto or Canada. It had been mentioned as occurring in China in the press but no steps had been taken to prepare for it in Canada. The author entered Vancouver with the most cursory of immigration examinations—two words: "Welcome home!"

However, the consequences for Toronto were rapid and far-reaching. The press attention to tourism in Toronto came with the headline: "Major conference cancelled due to outbreak" (GM, April 3, 2004). This referred to the decision of the American Association for Cancer Research to cancel its annual meeting in Toronto. This event was expected to bring 28,000 delegates to the city. The cancellation occurred in spite of a letter from Toronto's medical Officer of Health that said that "the risk of SARS transmission among the general population is extremely low" (GM, April 3, 2004).

Space does not permit the detailed history of SARS in Toronto: suffice it to say that the incidences of SARS were largely contained to hospitals, and there were never people walking on the streets of Toronto with protective masks. There were two waves of SARS in Toronto: the initial wave, which resulted in a World Health Organization (WHO) Travel Advisory issued on April 22, 2003. This advisory was lifted on April 30, 2003, and on May 14, 2003, Toronto was removed from the list of areas with recent local transmission. However, on May 20, 2003, a patient infected others in a Toronto hospital and Toronto was put back on the WHO list of places with recent local transmission. Toronto was eventually removed from this list on July 2, 2003 (Tufts, 2004).

The crisis resulted in a sharp and substantial decline in tourism activity, massive cancellations of reservations and events, and layoffs in the tourism industry and related sectors. This came at a time when global and Toronto tourism were already struggling from other events—post-September 11 jitters and fears of global terrorism, enhanced security measures and their associated costs, economic uncertainty, and the war in Iraq all taking their toll. However, the consequences of SARS for Toronto were much greater than in the case of the terrorist attacks of September 11, 2001.

The Toronto hotel sector was immediately and significantly affected: average hotel occupancy rates fell to 46.6% in April 2003, in comparison with 68% a year earlier and there was an estimated decline of tourism expenditures of $503 million, or 28% in 2003 in comparison with 2002 (Tufts, 2004).

It should be mentioned, however, that the consequences were not confined to Toronto for two reasons:

1. Toronto is a major point of entry providing access to other locations in Canada; and
2. The Toronto situation was inappropriately generalized in some cases in international decision making to the whole of Canada, which is the largest country in the world in area!

Statistics Canada (quoted by Tufts, 2004) reported that during the second quarter (April–June) at the peak of the crisis, the tourism industry in Canada was affected in the following ways:

- International visitors declined 14%.
- Spending by international visitors declined 13%.
- The international travel deficit grew to over $1.1 billion.
- Tourism employment decreased by 2.4% (during a period when seasonal employment usually increases.

Although the consequences of SARS were far-reaching, no attempt is made in this chapter to document in detail the impacts of SARS on Toronto. It is concerned more with the steps that were taken to promote recovery rather than the causes of the problems. However, it is an artificial distinction, for the two are related. Obviously, for revival of tourism, it was necessary to assure potential visitors that the problem had been exaggerated, confined, and, ultimately, resolved.

Results of Newspaper Analysis

The results are presented in two phases: before and after the World Health Organization (WHO) travel advisory since this was a major event in SARS-struck Toronto. Not only did it have an impact on the willingness of both domestic and international visitors to travel to Toronto, it came as a great surprise, and was widely regarded as an inappropriate, even ridiculous international response, particularly as it came at a time when the number of incidences of SARS was in decline and no new cases had been reported for a week.

From the first incidence of SARS in Toronto, there were substantial price cuts and layoffs in an industry that was already ailing. They can be viewed as being part of a recovery strategy and as attempts to make establishments more competitive. They are not reported in detail here although they were widely reported in the media and these reports added to the malaise.

Pre-WHO Travel Advisory

As indicated above, the first action that was taken with respect to tourism was to release a letter from Toronto's Medical Officer of Health in an attempt to allay fears. This followed the cancellation of a major conference in early April.

On April 11 (GM), it was suggested in a report commissioned by the federal government that they should provide $650 million in tax relief as part of a federal program to cut fees (which would be passed on to passengers) to encourage flying. However, this was not a totally new initiative, for the airlines were already struggling prior to the onset of SARS. SARS simply made their plight more pressing.

On April 15 (GM), there was widespread criticism of the mayor and a call on him to show leadership (although some questioned his ability to provide this). This was done in response to travel advisories issued by several countries about visiting Toronto and the outright ban some North American companies introduced on employee travel to the city. The mayor returned from a trip to Florida and promised action, but, at the same time, said that there was little that could be done now to promote the city and that "It's not wise to throw good money after bad."

In Toronto, where there is a large Chinese community, SARS was viewed by many as being a Chinese disease. In order to support the Chinese community, senior federal politicians (as well as various political candidates), including the prime minister, were photographed eating in Chinatown in an attempt reduce fears that bordered on racism.

On April 16 (GM), although officials from federal, provincial, and municipal governments as well as the hotel and convention industry were meeting to devise a response strategy, there was considerable criticism of lack of political leadership.

The need to allay fears, instill confidence, launch promotions, and initiate a special event (to bring people together, thereby indicating Toronto as being attractive and safe) were suggested by industry spokespersons who lamented the lack of a figure comparable to Mayor Rudolph Giuliani who had taken up the cause for New York following September 11.

On April 19 (GM), a major article on SARS and tourism in Toronto was published under the headline "From Hog Town [a nickname for Toronto] to Ghost Town." Part of this article reported an interview with Mike Rogers, President of the advertising agency that had been hired to revive New York City following September 11. He urged that the worst thing that could be done was to dwell on the threat. Rather, he suggested that the goal should be to convey a spirit of optimism. SARS should not even be mentioned. Advertising should show people going about their lives and enjoying themselves—especially families with children—for if people are taking their children, it will be assumed to be safe for all.

On April 23, a WHO travel advisory was announced advising against nonessential travel to Toronto in spite of the fact that there had been no new case of SARS in Toronto for a week. Thus, a negative judgment was made against Toronto by an international agency, and publicized globally. Even though the appropriateness of their decision was widely questioned in Toronto, additional damage was done and was very difficult to control.

Post-WHO Travel Advisory

The WHO travel advisory came as a complete surprise in Toronto and added even greater urgency to an already problematic situation.

On April 23 (GM), the mayor of Toronto urged residents to go about their lives as normal and was quoted as saying: "It's not the disease that's doing the damage; it's public perceptions about SARS that's hurting Toronto's tourism industry, and it's getting worse." The next day, following a moment of silence in recognition of the 16 victims who had died in Toronto to that time, a 10-hour emergency city council meeting debated the role of federal, provincial, and city commitments to advertising budgets. Midway through the meeting, the mayor announced that the city would soon launch a $25 million marketing campaign (city $5 million, provincial and federal governments $10 million each). This announcement came amid a flurry of measures, including calls for the province to give the city the right to levy a hotel tax, which was approved by the city council at the end of the meeting. In addition, the Medical Officer of Health was praised for her handling of the episode and she indicated that she could give the "all clear" if there were no new cases of SARS in 10 days. Nevertheless, a headline on April 25 (GM) read "Worst Is Yet to Come for Toronto, Officials Warn."

Approximately 1 week later, on April 29, following the provision of new information by Canadian health officials, the travel advisory was rescinded. It had been 20 days since the last case of community transmission. Nevertheless, the president of Tourism Toronto, the city's travel and convention marketing agency, indicated that the damage had been done and that, as a minimum, it could take 2 years before tourism would fully recover. On the same day, Ontario pledged to put in place a 5-month tourism tax holiday, a $118 million worldwide advertising campaign, and a $10 billion business recovery strategy (GM, April 30, 2004).

On May 14 the WHO took Toronto off its list of places affected by SARS. On May 22 the situation was further complicated by a case of mad cow disease in Alberta, which heightened fears for the recovery of the tourism industry (GM, May 22, 2004).

On about May 14, a proposal to hold a blockbuster event in the form of an outdoor concert headlining the Rolling Stones came to prominence. The concert was a key component of a strategy endorsed by a broad-based recovery group called the Toronto Tourism Recovery Coalition. This would be the anchor event, which would potentially attract hundreds of thousands of fans, of a strategy to fill hotels, restaurants, and other Toronto attractions. There is some irony in this as some years before the group had been less welcome in the city and members had been charged for possession of narcotics! The debate over this event and the planning for and occurrence of the concert make up a complex story. Suffice it to say that the concert eventually took place on July 30, 2004, amid much fanfare and considerable controversy. A study of credit card expenditures between July 27 and August 2 found significant increases of approximately $75.2 million in expenditures compared to the previous week. Restaurants enjoyed the largest (59.2%) increase in business, followed by travel and entertainment (20.6%), photography and supply stores (18.5%), drug stores and pharmacies (14.1%), hotels (7.1%), car-rental agencies (5.1%), and retail (5.1%). However, the retail sector received $39.6 million of the increased revenue, followed by restaurants ($20.9 million) with hotels only receiving $1.9 million. The relatively small injection of expenditures into hotels suggests that the event attracted primarily a day-trip local and regional market rather than overnight visitors. Regardless, a rough calculation suggests that the federal and provincial governments would have recouped about $12 million in taxes from the extra spending, more than doubling the $5 million that the two governments laid out in funds to bring the Rolling Stones and other acts to Toronto (GM, August 7, 2004).

Also, a newly formed nonprofit corporation, the Toronto03 Alliance, using $1 million in seed money provided by Canada's five major banks, put forward a private sector plan called "Summer in the City" to promote cultural and sporting events in and around the city. Prominent entertainers agreed to record, free of charge, public service announcements for TV and radio. One of the most successful initiatives was the sale of cheap packages combining a hotel room, fine dining, and admission to a show at much reduced prices. These offerings were taken up rapidly, particularly in the local and regional markets, and they have been so successful that they continue today, more than 2 years later.

Recovery?

At the time of this writing in late 2004, SARS is occasionally mentioned in the business sections of the newspapers as the financial statements of companies are reported and commented upon (see, e.g., "Legacy Hotels Loses $9.2M in Final Quarter of Bad Year, *The Record,* January 28, 2004, C). Less than desired performance is often attributed to the challenging economic environment that has been experienced, including SARS. However, the bird flu is now receiving more attention than SARS and warnings against overreaction, referring to the implications of SARS to public health, can be found. For example, a former Ontario public health official said, "Creating panic over things like bird flu is counterproductive—and distracts us from the problems that kill more people than SARS ever did" (GM, February 2, 2004, A11).

The SARS scare, according to one source (GM, February 9, 2004) cost Toronto "$500 million in tourism revenue and thousands of jobs." The SARS outbreak exposed deficiencies in Toronto's fragmented and underfinanced tourism efforts, which have been viewed as being inferior and less well financed than those of competitors from the United States. The Greater Toronto Hotel Association, in a voluntary agreement among its members signed in January 2004, imposed a 3% hotel-room tax to be paid by visitors. Approximately 30,000 of 33,000 hotel rooms are covered by the levy. Also, on February 11, the Ontario Provincial Government, a new government that was not in power during the SARS crisis, announced an investment of $30 million to promote tourism in Ontario. Of this sum, $2.8 million was earmarked for northern Ontario. Toronto may be expected to be a major beneficiary of this outlay, even more than northern Ontario itself, since for some international tourists Toronto may be viewed as being a gateway to the Canadian north. This will be used to finance a new "branding exercise" as a tool for "genuine renewal" of tourism in the region. As a result, Tourism Toronto will be able to spend about $20 million in 2004 compared with $8 million in 2003. However, the pent-up demand and the rapidly increasing value of the Canadian dollar are encouraging some Canadians to look farther afield (GM, January 10, 2004, T2).

The market appears to be rebounding slowly. However, more travel may not translate immediately into a corresponding growth in profits for the travel industry, for customers have become used to bargains, and price-cutting efforts date back to 2001. Businesses that slashed prices must now cope with the problem of "raising these rock-bottom prices to more sustainable levels in the face of ongoing price resistance" (GM, January 10, 2004, T2).

Discussion and Conclusion

A number of questions and conclusions arise from the thumbnail sketch of Toronto's recovery strategies that have been presented above.

The flow of information and responsible reporting are crucial at all levels. While it is not the author's intention to belittle the risks of SARS or to undervalue the loss of life that occurred, it is a fact that greater losses of life occur regularly in Toronto from influenza or complications from common colds. While events and situations should be reported and the government of China has been widely criticized for withholding information, the Canadian media have been widely accused of scare-mongering and arousing fears where minimal risks, in fact, existed. For example, the cumulative number of cases of SARS in Toronto was frequently reported, not allowing for deaths or recoveries, so the number of persons available to transmit the disease was generally less than implied in the reports.

In an increasingly turbulent environment, when disaster preparation plans are commonplace for society as a whole, is it reasonable to expect the tourism industry to have its own disaster reduction and mitigation plans? In an increasingly turbulent environment, what is to be regarded as normal business risk for the private sector, what risks can be insured against, and when is government involvement to be expected? What should be the role of national, provincial, and municipal governments, and who should coordinate and manage their efforts? What is a reasonable response time?

It may be worth making a distinction between short-term (damage control) and long-term strategies. The Rolling Stones concert, as a single event, can be viewed as being a short-term strategy. On the other hand, the collaboration between hotels, restaurants, and attractions in putting together packages can be viewed as a longer-term strategy in product development. At the same time, the urgency of embarking upon and providing additional money for new marketing strategies, under discussion when SARS occurred, was given greater urgency and decisions were probably brought forward.

Finally, although international visitors are often relatively high-spending, it is often forgotten that they cost more to attract and are more fickle than the local or regional market. Some market segments are more risk-averse than others and they need to be identified. As distance increases, so does the number of intervening opportunities, and the quality of information tends to diminish. A risk-reduction strategy can be served by ensuring that more reliable local and regional markets are not neglected in the pursuit of more fickle distant markets. Even in internationally prominent destinations, the regional market usually comprises a large proportion of the effective market. After all, if local people will not come, why should international visitors be attracted to visit?

Concept Definitions

Collaboration/partnerships Stakeholders may combine their knowledge, capital, and other resources to better achieve mutual interests in perceived opportunities or to respond to common threats. Sometimes collaboration may occur between similar stakeholders, such as two hotels, or it may be between partners with very different characteristics, such as a government department, a local tourism authority, and a number of tourism attractions.

Recovery strategies (short-term and long-term) Recovery strategies are steps that are taken to minimize the adverse consequences of an event and to restore functions to former levels as quickly as possible. They consist of immediate actions taken following the event as well as initiatives that are taken on a longer time frame.

Risk management Risk management comprises strategies that are put in place ideally in advance of but sometimes during adverse circumstances in order to reduce the negative consequences of such events and to speed recovery.

SARS Severe acute respiratory syndrome is a form of pneumonia that was initially observed in southern China in February 2003 and spread rapidly to other parts of the world, killing hundreds of people (although not as many as the flu and many other diseases), and greatly disrupting tourism, especially in locations where infected individuals were identified.

Travel advisory A travel advisory is an announcement, usually by governmental authorities, warning potential travelers of situations that may have implications for the ability to travel safely.

Turbulent environment A turbulent environment is one in which events that impinge upon an organization and affect its operations substantially can be expected to occur, although their nature and timing are difficult to predict.

Review Questions

1. When viewed from the perspective of the tourism industry, to what extent do you think it is useful to make a distinction between natural and human-caused extreme events?
2. Make a list of extreme events and classify them according to their likely magnitude (loss of life and damage to property), frequency (how often they are likely to occur), and duration (how long they last). What implications do your answers have for the tourism industry?
3. Consider the proposition that perceptions are as important, if not more important, than reality in the context of destination images.
4. What is meant by a risk-reduction strategy? How would you go about reducing the risks faced by the tourism industry in a destination of your choice?
5. Acquire a copy of your local newspaper, review the lead stories that are reported, and evaluate their implications for tourism.
6. What are the advantages and disadvantages of organizing a special event as a major component of a response of the tourism industry to an extreme event?
7. What are travel advisories and who should be responsible for issuing them? What are their effects?
8. To what extent are issues of public health important to tourists and tourism destinations? How might tourists prepare to reduce their likelihood of experiencing ill health while on holiday?
9. Review the Toronto case study and identify the individuals and organizations that were affected and took action to address the consequences of SARS for tourism.

References

The Globe and Mail. Toronto, various dates.

The Record. Kitchener, various dates.

Tufts, S. (2004). *The Impacts of the Severe Acute Respiratory Syndrome Crisis on Cultural Events and Organizations in Ontario, Final Report.* Toronto: Ontario Region of the Department of Canadian Heritage, Toronto.

9

Risk Management for Australian Commercial Adventure Tourism Operations

Damian Morgan and Martin Fluker

Learning Objectives[1]

- To explain why people seek both safety and excitement in adventurous tourist activities.
- To highlight the difference between real and perceived risks as they relate to adventure tourism.
- To identify and discuss the three core elements of the adventure tourism experience: the participant; the management practice of the operator; and the features of the setting.
- To describe the role of the legal system, industry accreditation, industry standards, and insurance with respect to the Australian adventure tourism industry.
- To list the principles of managing crisis situations in adventure tourism operations.
- To synthesize the elements relevant to the total commercial adventure tourism operating environment.

Society's awareness of risks posed by environmental, industrial, and biological hazards has never been greater (Kolluru, 1996; Smith 2001). This awareness reflects the fact that most people want to live and flourish in a *safe* society, protected from danger and injury. On the face of it, the rapid growth of Australia's adventure tourism industry over the last three decades seemingly counters the development of the safe society; this is because adventure tourism satisfies a tourist's desire to engage in risk-taking behaviors. But do tourists really wish to face a real and likely possibility of injury occurring to them? We argue here that most adventure tourists seek simultaneously safe and exciting experiences. Adventure tourism operators provide suitable adventure experiences by managing the range of inherent hazards and risk.

We know that many clients engage in commercial adventure tourism activities to experience thrills, arousal, and excitement (Fluker and Turner, 2000; Hall and McArthur, 1994). A participant's experience of positive arousal arises through the uncertainty created by an activity's inbuilt physical and social challenges (Morgan 2000). These challenges, especially those of a physical nature, are inextricably linked to the inherent risks provided by the natural environment (Brannan, Condello, Stuckum, Vissers, and Priest, 1992; Cheron and Ritchie, 1982). To ensure the anticipated experience for adventure tourism participants, the inherent risks must be managed at an appropriate level. With too little risk, the customer can find the experience dull and boring; too much risk and the operator and clients may confront an adverse or emergency situation (Morgan, 2000).

The Interlaken canyon disaster in Switzerland provides a pertinent example of tourist adventure gone wrong. As reported by Le Quesne (1999), on the 27th of July 1999 a group of 44 adventure tourists and 8 guides were rappelling and body-rafting down a 400-meter stretch of rapids and waterfalls. Heavy rainfall caused the banks of an upstream creek to falter, releasing "a 6 m high wall of mud brown water" down the watercourse. Twenty-one people died in this adventure activity.

Newspapers reports of the Interlaken tragedy speculated that the activity's guides ignored early warning signs of danger (e.g., Mann, 1999). Experienced river guides, not involved with the adventure company in question, also expressed serious concerns about the operation. It was suggested that company river guides were pressured to put profits before safety and lacked adequate knowledge of the local conditions (e.g., Nicholson, 1999). The Swiss court judgment supported this view of events and convicted six company managers and senior guides of manslaughter through culpable negligence (Bita, 2001a). The report of the trial by Bita (2001b) highlights a number of important considerations for risk management in commercial adventure tourism operations:

> Judge Thomas Zbinden found that most of the participants didn't know what canyoning was, let alone have any experience of it. The guides' supervisors ignored signs of a thunderstorm brewing in the Valley, he said, even though storms had been predicted in the local newspaper and on radio. They [the canyoning tour operators] had time to cancel the trip but went on regardless. The junior guides were not properly trained, the judge concluded, because they had not been instructed on how to read all the danger signs of an impending flood.

Professor of Tourism at Queensland University's Centre for Tourism and Risk Management, Dr. Jeff Wilks, quoted in *The Australian* (Strahan, 2001, p. 5), summed up the implications of this unfortunate event for Australia's adventure tour operations in stating, "due diligence is all about monitoring your staff and situation, and the message from Interlaken is to go back and have a look at your staff, their experience and bring your staff up to speed about what is happening."

The purpose of this chapter is to present an overview of the adventure tourism industry's risk management operating environment within Australia. We draw elements discussed in this overview into a practical model encapsulating relevant aspects of the risk management operating environment.

Commercial Adventure Tourism Activities

Adventure travel and adventure tourism are often discussed as the same phenomenon. Adventure travel is a form of tourism involving an ongoing and self-organized experience (Webber, 2001). The time required for adventure travel may run into weeks or even months. We define adventure tourism from an activity perspective. Thus, participants in adventure tourism activities will be drawn from adventure travelers as well as package tourists and perhaps local residents. Examples of adventure tourism activities include commercial operations in white-water rafting, sea kayaking, horseback riding, mountain biking, cross-country skiing, and rock climbing. The definition of commercial adventure tourism activities, adapted from an earlier definition offered by Hall (1992), is: "A broad spectrum of commercialized recreational operations that facilitate tourists' deliberate engagements with elements of risk emanating from the natural environment. The characteristics of the participant, the features of the setting, and operator management of the activity combine to influence the tourist experience of this engagement."

Adventure tourists are required to pay money to engage in commercial adventure tourism activities. For this payment, the adventure tourism operator provides specialized skills, knowledge, equipment, and access to a suitable location. Taking a legal perspective, the adventure experience forms a contractual arrangement between participant and operator, obligating the parties to legal rights and duties (Cordato, 1999). This is important, as adventure tourists will often be exposed to risks not routinely experienced in their everyday lives. We take up legal issues later in this chapter; the major point stated here being that the client's risk exposure necessitates adventure operators to recognize and manage hazards.

The Commercial Adventure Tourism Industry

The importance of the adventure tourism industry is recognized by the wider tourism industry. Commercial adventure tourism provides significant financial benefits for regional economies as these nature-based activities normally take place in rural and wilderness settings (Tourism Victoria and VTOA, 2002). In Australia, these benefits are reinforced where growth rates in international tourist participation in adventurous activities such as scuba diving, snorkeling, and white-water rafting have outstripped overall international visitor growth rates (Blamey and Hatch, 1998). A recent newspaper report claimed that adventure tourism is the fastest growing segment of the Australian tourist market, and estimated its worth at somewhere between $50 million and $500 million (Elias, 1999).

The commercial opportunities will continue to attract new adventure tourism operators. New operators will often have to meet government requirements including those relating to guide qualifications, operating regulations, and site permits (Cloutier, 1998, p. 32). Many operators will also seek accreditation through industry-based bodies that can provide a range of operational benefits including training and access to lower insurance premiums (e.g., VTOA, 2001a). These requirements assist the development of suitable standards and practices across the industry. Nevertheless, the nature of adventures means that operators and clients face a range of natural hazards found in the environment. The potential consequences for loss and injury are considered in the following section.

Injury Research in Commercial Adventure Tourism

It is not surprising that adventure tourists suffer injury in commercial adventures, given that these activities occur in natural settings characterized by inherent physical risks (Bentley and Page, 2001). Injuries have been reported in a range of commercial adventure tourism operations including scuba diving, rafting, and horse trekking.

The injury patterns are somewhat imprecise because of the difficulty of determining specific injury (i.e., morbidity and mortality) rates in adventure tourism activities (Bentley and Page, 2001; Langley and Charmers, 1999). For example, not all injuries will be severe enough to require medical treatment. Even where an injury is treated at a local surgery or in a hospital, records may not always provide specific information regarding the circumstances of injury (e.g., whether the injury occurred during a commercial adventure activity) or the person (e.g., whether the victim was a short stay tourist or a permanent resident). Bentley and Page (2001) acknowledge other data deficiencies including the lack of specific exposure (to physical risk) data required to determine the relative risk of participation.

Nevertheless, adventure tourist injury rates for specific activities have been reported in Australia and elsewhere (e.g., Bentley, Meyer, Page, and Chalmers, 2001; Bentley and Page, 2001; Wilks and Coory, 2000; Wilks and Davis, 2000). Considering this data, serious injuries in Australia's adventure tourism industry are uncommon events in most activities. Nevertheless, incidents such as the Interlaken tragedy and the 1998 disappearance of the Lonergans during a dive on the Great Barrier Reef (see Wilks and Davis, 2000), highlight the imperative need for operators to manage the potential physical risks posed by adventure tourism activities.

Further development of reporting methods and industry-specific definitions will provide the opportunity for future injury epidemiology studies in adventure tourism. Although present research is limited, common risk factors include the unfamiliar nature of the activities undertaken (Wilks and Atherton, 1994), errors of judgment made by the employees of operators (e.g., Maritime Safety Authority, 1994), and possibly a greater propensity for risk taking when away from the home environment (e.g., Carey and Aitken, 1996). The extent and cost of injury across the context of all adventure tourism activities remains an important area for future investigation.

Injury data provide one avenue for understanding the range of hazards manifest in adventure activities. Through a better understanding the elements of adventure, we can make more astute judgments about the source of hazards and potential risks.

Elements of Adventure

The definition used for commercial adventure tourism specifies that the client experience is determined in combination by the participant characteristics, the management practice of the operator, and the features of the setting. The combination of these elements would normally produce positive outcomes for all those involved. In some cases, however, the combination can also produce undesired outcomes ranging from mishap-induced fear or embarrassment, to more serious incidents causing severe injury or death. The role of each of these elements in the adventure is now discussed, with an emphasis on features that place a client at greater risk relative to other adventure participants.

The Adventure Tourism Participant

The participant brings a range of previous experience and expectations to the adventure. Our research of rafters has supported the notion that clients' fear perceptions, induced by physical hazards, and control over those fears (both before and during the activity), are central to the resultant adventure experience (Morgan, 2000). Other research has shown that adventure clients rate operator safety standards as the most important feature of an adventure activity (e.g., Hall and McArthur, 1994).

The activity guide plays a critical role in manipulating clients' perceptions of safety, fear, and control. The guide is critical to clients' experiences of adventure. From our experience working in and researching this industry, adventure guides manage and plan the trips to minimize the *real* risks while delivering to clients the essential elements of a core adventure experience. Through this process, the adventure participants may not even be aware of the potential risks and the method of their management. The following example outlines the adventure process in a white-water rafting activity.

Adventure tourism activities typically begin with a safety talk to the client group. This will normally include an explanation of the adventure and use of equipment. Following the talk, and depending on the nature and duration of the activity, the adventure operator may give experienced or competent clients more scope and responsibility when dealing with the inherent risks. Allowing competent participants to exercise more control over the activity will enhance their experience of perceived risk (and the adventure experience) without necessarily compromising client safety. For example, clients that have repeated the same adventure a number of times do not always need to undertake basic training in the standard tasks of the activity (the standard tasks in a white-water rafting activity include holding on, back paddling, and assuming the white-water floating position). The limited activity time can instead be used to encourage more experienced clients to build on the standard tasks through developing higher level rafting skills. This might include navigating a raft through a particular rapid by initially choosing a line and then calling instructions to other paddlers in the raft, this skill development occurring under the watchful eye of the trained river guide.

In contrast, novice clients will hold only vague expectations of what skills might be involved in an adventure activity. Novice white-water rafting participants, for example, may not even expect to put in any effort, apart from occasionally holding on to the raft. This underlines the importance of novices becoming used to the activity and equipment before becoming subject to risks beyond their perceived control. For example, novice rafters usually have paddling skills practice with their individual rafting guide before or shortly after launching. Successfully training novice paddlers with minimal prior knowledge can become extremely important during the adventure. Training not only enhances the client's experience of the activity, but enhances safety as well; the guide may need to rely on the client's compliance and training in situations of real risk exposure.

For example, the rafting guide would aim ideally to be able to guide the raft down a river with a minimum of reliance on the crew; however, in some settings often precise and coordinated paddling is required from all crew members in the raft to avoid hazardous circumstances such as those presented by large stopper waves. Capsizing on a particularly hazardous rapid might create strong psychological distress within a client (and this condition can also contribute to potential physical danger).

Interestingly, a lack of awareness of the true potential risk in an adventure activity can result in participants holding very different risk perceptions compared to those held by guides and operators. For adventure clients, their perceived risk becomes the defining feature of the adventure experience (Morgan, 2000). Clients' expectations of risk can arise through a number of causes including the likelihood of physical injury, the possibility of not receiving value for money, or the potential advent of psychological distresses through social discomfiture (Cheron and Ritchie, 1982). Notwithstanding this expectation of risk, clients' perceived risks during the adventure have been inextricably linked with their fears, awareness of dangers, anxiousness, and feelings of personal control (Morgan, 2002).

A range of preexisting factors will also influence client risk perceptions during adventure tourism activities. These include previous experience in the situation, personality, age, gender, and culture (Kasperson and Dow, 1993). Participants' level of control, mood, personality, and group dynamics also influences risk perceptions. To summarize, individuals tend to perceive less risk in behavior that is voluntary, under personal control, or undertaken as part of a group. Adventure tourists typically undertake activities voluntarily and as part of a group. The level of a participant's personal control in the activity, however, will vary between individuals and activities.

Individuals who carry accurate perceptions of the activity's risk would be expected to pay close attention to information and training provided by the operator. Where participant perceptions of risk are flawed or biased or if critical information is absent, participants may not be prepared for the risks they encounter. This occurrence becomes more likely in activities with inadequate management practices or where atypical environmental events in the setting have not been predicted. These aspects are discussed in the following sections.

Management Practices of the Adventure Operator

The second element determining the client's adventure experience is operator management practices. Clients will, to a large extent, rely on the care of the operator and staff to predict and manage the activity's inherent risks. The responsibility for risk management is therefore shifted from the client to those in charge of the activity. This brings into focus the need for competent and trained staff skilled in both the activity and in people management.

Facilitating activity management requires operators and delegated guides trained in both hard and soft skills. Hard skills refer to those areas in which a guide requires technical competence. This might include, for example, skills in craft control, rescue, first aid training, navigation, and use of specialized equipment. These skills are relatively easy to measure directly or may become assumed to be acquired competencies through a designated level of experience (e.g., 150 trips on a specified river). Guides will use these technical skills to provide their clients with suitable training and equipment for the adventure.

Soft skills refer to interpersonal qualities used by the guide to manage customer experiences during the adventure activity. These skills are important to control client moods and anxieties where little time for training is possible or where unexpected circumstances arise. Adept guides read their clients' moods and personality, and based on this information, engage in appropriate communication strategies. This will include using clients' names, giving them confidence through positive feedback, and routinely checking how they are feeling during their adventure.

In addition to employing suitable staff, other practices are required for risk management. These practices will vary depending on the nature of the activity and can include appropriate staff-to-client ratios, trip planning, safety communication, first aid and emergency preparation, clothing and equipment maintenance, hazard indicators, and information recording procedures (see Cloutier, 2000, for a comprehensive overview from a Canadian perspective).

The Adventure Tourism Setting

The final element contributing to the adventure experience is the activity setting. Commercial adventure tourists will gain access to the activity's natural settings through the operator's expertise and supply of equipment. Many of these settings, such as inland rivers, the open oceans, or rugged mountain ranges, will undergo constant change. In facilitating the adventure, adventure tourism operators must be aware of specific hazards in the setting and how changes in conditions can exacerbate these hazards. Failure of operators to regard the dynamics of the environment presents one area where the best hazard minimization systems can fail. This situation can occur through operators taking a mindless approach to the examination of environmental hazards.

For Langer (1989), mindlessness is a state of consciousness where individuals process information using existing categories derived from past experience. The notion of mindlessness may have serious implications for adventure tourism operators given that this condition, as Langer contends, is often caused through repetition of behavior. Adventure tourism instructors often deal with high volumes of people. This can result in mindlessly categorizing participants. As one consequence, participants perceiving greater than average risk, or whose competency is below the average, might become subject to risk exceeding their capacity for control. Similarly, guides may mindlessly categorize environmental hazards based on their past experiences. For example, white-water rafting guides may not be aware of increased risk posed by a river that is rarely in flood. In these circumstances, following the usual course might not be the safest path in the changed conditions.

Creating mindful tour guides is an important safety consideration (Moscardo, 1997). This would include techniques and training that requires guides to suggest appropriate responses to unusual or unexpected situations. The causes and effects of mindless categorization of clients and the environment by river guides remains an important area for future research into adventure tourism activities.

Legal Considerations, Accreditation, Industry Standards, and Insurance

Where undesirable incidents result in damage claims, the Australian legal system will typically attempt to apportion blame to those found responsible and award compensation to those parties suffering loss or injury. These cases can often revolve around a breach of duty of care where the risk should have been reasonably foreseeable (Wilks and Davis, 2000). Further, the adventure tourism operations are subject to mandatory legal requirements pertaining to applicable jurisdictions; for instance, in Australia operators are subject to common law, trade practice, and health and safety legislation, and in many cases, government licensing requirements.

Specific legislation (e.g., the Victorian Government [2001] Maritime Act, 1988) will in some instances specify standards of operation. These legislative requirements will work to determine the minimum standards of entry for operators. Similarly, land managers may not permit an operator access to a location without meeting minimum requirements. For example, Parks Victoria (2001) requires operators to provide evidence of public liability and in some cases details of operator accreditation.

As a more general strategy to manage the adventure tourism industry, many countries are now implementing voluntary safety codes of practice. These safety codes typically encompass key legal, environmental, safety, and customer service principles. This approach may be the most suitable public policy option, particularly where accidents are recognized to be part of the experience. This view focuses blame on the individual involved rather than a governing authority (Johnston, 1989). Bentley, Page, and Laird's (2001, p. 239) study pointed to the existence of support for this notion in New Zealand's adventure tourism industry. Based on operator reports of injuries, their study identified that clients' "failure to attend to and follow instructions" was a prime cause of injury. Public policy should therefore emphasize the need for safety management to begin with an education and training foundation directed towards the individual level (i.e., employees and clients).

A number of more formalized accreditation systems relating to client safety operate throughout Australia. Although not generally enforceable, these systems are often supported through industry-based standards or codes of practice where adherence may be cited as evidence of an operator meeting legislative obligations. For example, the Queensland Government's (2001) industry code of practice details how horseback riding establishments can meet the obligations of the Queensland's Workplace Health and Safety Act.

Industry accreditation may also require certification of an operator's employees. Certification for specific activities will normally be based upon an instructor's hard skills and activity experience. For example, instructor certification can be gained through private organizations such as the Professional Association of Diving Instructors (PADI, 2001), or through competency standards obtained through the national vocational education and training (VET) system (ORCA, 2001).

Insurance cover for operators allows some of the risk that they carry to be transferred to another party. For example, adventure tourism operators (and their employees) require public liability insurance (PLI) to protect them from claims of third parties (e.g., clients) resulting from loss or injury often due to a breach of the duty of care (Liability Insurance Taskforce, 2002). Risk management strategies assist in both recognizing and minimizing the risk retained by the operator and in identifying that to be transferred to the insurer.

Rising PLI premiums have been recently identified as driving some adventure tourism operators from the industry. The reasons for premium increases are complex and beyond the scope of this work (see Bartholomeusz [2002] and VTOA [2001b] for a brief analysis of some relevant factors). Aside from passing on higher charges to customers, a number of suggestions have been proposed to address this problem. These include agreements made with clients not to sue if injured (e.g., through legally binding waivers), caps on insurance payouts, and legal contesting of insurance claims to develop case histories (Kemp, 2002). Whatever solution is found, adventure tourism operators will nevertheless continue to require risk management planning, including preparation for potential crises.

Managing Crises in the Adventure Tourism Industry

Accreditation and legislative requirements will not prevent all crises from arising in commercial adventure tourism operations. Moreover, insurance coverage should not take the place of appropriate planning for risk and crisis management. Crisis management requires operation managers to recognize the crisis potential and so put in place specific strategies to deal with these situations. Augustine (2000) suggests that crisis management should be addressed in six stages: avoiding the crisis, preparing to manage the crisis, recognizing the crisis, containing the crisis, resolving the crisis, and profiting from the crisis. In essence, this model describes a methodical approach to controlling crises with an emphasis on initial avoidance. Although the first stage, crisis avoidance, is "the least costly and simplest way to control a potential crisis" (Augustine, 2000, p. 8), where crises do arise, the next five steps in crisis management will have an overall aim to minimize all ensuing negative impacts.

For stage two, preparing to manage the crisis requires operators to adopt the mindset that renders crises as inevitable aspects of adventure operations. This view demands that plans and strategies are developed to address crises. Presumably, most adventure tourism operators across the industry would have these kinds of strategies in place for the management of foreseeable events. These strategies can include regular crisis simulations, worst-case scenario communication, protecting and directing endangered clients, head counts, coordination with emergency response agencies, rescue techniques, establishment of employees' crisis roles, first aid, transport procedures, standardized reporting procedures, and public relations. Augustine (2000, p. 14) also suggests that crisis management should consider "second order effects." For example, having survived a severe storm, a sea kayaking party may find themselves without communication. Here, emergency supplies of food and blankets may be critical for life support until search and rescue teams can locate the party.

Stage three involves recognizing the crisis. Often, early recognition will only come to those with extensive experience in the activity and in local conditions. As described in the earlier discussion of the Interlaken tragedy, local townspeople claimed to have identified the approaching crisis with their warnings going unheeded. To provide early crisis warning, objective environmental indicators of potential hazards should be identified where possible. These indicators, for example, may relate to certain weather patterns, river levels, or snow conditions. Where conditions cause indicators to be exceeded, specific actions would be automatically triggered (e.g., canceling an adventure) removing this decision responsibility from individual employees.

Containing the crisis (stage four) is where "tough decisions have to be made and made fast" (Augustine, 2000, p. 20). These situation-specific decisions have the overriding aim of preventing death and further injury. The crisis management plan will be invoked and subsequently a company representative should establish links with local authorities and media. The successful crisis resolution (stage five) will be a result of careful crisis planning and management. The final outcome can also depend to some extent on luck (whether good or bad) and this element should be recognized in the final stage, profiting from the crisis. This stage involves assessing the chain of events leading to the crisis and determining the success or otherwise of crisis management planning. The *profit* or benefits arising from the crisis surround the communication of the circumstances to the industry so that other operators can effectively avoid or manage similar crises in the future. Lastly, to avoid financial losses, operators should carry adequate insurance for the protection of their clients and themselves (Wilks and Davis, 2000).

The advent of habitual crises in adventure tourism will engender potential and current consumers of adventure tourism with suspicious attitudes towards the industry. Managing crises in a responsible manner requires that operators show due "concern for its customers and commitment to corporate ethical standards" (Augustine, 2000, p. 28) to regain a degree of confidence and trust. This emphasizes the need for objective standards to be implemented within (and by) the industry and supported through appropriate public policies. Fundamental to the system, these standards should be communicated clearly to the adventure tourism public.

Linking Aspects within the Commercial Adventure Tourism Operating Environment

The objective of this chapter was to provide a practical model that encompasses aspects within the risk management operating environment. Figure 1 draws together the aspects relevant to commercial adventure tourism that have been discussed in this chapter. The figure is intended to represent features common to the majority of operations in the commercial adventure tourism industry. Of course, the industry is highly diverse in the types of activities offered, the nature of the physical settings, type of clothing and equipment, the activity duration, and the target markets. This diversity should be considered when designing risk management strategies specific to different categories of adventurous activities and also be informed through the seven-step process of risk management described by Standards Australia (1999). Furthermore, the scope of this chapter does not permit a full evaluation of all the aspects contained in Figure 1 (e.g., marketing). Interested readers are directed to the listed references at the close of this chapter (e.g., market aspects relevant to the industry are discussed by Fluker and Turner, 2000).

In Figure 1, the aspects relevant to commercial adventure tourism are partitioned into discrete zones, loosely based on Covey's (1990) circles of influence and concern. The individual adventure operator does not have direct responsibility for aspects situated in the two *zones of concern* (Figure 1). Rather, this is the concern of legislators (in the legal environment), industry bodies (for accreditation), the

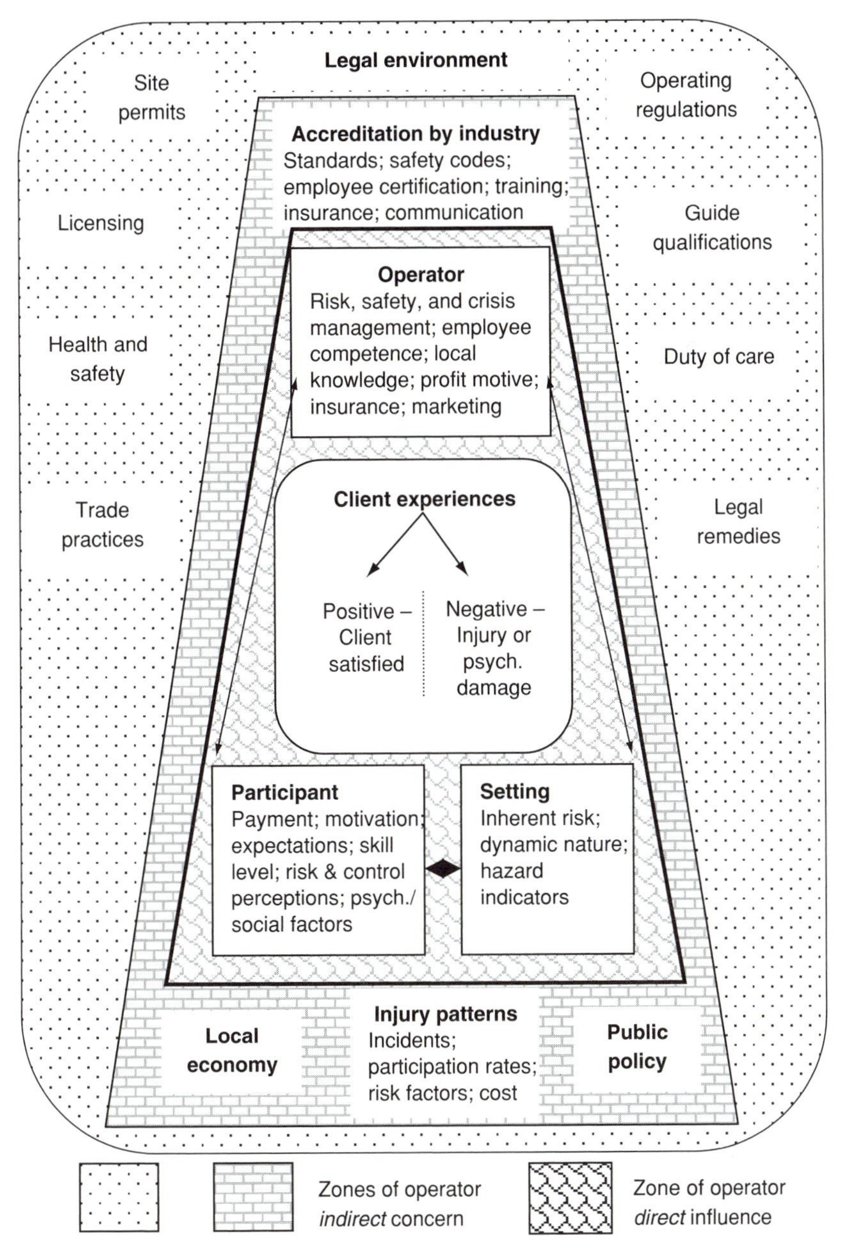

Figure 1 Commercial adventure tourism operating environment.

health system and health researchers (to identify injury patterns), and the wider community (in the public policy process).

Although legal responsibilities rest elsewhere, operators must nevertheless adhere to mandatory requirements of legislation and common law (or face the legal consequences). Many of these legal requirements that concern commercial adventure operators are listed in the outer zone of Figure 1. Similarly, operators should remain cognizant of non-mandatory aspects represented in the inner zone of concern (Figure 1). Indeed, operators will benefit through accreditation standards and practices and also gain a better understanding of the operation of specific risk factors in adventure activities through injury reporting and research.

The three elements that combine in providing the client's experience (the operator, the participant, and the setting) are within the operator's zone of influence (Figure 1). The operator carries direct responsibility for managing these aspects to provide appropriate client outcomes (depicted in Figure 1 as either a positive or a negative client experience).

Fortunately, client outcomes from adventure tourism experiences will usually be positive. Improvements to product quality and professional delivery will ensure that clients will continue to experience satisfying adventures into the future. This predicted trend is supported through the continued efforts made by industry bodies to improve operator practices (with some of these initiatives discussed earlier in this chapter). Nevertheless, recognition by operators and industry representatives of the aspects and attendant links within the risk management operating environment will aid in the delivery of suitable adventure products.

Research Directions for Commercial Adventure Tourism

There are many areas for research that will promote a better understanding of the adventure tourism industry. Risk management requires knowledge across a range of aspects within the commercial adventure tourism operating environment. Multidisciplinary research projects should focus on these specific aspects and their links to the adventure tourism industry.

A number of research areas have been suggested earlier including the role of suspected risk factors for injury and the causes and effects of mindless operations of activities. More broadly, research into the effectiveness of accreditation programs and organizations in determining and communicating suitable industry standards is required to benchmark current operator practices. A better understanding of the relationship between marketing practices and the client experience of adventure would also assist the adventure tourism industry and build on previous research (e.g., Wilks, Atherton, and Cavanagh, 1994).

Conclusion

Adventure tourism activities carry the onus for managers and guides to provide an experience for their clients within an acceptable margin of safety. The physical risks that do exist should be clearly communicated to clients before they commit to the adventure. In this respect, commercial adventure tourism operations will become increasingly proficient and professional. Nonetheless, clients will be simultaneously seeking new and more challenging adventure experiences. To meet this demand, structured management frameworks underpinned by current legislative and accreditation requirements will not only improve safety standards but work to effectively match the tourist demand to industry supply at appropriate risk levels and in suitable locations. This practice will provide long-term benefits for adventure operators and their clients.

Concept Definitions

Adventure tourism A broad spectrum of commercialized recreational operations that facilitate tourists' deliberate engagements with elements of risk emanating from the natural environment (Hall, 1992).

The participant The characteristics of the participant, including their past experience in the activity, form the first of three core elements which significantly determine the outcome of an adventure tourism experience.

Management Management techniques employed by the adventure tourism operator, such as ensuring that their guides hold the appropriate first aid qualifications and having crisis management plans in place, make up the second of three core elements that affect the outcome of the adventure tourism experience.

The setting The unpredictable and dynamic nature of most adventure tourism settings is the third core element of the adventure tourism experience.

Commercial adventure tourism operating environment The combination of legal, social, environmental, managerial, and operational elements of concern or influence in adventure activity operation.

Note

[1] This chapter was previously published in slightly different form as: Morgan, D., and Fluker, M. (2003). *The Journal of Hospitality and Tourism Management,* 10(1), 46–59.

Review Questions

1. What is the most important lesson to be learned from the 1999 Interlaken canyoning disaster?
2. Discuss the possible risks, from both the operator and client perspective, inherent to the following activities: white-water rafting; bungee jumping; and horse trekking.
3. Explain the potential implications for an adventure tour operator holding an unrealistic understanding of their client's perceptions of the experience.
4. Which factors in Figure 1 (the commercial adventure tourism operating environment) does the operator have the most control over? Which factors does the operator have the least control over? Justify your answer.
5. Accreditation can assist adventure tourism operators in meeting a designated level of safety. Besides legislation, what strategies can encourage adventure tourism operators to become accredited?

6. Imagine that you are the office-based manager of a company that organizes multi-day bushwalking tours in remote locations. You have just received a satellite phone message from one of your guides in the field. She told you that ""a landslide bringing down tons of stones, mud, and rubble has just destroyed the campsite. Between six and ten of the party are yet to be accounted for." Describe the steps that you would take to manage this crisis.

References

Augustine, N. R. (2000). Managing the crisis you tried to prevent, in *Harvard Business Review on Crisis Management*. Boston: Harvard Business School Press.

Bartholomeusz, S. (2002, March 30). Insurers can't help bad luck, only bad management. *The Age,* pp. 1, 2.

Bentley, T. A., and Page, S. J. (2001). Scoping the extent of adventure tourism accidents. *Annals of Tourism Research,* 28(3), 705–726.

Bentley, T. A., Page, S. J., and Laird, I. S. (2001). Safety in New Zealand's adventure tourism industry: The client accident experience of adventure tourism operators. *Journal of Travel Medicine,* 7(5), 239–245.

Bentley, T., Meyer, D., Page, S., and Chalmers, D. (2001). Recreational tourism injuries among visitors to New Zealand: An exploratory analysis using hospital discharge data. *Tourism Management,* 22, 373–381.

Bita, N. (2001a, 13 December). Appeal to Interlaken families. *The Australian,* p. 5.

Bita, N. (2001b, 13 December). Verdict brings peace to some, warnings for others. *The Australian,* p. 13.

Blamey, R., and Hatch, D. (1998). *Profiles and motivations of nature-based tourists visiting Australia*. Canberra, Australia: Bureau of Tourism Research.

Brannan, L., Condello, C., Stuckum, N., Vissers, N., and Priest, S. (1992). Public perceptions of risk in recreational activities. *Journal of Applied Recreation Research,* 17(1), 144–157.

Carey, M. J., and Aitken, M. E. (1996). Motorbike injuries in Bermuda: A risk for tourists. *Annals of Emergency Medicine,* 28(4), 424–429.

Cheron, E. J., and Ritchie, J. R. B. (1982). Leisure activities and perceived risk. *Journal of Leisure Research,* 14(2), 139–154.

Cloutier, R. (1998). *The Business of Adventure: Developing a Business in Adventure Tourism*. Kamloops, British Columbia Canada: Bhudak Consultants.

Cloutier, R. (2000). *Legal Liability and Risk Management in Adventure Tourism*. Kamloops, British Columbia Canada: Bhudak Consultants.

Cordato, A. J. (1999). *Australian Travel and Tourism Law*. Sydney: Butterworths.

Covey, S. R. (1990). *The 7 Habits of Highly Effective People: Restoring the Character Ethic*. Melbourne: The Business Library.

Elias, D. (1999, August 1). Tourism chasing the rush. *The Age,* p. 16.

Fluker, M., and Turner, L. (2000). Needs, motivations, and expectations of a whitewater rafting experience. *Journal of Travel Research,* 38(4), 380–389.

Hall, C. M. (1992). Adventure, sport and health tourism, in C. M. Hall and B. Weiler (eds.), *Special Interest Tourism*. London: Belhaven Press, pp. 141–158.

Hall, C. M., and McArthur, S. (1994). Commercial white-water rafting in Australia, in D. Mercer (ed.), *New Viewpoints in Australian Outdoor Recreation Research and Planning*. Melbourne: Hepper-Marriott, pp. 109–118.

Johnston, M. E. (1989). *Peak Experiences: Challenge and Danger in Mountain Recreation in New Zealand.* Unpublished doctoral dissertation, University of Canterbury, Christchurch, New Zealand.

Kasperson, R. E., and Dow, K. (1993). Hazard perception and geography, in T. Garling and R. G. Golledge (eds.), *Behaviour and Environment: Geographical and Psychological Approaches.* The Netherlands: Elsevier Science, pp. 193–222.

Kemp, S. (2002, January 19). Insurance threat to business. *The Age,* p. 3.

Kolluru, R. V. (1996). Public health and safety risks: Prevention and other initiatives, in R. V. Kolluru, S. M. Bartell, R. M. Pitblado, and R. Scott Stricoff (eds.), *Risk Assessment and Management Handbook.* New York: McGraw-Hill, s. 5.1–5.61.

Langer, E. J. (1989). *Mindfulness: Choice and Control in Everyday Life.* London: Harvill.

Langley, J. D., and Charmers, D. J. (1999). Coding the circumstances of injury: ICD-10. A step forward or backwards? *Injury Prevention,* 5(4), 247–253.

Le Quesne, N. (1999). Switzerland's river of tears. *Time South Pacific,* 32, 40.

Liability Insurance Taskforce. (2002). *Report: Liability Insurance Taskforce, February 2002.* Brisbane, Australia: Queensland Government.

Mann, S. (1999, 30 July). How "a lousy little creek" devoured 19 adventurers. *The Age,* P. 1.

Maritime Safety Authority. (1994). *Accident Report and Investigation Summary* (Report No. 94 487). Wellington, New Zealand: Author.

Morgan, D. J. (2000). Adventure tourism activities in New Zealand: Perceptions and management of client risk. *Tourism Recreation Research,* 25(3), 79–89.

Morgan, D. J. (2002). Risk, competence and adventure tourists: Applying the adventure experience paradigm to white-water rafters. *Leisure,* 26(1–2) . 107–127.

Moscardo, G. (1997). Making mindful managers: Evaluating methods for teaching problem solving skills for tourism management. *Journal of Tourism Studies,* 8(1), 16–24.

Nicholson, B. (1999, 31 July). Risks taken for profit, say guide. *The Age,* p. 3.

ORCA. (2001). Training and accreditation. *Outdoor Recreation Council of Australia.* [Available at http://www.ausport.gov.au/orca/tanda.html, accessed 24 June 2001.]

PADI. (2001). Scuba diving professional. *Professional Association of Diving Instructors.* [Available at http://www.padi.com/courses/pro/default.asp, accessed 24 June 2001.]

Parks Victoria. (2001). Information for prospective tour operators. *Parks Victoria.* [Available at http://www.parkweb.vic.gov.au/, accessed 25 June 2001.]

Queensland Government. (2001). *Horse Riding Schools, Trail Riding Establishments and Horse Hiring Establishments Industry Code of Practice 2002.* [Available at http://www.detir.qld.gov.au/hs/icp/icp012.pdf, accessed 25 June 2001.]

Smith, K. (2001). *Environmental Hazards: Assessing Risk and Reducing Disasters* (3rd ed.). London: Routledge.

Standards Australia. (1999). *Risk Management (AS/NZS 4360:1999).* Strathfield, Australia: Standards Association of Australia.

Strahan, N. (2001, 17 December). Tour firms warned on canyon findings. *The Australian,* p. 5.

Tourism Victoria and VTOA. (2002). *Victoria's Adventure Tourism Action Plan 2002–2004* (available from the Victorian Tourism Operators Association).

Victorian Government. (2001). *Marine Act 1988 (Act No. 52/1988).* [Available at http://www.dms.dpc.vic.gov.au/, accessed 26 June, 2001.]

VTOA. (2001a). Tourism accreditation—An overview. *Victorian Tourism Operators Association*. [Available at http://www.vtoa.asn.au/accred/index.html, accessed 26 June 2001.]

VTOA. (2001b). Insurance costs—Why are they escalating? *Tourism News,* 15(6), 2.

Webber, K. (2001). Outdoor adventure tourism: A review of research approaches. *Annals of Tourism Research,* 28(2), 360–377.

Wilks, J., and Atherton, T. (1994). Health and safety in Australian marine tourism: A social, medical and legal appraisal. *The Journal of Tourism Studies,* 5(2), 2–16.

Wilks, J., Atherton, T., and Cavanagh, P. (1994). Adventure tourism brochures: An analysis of legal content. *Australian Journal of Hospitality Management,* 1, 47–53.

Wilks, J., and Coory, M. (2000). Overseas visitors admitted to Queensland hospitals for water-related injuries. *Medical Journal of Australia,* 173(5), 244–246.

Wilks, J., and Davis, R. J. (2000). Risk management for scuba diving operators on Australia's Great Barrier Reef. *Tourism Management,* 21, 591–599.

10

The Effect of Disaster on Peripheral Tourism Places and the Disaffection of Prospective Visitors

Wilson Irvine and Alistair R. Anderson

Learning Objectives

- To develop an understanding of the concept of image.
- To show how image influences tourists in the process of choosing a holiday destination.
- To understand what factors peripheral tourism areas depend on in order to attract tourists.
- To be able to explain why tourism areas are vulnerable to any change in perception.
- To appreciate the effects of different factors created by disaster on different business locations and types.
- To explain why peripheral tourism areas are likely to become more dependent on remaining attractive.
- To be able to analyze how and why providers' perception of the effects of disaster may be more severe than the actual effects.

Introduction

The purpose of this chapter is to explore the impacts of foot and mouth disease on peripheral tourist destinations. Our theoretical orientation is that tourists are attracted to the otherness of peripheral places; that while such places stand as different, this conception of difference can rapidly shift from attraction to repulsion. Hence peripheral tourism is very vulnerable to shifting perceptions of safety and risk. In particular, the powerful lens of media portrayal impacts strongly and indiscriminately on peripheral places. We first outline our conceptual framework, then report on the role of tourism in the periphery of Scotland, in the Grampian region, and in the northern periphery of England, the Cumbria region. To develop our understanding we studied two different areas; Grampian, which was only indirectly affected, and Cumbria, which was directly affected by the presence of the disease. Our empirical data, collected by survey shows a very uneven effect: by region as we had expected, but also by type of tourist establishment. Finally we discuss our findings in the light of our theoretical model.

The Attraction of Places

Although Mathieson and Wall (1982) note that there is little agreement about the importance of any specific factor to motivate people to visit particular places Tiefenbacher and colleagues and Galloway (Tiefenbacher et al. 2000; Galloway, 2002) propose two types of motivation, push and pull factors. Crompton (1979b) and Goossens (2000) identified that push factors are broadly associated with demographic attributes and psychological variables such as need and personal values. Pull factors are seen as those external to the individual and are aroused by the destination. Dann (1981) points out how many researchers focus on the pull factors, since these represent the specific attractions of place. But Goossens (2000) suggests that both sets of factors should be considered, since each is one side of the motivational coin. Emotion is seen to be the connecting link, because tourists are pushed by their emotional needs and pulled by the emotional benefits. Leisure is thus seen as a positive and subjective experience, in particular, that emotion plays a major role in hedonistic consumption. In terms of destination pull factors, there is broad agreement about the influence of tourism image on the behavior of individuals (Mansfeld, 1982; Ashworth and Goodall, 1990).

The tourism image is often defined as an individual's overall perception or total set of impressions about a place (Hunt, 1998; Phelps, 1999; Fakeye and Crompton, 1997), or as the mental portrayal of a destination (Crompton, 1979a; Alhemoud and Armstrong, 1996). Such images are the manifestations of the social construction of a place. In other words, as Carter (1998) puts it, the symbolic contributes to the sense of place. The image of a destination consists, therefore, of the subjective interpretation of reality made by the tourist. Zukin (1995) draws attention to the interrelationship of the material and the symbolic. She argues that the production of particular space has a synergy with its production of symbols. This echoes the postmodern analysis of Harvey (1989) about the commoditization of symbol (Lash and Urry, 1994). There is now considerable evidence (Gartner, 2000; Kent, 1990; Crompton and Ankomah, 1993) about the influence of the tourism image on the choice of holiday destination. This means that places with stronger positive images will have a higher probability of being included and chosen in the process of decision making (Alhemoud and Armstrong, 1996; Bigne et al., 2001). Pike (2002) recently reviewed 142 academic papers about image. One key element of his extensive review was that images were either favorable or not. Gunn (1972) discusses the image formation of destinations and makes a distinction between induced and organic components.

Induced images are created by the strategic promotion of place. Organic components, in contrast, are not directly created, but formed by channels of mass media. Gunn contends such organic images are argued to be more influential. This supports the idea of the power of a social construction of place. Schutz (1972) talks about the stocks of knowledge of a phenomenon acquired inter-subjectively by individuals; formed through organic representations produced by media, education, government, and other institutions. Tiefenbacher et al. (2000) argue that such perceptions are generated by advertisement, movies, and word of mouth. Among a range of proposed factors they suggest that "keeping up with the Joneses" is important. Thus group perceptions of a place are an influence. Reid and Reid (1993) make a similar point, that positive images are shared and also lead others to visit the location.

Image will therefore influence a tourist in the process of choosing a place to stay (Bigne et al., 2001) and image, and its influence, is likely to be constructed prior to the actual experience of the place. Thus influence begins at the stage of choosing the holiday destination, and consequently destination choice cannot be explained exclusively in terms of the objective environment (Johnson and Thomas, 1995). As Gallanti-Moutafi (1999) notes, tourists embark on their journeys with already formed images, largely the product of popular cultural representations. Places are transformed into a tourist site through the system of symbolic and structural processes. Tourists "read" these signs and judge their aesthetic appropriateness. Stewart et al. (1998) stress how interpretation of place provides a better framework for understanding perceptions of place than merely asking visitors to recount "facts." Moscardo (1999) suggests this is because people cannot process all the available information. Conversely, Owen et al. (1999) suggest that because of a lack of detailed information, prospective tourists will place greater reliance on long established impressions and possibly stereotypical impressions. Mathieson and Wall (1982) suggest, in terms of push factors, that the motivation to visit a place may depend upon the perception of the value in visiting that destination. Thus images of place are broad conceptions, loosely formed and probably based on the assimilation of diverse and incomplete information. For example, Dann (1996, p. 79) shows how representations of destinations rely on cultural stereotypes and received images, "which remain to be confirmed or invalidated by experience." Images of place and the consequent choice of destination are therefore an individual subjective interpretation, but formed from social and shared representations selected from our economy of space and sign (Lash and Urry, 1994).

Consequently we may argue that tourism consists of a demarcation of both space and time. For time, as Baudrillard (1998) points out, it is a leisure time, differentiated from work time and caught up in the consumption of signs and experiences. For space, tourism is about created leisure space, places that are first signed as appropriate (Urry, 1990) and then consumed. "A new, or renewed importance attaches to place . . . even when these are imagined or invented" (Kumar, 1995, p. 123). Tourism thus creates specific social space (Meethan, 2001). Yet this specificity of place is also caught up in the headlong dash of space-time compression, what Harvey calls the annihilation of space through time. As Harvey (1989, p. 293) puts it so well,

> Mass television ownership coupled with satellite communication make it possible to experience a rush of images almost simultaneously, collapsing the world's spaces into a series of images on a television screen . . . mass tourism, films made in spectacular locations, make a wide range of simulated or vicarious experiences of what the world contains available to many people.

In this fission and fusion of the local and the global, the global spread of tourism depends upon the specificity of place, but the processes of globalization bring about a greater range of wider couplings, thus demonstrating that globalism pulls two ways. One specific arena of this global local acting out is the periphery, where the otherness of image places a vital role in attracting tourists.

Peripheral Places

The notion of peripherality is also a matter of perception. Brown and Hall (1999, p. 9) argue that a place that is remote and difficult to reach may be perceived by tourists to have certain qualities symptomatic of its situation, such as natural beauty, quaintness, and otherness. Such places are seen as "authentic" (Urry, 1990), rich in symbolic representations of the unspoiled, the pristine, and the traditional. Urry (1995) also makes a powerful case to show that it is this otherness which creates attraction. Thus as Blomgren and Sörensen (1998) propose, the attractiveness of a periphery relies on the subjective interpretation of such symbols. Anderson (2000) argues that peripheral spaces have moved from outlying production zones to become areas that are consumed in their own right. He argues that it is their very otherness, non-industrial nature, distance, and an absence of core activities that creates value in the consumer's eye. Moreover, it is those very qualities of otherness that are consumed (2000, p. 102): "the periphery is the ideal zone for the production of aestheticized cultural goods." Brown and Hall (1999) describe a peripheral area as one that suffers geographic isolation, being distant from core spheres of activity, and having poor access to and from markets. Such areas, they claim, are characterized as economically marginalized with much of the business activity confined to microbusiness. But as Wanhill (1997) notes, the European Union's Maastricht Treaty acknowledged that tourism could reduce regional disparities. Taken together, we see the importance of tourism for the peripheral place, highly dependent on the "difference" of image from the core, but we also see how it appears to depend on a positive image.

The Vulnerability of Subjective Interpretations

One problem linked to image and the motivation to visit is the fragility of the symbolic otherness. Pearce (1982) considers appropriate images as transitory, but ones that are insulated from danger. Meethan (2001) talks of trust in a destination; trust in it measuring up to its image. He makes the salient point that the elimination of risk and issues of safety appear as prime factors in choice of destination. Cavlek (2002) points out that peace, safety, and security are the primary conditions for the tourism development of a destination. He also notes (2002, p. 479) how "nothing can force them to spend a holiday in a place they perceive as insecure." Indeed, Sonmez and Graef (1998, p. 120) argue that if the destination choice is narrowed down to two alternatives that promise similar benefits, the "one that is safe from threat is likely to be chosen." Pearce (1988) suggests that concern for personal security is a major factor in the decision-making process through which individuals make their travel choices (Sonmez, 1998). Although Galloway's (2002) paper explores sensation seeking as an explanatory factor in motivation and Elsrud

(1999) discusses "risk taking" as an attraction in backpacking holidays, these are special instances when risk appears to enhance image. In this way, different groups may have different perceptions or even different social constructs of thrills and danger (Carter, 1998).

In any case we know that images are very incomplete representations. Indeed they may be fundamentally inaccurate. For example, Cavlek (2002) reports that during the Indonesian crisis, tourism to Bali was not affected. This was because of the general lack of awareness that Bali was part of Indonesia. Similarly the Greek island of Kos was badly affected by the misinformed associating it with Kosovo! Drabek (2000) notes how the effects of crisis ripples out to areas where no such problem exists (Cavlek, 2002). Yet as Faulkener and Vikuluv (2001) noted, crises have become integral to business activity, but tourism suffers more than any other. Faulkner and Vikuluv (2001) propose that all destinations face the prospect of either a natural or a human-induced disaster. In particular, Cavlek (2002) suggests that government warnings to potential tourists always have strong psychological effects, thus creating a major impediment to selling holidays, even to parts of the country still entirely safe.

Thus far we have explored the importance of image in motivating tourism. We have demonstrated that the otherness of peripherality is a key mechanism for attracting tourists. This otherness, we have argued, is an incomplete social construction, driven by, and created within, the forces of globalization but dependent upon a positive impression of local place. We have also noted how perceptions of risk, real or imagined, like the images themselves, may act to reverse the attraction and turn it into a repelling force. We now continue to explore the vulnerability of specific peripheral tourist places.

Tourism in Scotland

Tourism is Scotland's most important industry, injecting £2.5 billion into the economy annually (the web site www.scotexchange.net, Media Office, 2003). It is the fourth biggest employer, employing 193,000, some 8% of the workforce. In 1995 the UK ranked fourth in the top ten tourist destinations in Europe, with 23.7 million arrivals (De Vaal, 1997). However, inbound tourist statistics show that only 1.9 million of these United Kingdom visitors came to Scotland, with that figure dropping to 1.5 million in 2001 (the web site www.staruk.org.uk, Key Facts, 2002).

This decrease is blamed on the effects of the September 11 terrorist attack and on foot and mouth disease. As Bowditch (2002, p. 2) noted, "The tourism industry hit by foot and mouth disease, a strong pound, the effects of terrorism and years of poor management, desperately needs a boost." 2002 saw increased visits to Scotland by UK residents, with visits being up by 10% compared to 2001 (McKay, 2003). Thus we see that the "disaster" of foot and mouth disease had a very real impact on tourist numbers. The Tourism Attitudes Survey (Know Your Market, 1999) indicated that 81% of all visitors came to Scotland for its beautiful scenery, much of which by its very nature is peripheral. Destination attractions include Scotland's most dramatic and atmospheric places, from Culloden Moor to Glencoe for the Brave Heart or the Loch Ness Monster experience (Visit Scotland, 2002). Thus otherness is actively promoted: "Nowhere else will awaken your senses in quite the same way from loch side or seacoast to wild summit or vibrant cities—dramatic landscapes will inspire" (Visit Scotland and British Tourist Authority, 2002, p. 1). Other tourist literature promotes "distinctive Highland culture, colorful history, built heritage, and opportunities for quality recreation within our natural environment" (The Highlands of Scotland Tourist Board, 1997, p. 1).

Not all agree; Michael O'Leary, chief executive of the Irish airline Ryan Air, argued that Scottish tourism was "stuck in the 50s, advertising Haggis and Castles to the Europeans" (Robertson, 2001, p. 1).

Tourism in Peripheral Grampian

Grampian is the northeast shoulder of Scotland with a tourist product primarily focused on scenery and castles. Scenery, heritage, and history play a major part in tourist attraction; seeing historic house and castles is important for eight out of ten visitors. Grampian's attractions currently range from outdoor activities and natural and built heritage to adventure and theme parks (Aberdeen City Council, 2002). However Aberdeen and Grampian visitor numbers fell by 13.1% from 140,743 in 2001 to 122,255 during the same period in 2002 (the web site www.scotexchange.net, Media Office, 2003). This reduction was confirmed by the local tourist board, at –12.8%. Researchers have spent much time debating Grampian's problems of seasonality and peripherality and analyzing what disadvantage, if any, is placed on the area because of these factors. Peripherality has been viewed as the biggest problem, being held responsible for the increasing amount of difficulties being experienced within the industry (Baum, 1996), and is most often viewed as the most consistent policy issue within cold-climate areas. A peripheral area is seen as an area of remote geographical isolation that is far away from central areas of activity, with poor infrastructure, meaning access is difficult (Brown and Hall, 2000). This problem is especially evident in Grampian where the majority of the region is isolated from major cities. Table 1 shows key differences between core and peripheral areas.

Table 1 Core and Peripheral Areas

Core	*Peripheral*
High levels of economic vitality and a diverse economic base.	Low levels of economic vitality and dependent on traditional industries.
Metropolitan in character. Rising population through inmigration with a relatively young age structure.	**More rural and remote—often with high scenic values. Population falling through outmigration, with an aging structure.**
Innovative, pioneering, and enjoys good inflows.	Reliant on imported technologies and ideas, and suffers from poor information flows.
Focus on major political, economic, and social decisions.	**Remote from decision making leading to a sense of alienation and lack of power.**
Good infrastructure and amenities.	**Poor infrastructure and amenities.**

Source: Brown and Hall (2000, p. 9).

The highlighted sections in Table 1 show key factors affecting the Grampian region. Aberdeen city is the core city, bringing some new people and business to the area. Nonetheless, the peripheral hinterland surrounding the city is typical of peripheral places. The 2001 Census highlighted this, showing only 15% of the population was under 15 years old, whereas over 18% of the population are 65+ (the web site Aberdeen.net, Aberdeen City Council 2002). Aberdeen, however, still makes up a large percentage of the area, with a population of 212,215, which is 4.2% of the population of Scotland, although this figure is falling, due to younger generations moving away. Statistics show that in 1991 the population was 211,910, dropping by 3,040 in 2001, clearly defining problems of an aging society. The northeast of Scotland is undoubtedly dependent on the oil industry and contains a city and a number of smaller towns mainly reliant on oil and some traditional industries. However, it is an area with a mainly peripheral structure with poor roads and a large rural community not dissimilar to Cumbria. It also contains some unique tourist attractions and wonderful scenic beauty comparable to the northeast of England.

Tourism in Cumbria

Cumbria is located in the north of England and is both peripheral and highly dependent upon tourism. Unlike Grampian, where there were no reported cases of foot and mouth disease, Cumbria was directly affected by the presence of the disease. Cumbria includes the Lake District National Park and the Hadrian's Wall World Heritage Site and has recently been awarded Green Globe Destination Status (see the web site www.golakes.co.uk, Cumbrian Tourist Board, 2003). It is a relatively remote area composed of sparsely populated sectors with some minor concentrations of population. The Lake District itself offers a rural experience with a plethora of cultural events, museums, and galleries. It is dominated by small owner-managed businesses. However, it has been directly affected by foot and mouth and other threats since 2001 and there has been a marked decrease in visitor numbers as demonstrated in Table 2. Table 2 shows the drop in numbers experienced by tourism businesses due to the foot and mouth epidemic. The trips refer to the summer months of June to September 2000–2002.

Table 2 demonstrates that, of the four representative English locations, Cumbria was the worst affected and experienced the greatest percent reduction in visitor numbers, with little evidence of recovery in 2002.

Table 2 Tourism Trips in England

Summer Trips	*All Tourism Trips 2000 (million)*	*All Tourism Trips 2001 (million)*	*All Tourism Trips 2002 (million)*
England	64.9	55.5	63.0
Cumbria	2.3	1.8	1.8
Yorkshire	6.0	5.4	5.8
London	9.1	7.7	8.0

Source: Adapted from the web site www.staruk.org.uk (International Passenger Survey, 2003).

Foot and Mouth Disease

Foot and mouth disease is one of the most contagious animal diseases. Although most affected adult animals will recover within two weeks, the drop in yields could have enormous economic impact. It has few effects on humans. Nonetheless, the UK government policy of slaughtering affected or at-risk herds had an enormous impact on Britain's countryside. The first cases of foot and mouth disease (since 1967) were confirmed on February 10, 2001. Within two weeks the disease had spread to a large number of cases. After peaking in April/May numbers tailed off to October 2001. As Anderson (House of Commons, 2003, p. 5) noted, a total of 2,026 cases of the disease were identified and a total of over 4 million animals were culled during the crisis. Media attention during the crisis focused dramatically on the agricultural community, showing the destruction of livestock and the closure of farms across the county. However, the vast majority of this was brought about without confirmation of the disease in that area. Ireland and Vetier (2002, p. 5) detail the steps taken when farms were not directly affected, but were unfortunate enough to be in the cull area, or have links with foot and mouth cases. "A quarantine ban was established on farms with trace former connections to the confirmed case and animals killed under the classification of 'slaughter on suspicion.'" However, there were also direct affects felt in one way or another by most other industries, in particular tourism that is dependent on access to accommodation and associated tourist facilities.

The two study areas affected by the disease were Cumbria and the Grampian region, both peripheral in the sense that they are not associated with any major population areas and the accompanying activity and tourism attractions that are principally in the countryside. Cumbria was affected directly with a large number of cases, but Grampian only indirectly. Grampian did not have one single case during the outbreak and was more than 150 miles from the nearest case in the south-west of Scotland.

The Impacts of Foot and Mouth in Cumbria

Cumbria's main industries are agriculture and tourism and tourism was affected just as badly as agriculture. As Ireland and Vetier (2002, p. 6) put it, it is "evident that demand failure among tourists has a severe impact on the British tourism industry." The BBC News web site (April 2001) dramatically described the devastation and fear of the unknown future for the farmers of Cumbria saying, "Cumbria is holding its breath. Not just in dread of future outbreaks, but also because of the smell of the burial sites." Television dramatized the extreme actions taken by the government and the effect on peoples' lives. Tourism in Cumbria particularly suffered when the Government closed the countryside down. The Anderson report (House of Commons, 2003, p. 3) noted the closure of many footpaths, "the instrument to close footpaths and bridleways were necessary not only in infected areas but also outside them." But many of the tourism businesses within Cumbria rely on the footpaths and surrounding areas to be open for them to survive. One consequence was the difficulty in gaining access to many of the small villages within Cumbria when these roadways are closed. Television coverage of the foot and mouth epidemic detailed every case and scare within Cumbria. The television coverage scared many potential tourists away from the countryside; many areas that had no contact with the outbreak suffered because of the media messages given. Ireland and Vetier (2002, p. 1) argued that "exaggerated media reporting of a crisis can be as damaging as inept Government policy." Many of the tourism businesses within Cumbria closed because of the dramatic reduction of visitors within the area. After the epidemic was over tourism organizations within Cumbria began to try to rebuild the businesses by extra advertising and property upgrading. Although considerable efforts were made, tourism numbers were still poor compared to previous years.

The effects within Cumbria devastated both tourism and agricultural businesses. Many of the businesses exist in remote towns and villages spread out within the district and many directly rely on tourists drawn to their areas by the impressive wildlife. Thus the closure of footpaths and roads connecting the remote areas of Cumbria to the tourists meant that most of this wildlife could not be reached. This reliance on footpaths and road connections for the remote tourism businesses caused many of the problems when the foot and mouth epidemic struck. It caused a decrease in tourists so great that all tourism businesses within the area were affected. A large number of the Cumbrian attractions were shut down for at least 3 months. Many of these never reopened. In addition to the closure of businesses, the loss in tourism numbers reduced turnover within the area; many people lost their jobs because businesses couldn't afford to support themselves, let alone pay wages, nearly all business investment stopped. Table 3 shows the drop in employment numbers in Cumbria between 2000 and 2001. The businesses include hotels, restaurants, tour operators, and cultural activities, all of which rely greatly on tourism within small rural areas. Job losses of more than 25% were experienced.

It is clear from the foregoing that foot and mouth impacted badly on both peripheral tourist areas. The impact in Cumbria was most obviously an effect of the physical presence of the disease. No doubt this effect was created by the sensational, stark, and unpleasant images of animals being culled and their bodies dramatically burned. However, Grampian was remote from the direct effects, but though somewhat less, still badly affected by foot and mouth. We turn now to consider, in detail, which businesses were most affected and analyze these findings to try to establish why the impact was so great on a place not directly affected by the disease.

Table 3 Employment in Cumbria

	2000 Jobs	*2001 Jobs*
Tour operator	655	549
Hotel	9,633	6,843
Restaurant	5,816	4,409
Cultural activity	850	720

Source: Adapted from the web site www.staruk.org.uk (International Passenger Survey, 2003).

Methodology

Our sample frames were drawn from tourist businesses in Grampian and Cumbria. The Grampian sample of 180 businesses was drawn from a sample frame provided by Dunn and Bradstreet. The Cumbrian sample, 170 businesses, was selected by choosing one in five from a list taken from the official Cumbria Tourist Board Guide (2002). The Grampian sample, the main locus of our study, was surveyed twice. The first survey, Survey A, was carried out in April 2001 at the height of the outbreak and had 85 responses (47%). The second survey, Survey B, was carried out in March 2003 and had 60 responses (33%). Eighteen other serveys were returned uncompleted. Either the business was no longer operating from the address or was under new ownership. These surveys were intended to provide data to allow us to gauge and compare the anticipated with the real effects. The Cumbrian survey, Survey C, was carried out in February 2003 and contained a number of identical or similar questions. This survey had a response of 70 giving a 39% response rate. Questions were asked about both the expected and the actual effects of the disease. In all of the surveys many of the questions were open ended to allow respondents to enlarge on the data.

Data Analysis

The data were analyzed using descriptive statistics to analyze single variables and simple non-parametric tests were used to compare variables and significance of normally distributed results. The tests included frequency and cross-tabulation analyses. The cross-tabulation analysis (Pearson chi-square test) was used to check significance within the normally distributed results. A number of variables were recoded so that results were considered important and significant. Significance was tested at a 90% confidence level (the majority of tests proved significant and are all represented). All of the tests were carried out after the variables were coded onto SPSS (Statistical Package for Social Sciences) and entered and run in order to test the confidence of the data. A large number of the variables had open-ended responses, which were grouped using a Pragmatic Content Analysis in order to collate the similar responses and include them as part of the descriptive analysis. A number of tables were constructed at appropriate stages to describe the results.

To provide a comparative framework, tourism providers were recoded into two types of business organizations. First were Hotel, Guest House, and Bed and Breakfast providers; and second were "other" providers that covered a diversity of organizations from Caravan sites and Golf Courses to specialty equipment or other service providers in both areas. The characteristics of the different surveys are shown in Table 4.

The Grampian Surveys, Data, and Discussion

Survey A took place in Grampian at the height of the disease and could be expected to reflect the worst expectations of the impact. We also expected these prognoses to reflect the general gloom created by the vivid and dramatic media portrayal. The results shown in Table 5 confirm our expectations and show the extent of business reduction anticipated.

Table 4 General Characteristics of the Samples

Survey	*Business Type*	*N*	*Percent of Total (%)*	*Professional Body Membership (%)*	*Customer Type: Tourist (%)*
A. Grampian, Apr 01	1. Hotel/GH/B and B	58	69	65	36
(Total N = 180)	2. Other	27	31	34	50
B. Grampian, Apr 03	1. Hotel/GH/B and B	42	70	68	34
(Total N = 180)	2. Other	18	30	67	17
C. Cumbria, Feb 03	1. Hotel/GH/B and B	46	66	66	93
(Total N =170)	2. Other	24	34	58	75

Table 5 The Impact of the Disease (All %): Survey A

Business Type	*Cancellations Anticipated*	*Volume Decrease*	*Profit*	*Staff Cuts*	*Closure*	*Length of Downturn: Greater than Weeks (Wreater than a Year)*
Type 1	56	70	67	28	14	86 (36)
Type 2	55	56	59	22	26	48 (18)

The overall view was very pessimistic. More than half our respondents anticipated cancellations and large decreases in business volume and profits. Some 25% expected to have to lose staff and a significant number anticipated closure of their business. Most appeared to expect the impact to last for some considerable time. Taken by type of business, Type 1, that is, those most likely to be dependent on visitors, we see a very large impact on volume and the duration of the effects.

Survey B (Table 6), conducted after the outbreak, was able to measure the real impact. It shows that while the impact was large, it was not as bad as had been anticipated. It is worth noting that 10% of our original sample had gone away. This could be partially attributed to the impact or it could simply be normal business churn. Table 5 demonstrates that cancellations were worse than anticipated for Type 1 businesses and at 64% of bookings, reflect a major loss of business. Nonetheless, we note that actual volume decrease was "only" 53%, suggesting that some replacement visitors

Table 6 The Impact of the Disease (All %): Survey B

Business Type	*Cancellations*	*Volume Decrease*	*Profit*	*Staff cuts*	*Closure*	*Length of Downturn: Greater than Weeks (Greater than a Year)*
Type 1	64	53	47	14	05	47(26)
Type 2	28	23	22	06	05	23(12)

were found. Again the worst impact was on accommodation types of businesses, with 47% lasting for more than weeks and 26% being affected for more than a year.

The Cumbria Survey

The Cumbrian survey (Table 7) is a snapshot of data collected two years after the epidemic. Since Cumbria was physically affected as an area where the disease was present, the data provides us with some comparison about perceptions and impacts. Table 6 shows a dramatic reduction in visitor numbers with a 98% and 91% reduction in each type of business. This resulted in a marked decrease in business with 96% and 90% for Type 1 and Type 2, respectively, indicating a "loss of business." When asked about specific percentages of "loss of business," about 20% in both types of businesses affected identified an actual loss of business of "more than 50%," with approximately 80% of those affected in each category identifying a loss of between 1 and 50%. Staff cuts were highest in the "other" types of businesses (this differs from both Grampian surveys). A very large percentage in both types experienced the effects for more than a year with a large number still suffering.

Table 7 The Impact of the Disease (All %): Survey C

Business Type	*Visitor Numbers (decrease %)*	*Volume Decrease*	*Staff Cuts*	*Length of Downturn: Greater than Weeks (Greater than a Year)*
Type 1	98	63	22	48 (30)
Type 2	98	63	2	50 (25)

The Effects of the Foot and Mouth Disease on Core and Peripheral Areas within Grampian (Survey B, Longitudinal Survey)

Table 8 clearly articulates that Type 1 organizations in this survey were larger and depended less on seasonal business than did Type 2 organizations. This means that the mainly accommodation providers were less seasonal and larger, and surpris-

Table 8 Other Characteristics of Businesses in Grampian: Survey B

Business Type	*Seasonal (%)*	*Size > 10 Staff (%)*	*Peripheral (%)*	*Impact, Type of Volume Decrease (%)*
Type 1	17	46	55	53
Type 2	33 (.1)	08 (.2)	23 (.010)	23 (.076)

The figures in brackets represent the confidence levels of the data

ingly more were situated in peripheral areas: villages and remote areas. The accommodation providers, predominantly located in peripheral areas, had a much greater decrease in volume of 53%.

Table 9 compares core and peripheral businesses (peripheral businesses being smaller and more seasonal), throughout the region. It seems clear that peripheral businesses situated in Grampian experienced a greater decrease than the core businesses situated in the city and towns. Sixty-three percent of the respondents were situated in the core and 37% in peripheral locations. There was a much greater effect on profitability in the peripheral locations with 59% experiencing some type of decrease in profitability, including a sizeable group that experienced a "large decrease" in profitability (25%). There was much the same picture for impact on volume where the peripheral businesses clearly suffered most.

Table 9 Peripherality and Effects: Survey B

Situation	*Percent of Total Sample*	*Number of Businesses With More Than Two Full Time and Part Time Staff (%)*	*Non-Seasonal (%)*	*Impact, Type of Profitability Decrease (%)*	*Impact, Type of Volume Decrease (%)*
Core	63	63	79	28	35
Peripheral	37	35 (.8)	73 (.5)	59(25) (.1)	64 (.1)

"The Figures in brackets represent the confidence levels of the data"

From these data, we conclude that the effects of foot and mouth disease on tourism businesses were considerable. The Grampian longitudinal studies indicate that although bad, the effects were not quite as bad as anticipated; this demonstrates that even the tourism providers had pessimistic perceptions of the seriousness of the disaster and the related effects on their businesses. In both areas the impact was immediate, manifest in dramatic drops in volume of business and profitability and reductions of staff numbers. It was also long-term, a large number of businesses taking almost a year to recover. In some cases, though we cannot be certain how many, the businesses actually closed. There were also some unexpected findings. We found that caravan sites in Grampian had an increase in business volume. Since the opposite is true of Cumbria, we deduce that visitors had deserted caravan parks in the affected areas and were drawn to new areas over the period. We also found some remarkable instances where substitute products were used. These included the use of geese instead of sheep at a sheep visitor attraction. There was some evidence of specific spikes of business activity, probably related to a "Dunkirk" spirit and a campaign to support domestic businesses and special marketing initiatives made at the time. Peripheral businesses were more seasonal and smaller, and clearly suffered the most in the disease situation with more negative effects on profitability and volume than businesses situated in core areas. The overall effects confirm the perception of lack of security and safety in these areas (Cavlek, 2002), and these effects have rippled out into the non-affected area (Grampian) as identified by Drabek (2000).

Conclusions

The data and analysis appear to support our original argument, that the attractions of otherness are fickle. Peripheral tourist areas, which depend on their portrayal as appropriate places for visitors, are vulnerable to any change in perception. As our data demonstrates, the impact of any circumstance that detracts from the attraction has serious economic consequences. Lending strength to our case about perception, rather than reality, is the comparison between Grampian and Cumbria. Both are peripheral places and are highly dependent upon tourism; both are rural scenic places, so that the portrayal of otherness is symbolically dependent upon an Arcadian image. This rural otherness is a contrast to the urban, but is also bucolic, replete with the benign of rural life. Unsurprisingly, the confrontation to this imagery with media pictures of smoking cattle funeral pyres resulted in repelling visitors.

However, the contrast between the areas with and without the presence of the disease is significant. Cumbria was very badly affected, but Grampian experienced no cases of foot and mouth. Cumbria was effectively closed to visitors, but Grampian was only marginally physically affected. Yet, broadly speaking, the impact on tourism was similar, though the effects were admittedly worse in Cumbria. This seems to confirm that perceptions, rather than facts or real circumstances, create the disastrous effects of catastrophe. Within Grampian the businesses situated in peripheral areas suffered most.

There are some serious implications for the economics of peripheral places in these findings. We know that for such places a designation of difference, the otherness of such places, is a tourism attractor. We know that peripheral places will continue to suffer from the centripetal forces drawing income into urban cores. Consequently we realize that peripheral places are likely to become more, rather than less, dependent on remaining attractive. Globalization seems to suggest that the importance of local place is likely to be on one hand reduced in international convergence. On the other hand, the distinctiveness of some peripheral places may become greater, simply in contrast to the convergence of others. Moreover, the massification of communication in globalization will exaggerate the qualities of peripherality. It may enhance, but, as in the case of catastrophe, it may repel. Thus peripheral places are becoming increasingly vulnerable to the fickleness of attraction. Mere facts, information alone, are unlikely to ameliorate the impact of catastrophe. Tourism decisions seem to be made in the heart, not in the head.

Action must be considered by agencies to mitigate against these perceptions about peripheral areas. These actions must be proactive through educating tourists or at the very least initiated immediately when any threat is anticipated, as the effects on peripheral areas are real and serious.

Concept Definitions

Peripheral area An area that is geographically isolated and lacks industrial milieu and related infrastructure.
Image The perception of a place, both tangible and intangible.
Disaster A real effect that brings initial negative factors to those places and people affected.

Review Questions

1. What are the factors that determine image?
2. How does image influence tourists in the process of choosing a holiday destination?
3. What factors do peripheral tourism areas depend on in order to attract tourists?
4. Why are tourism areas vulnerable to any change in perception by prospective customers?
5. Describe the real and anticipated general effects of disaster.
6. What are the effects of the different factors created by disaster?
7. Why are peripheral tourism areas more likely to become more dependent on remaining attractive in the near future?
8. Discuss why the provider's perception of the effects of disaster is normally more severe than the actual effects.

References

Alhemoud, A. M., and Armstrong, E. G. (1996). Image of tourism attractions in Kuwait. *Journal of Travel Research,* 34(4), 76–80.

Anderson, A. R. (2000). Paradox in the periphery. An entrepreneurial reconstruction? *Entrepreneurship and Regional Development,* 12, 91–109.

Ashworth, G., and Goodall, B. (1990). Tourist images: Marketing considerations, in B. Goodall and G. Ashworth (eds.), *Marketing in the Tourism Industry: The Promotion of Destination Regions*. London: Routledge, pp. 213–238.

Baudrillard, J. (1998). *The Consumer Society*. London: Sage.

Baum, T. (1996). Images of tourism past and present. *International Journal of Contemporary Hospitality Management,* 8(4), 25–30.

Bigne, J. E., Sanchez, M. I., and San, J. (2001). Tourism image, evaluation variables and after purchase behaviour: Inter-relationship. *Tourism Management,* 22(6), 607–616.

Blomgren, K. B., and Sorensen, A. (1998). Peripherality—Factor or feature? Reflections on peripherality in tourist regions. *Progress in Tourism and Hospitality Research,* 4, 319–336.

Brown, F., and Hall, D. (1999). Introduction: The paradox of periphery, in F. Brown and D. Hall (eds.), *Case Studies of Tourism in Peripheral Areas*. Nexø: Research Centre of Bornholm, pp. 7–14.

Brown, F., and Hall, D. (eds.). (2000). *Tourism in Peripheral Areas*. Clevedon, UK: Channel View.

Carter, S. (1998). Tourists' and travelers' social construction of Africa and Asia as risky locations. *Tourism Management,* 19(4), 349–358.

Cavlek, N. (2002). Tour operators and destination safety. *Annals of Tourism Research,* 29(2), 478–496.

Crompton, J. L. (1979a). An assessment of the image of Mexico as a vacation destination and the influence of geographical location upon that image. *Journal of Travel Research,* 14(4), 18–23.

Crompton, J. L. (1979b). Motivations for pleasure vacations. *Annals of Tourism Research,* 6, 408–424.

Crompton, J. L., and Ankomah, P. K. (1993). Choice set propositions in destination decisions. *Annals of Tourism Research,* 20, 461–476.

Cumbrian Tourist Board, 2002, Cumbria. The Lake District Holidays & Breaks Guide, Peneith England.

Dann, G. M. S. (1981). Tourist motivation: An appraisal. *Annals of Tourism Research,* 8, 187–424.

Dann, G. (1996). The people of the tourist brochures, in T. Selwyn (ed.), *The Tourist Image: Myth and Myth Making in Tourism*. Chichester, UK: Wiley.

De Vaal, D. (1997). *A Survey of Continental European Visitor Attractions*. UK: Deloitte and Touche, pp. 12–40.

Drabek, T. (2000). *Emergency Management, Principles and Applications for Tourism, Hospitality and Travel Management*. Federal Emergency Management Agency. www.fema.gov/emi/edu/higher.htm.

Elsrud, T. (1999). *Risk creation in travelling; Risk taking as narrative and practice in backpacker culture*. Conference paper presented at the First International Conference on Consumption and Representation, University of Plymouth, UK, September.

Fakeye, P. C., and Crompton, J. L. (1997). Image differences between prospective, first time and repeat visitors to the Lower Rio Grande valley. *Journal of Travel Research,* 30(2), 10–16.

Faulkner, B., and Vikulov, S. (2001). Katherine, washed out one day, back on track the next: A post-mortem of a tourist disaster. *Tourism Management,* 22, 331–344.

Gallanti-Moutafi, V. (1999). The self and the other. *Annals of Tourist Research,* 27(1), 203–224.

Galloway, G. (2002). Psychographic segmentation of park visitor markets; evidence for sensation seeking. *Tourism Management,* 23, 581–596.

Gartner, W. C. (2000). Tourism image: Attribute measurement of state tourism products using multidimensional scaling techniques. *Journal of Travel Research,* 28(2), 16–20.

Goossens, C. (2000). Tourism information and pleasure motivation. *Annals of Tourism Research,* 27(2), 301–321.

Gunn, C. (1972). *Vacationscape: Designing Tourist Regions*. Austin, TX: Bureau of Business Research, University of Texas.

Harvey, D. (1989). *The Condition of Postmodernity*. Oxford, UK: Blackwell.

Hunt, J. D. (1998). Image as a factor in tourism development. *Journal of Travel Research,* 13(3), 1–7.

Ireland, M., and Vetier, L. (2002). *The reality and mythology of the foot and mouth disease crisis in Britain: Factors influencing tourist destination choice*. Working paper.

Johnson, P., and Thomas, B. (1995). The analysis of choice and demand in tourism, in P. Johnson and B. Thomas (eds.), *Choice and Demand in Tourism*. London: Mansell, pp. 1–12.

Kent, P. (1990). People, places and priorities: Opportunity sets and consumers holiday choice, in G. Ashworth and B. Goodall (eds.), *Marketing Tourism Places*. London: Routledge, pp. 42–62.

Kumar, K. (1995). *From Post-Industrial to Post-Modern Society*. Oxford, UK: Blackwell.

Lash, S., and Urry, J. (1994). *Economies of Signs and Space*. London: Sage.

Mansfeld, Y. (1982). From motivation to actual travel. *Annals of Tourism Research,* 19, 399–419.

Mathieson, A., and Wall, G. (1982). *Tourism: Economic, Physical and Social Impacts*. London: Longman.

Meethan, K. (2001). *Tourism in Global Society: Place, Culture and Consumption.* Hampshire, UK: Palgrave.

Moscardo, G. (1999). Mindful visitors: Heritage and tourism. *Annals of Tourism Research,* 23(2), 376–397.

Owen, R. E., Botterill, D., Emanuel, L., Foster, N., Gale, T., Nelson, C., and Selby, M. (1999). Perceptions from the periphery: The experience of Wales, in F. Brown and D. Hall (eds.), *Case Studies of Tourism in Peripheral Areas.* Nexø: Research Centre of Bornholm, pp. 15–48.

Pearce, P. L. (1982). *The Social Psychology of Tourist Behaviour.* Oxford, UK: Pergamon.

Pearce, P. L. (1988). *The Ulysses Factor: Evaluating Visitors in Tourist Settings.* New York: Springer.

Phelps, A. (1999). Holiday destination image—The problem of assessment: An example developed in Menorca. *Tourism Management,* 7(3), 168–180.

Pike, S. (2002). Destination image analysis—A review of 142 papers from 1973 to 2000. *Tourism Management,* 23, 541–548.

Reid, L., and Reid, S. (1993). Communicating tourism supplier services; Building repeat visitor relationships. *Journal of Travel and Tourism Marketing,* 2, 3–19.

Schutz, A. (1972). *The Phenomenology of the Social World.* London: Heinemann.

Sonmez, S. F., and Graefe, A. R. (1998). Influence of terrorism, risk on foreign tourism decisions. *Annals of Tourism Research,* 25(1), 112–144.

Sonmez, S. F. (1998). Tourism, terrorism, and political instability. *Annals of Tourism Research,* 25(2), 416–456.

Stewart, E. J., Hayward, B. M., and Devlin, P. J. (1998). The "place" of interpretation: A new approach to the evaluation of interpretation. *Tourism Management,* 19(3), 257–266.

Tiefenbacher, J. P., Day, F. A., and Walton, J. A. (2000). Attributes of repeat visitors to small tourist-orientated communities. *Social Science Journal,* 37(2), 299–308.

Urry, J. (1995). *Consuming Places.* London: Routledge.

Urry, J. (1990). *The Tourist Gaze.* London: Sage.

Wanhill, S. (1997). Peripheral area tourism—A European perspective. *Progress in Tourism and Hospitality Research,* 3, 47–70.

Zukin, S. (1995). *The Cultures of Cities.* Oxford, UK: Blackwell.

Newspapers

Bowditch, G. (2002, December 9). Trying to buy in tourists is selling Scotland short. *Sunday Times,* London.

McKay, D. (2003, January 13). Future of tourism to be debated at Abertay. *Press and Journal,* Aberdeen.

Robertson, J. (2001, July 29). Scots doing too little to win back US tourists. *Sunday Times*, London.

References Available Online

Aberdeen City Council. (2002). *Top City Attractions Pull in the Crowds.* www.aberdeen.net.uk/acc_data/news. Accessed February 2, 2003.

Cumbrian Tourist Board. (2003). *Facts About Cumbria*. www.golakes.co.uk. Accessed July 23, 2003.
International Passenger Survey. (2003). *Inbound Tourism Trends 1995–2002*. www.staruk.org.uk. Accesssed January 10, 2003.
Key Facts of Tourism for Cumbria 2000–2002. (2002). www.staruk.org.uk. Accessed July 20, 2003.
Know Your Market. (1999). *Tourism Attitudes Survey 1999*. www.scotexchange.net. Accessed January 12, 2003.
Media Office. (2003). *Scotland's Most Important Industry*. www.scotexchange.net. Accessed January 14, 2003.

Other Publications

Highlands of Scotland Tourist Board. (1997). *Annual Report 1996/97*, Strathpeffer, Scotland.
House of Commons. (2003). *Foot and Mouth Disease 2001: Lessons to Be Learned*. Inquiry Report HC888 (Anderson). The Stationery Office, London.
Visit Scotland and British Tourist Authority. (2002). *Scotland: Where to Go and What to See 2001*, Edinburgh.
Visit Scotland. (2002). *Moffat Center for Travel and Tourism, 2002 Scottish Visitor Attraction Barometer*, October 2002 Report, Visit Scotland.

11

Safety and Security Issues Affecting Inbound Tourism in the People's Republic of China

Zélia Breda and Carlos Costa

Learning Objectives

- To acknowledge that destinations are particularly vulnerable to political, economic, and social stability.
- To identify the main risk factors to tourism.
- To understand how these issues can damage destination image and impact tourism demand.
- To become familiar with tourism development in the People's Republic of China.
- To learn about the safety and security situation in China and examine which factors deter or might undermine tourism growth.

Introduction

Over the past few years, the tourism industry has been seriously undermined by the growing lack of safety and security. This factor has been identified as one of the five forces causing changes in the tourism sector in the new millennium. Crime, terrorism, food safety, health issues, and natural disasters are the main areas of concern. However, these issues only started to gain more visibility after the September 11 events. Terrorist attacks have also been experienced in other parts of the world and they are pushing the travel industry to deal with a major travel paradigm shift, which is based on the fact that tourism security is now a key concern for travelers. It is now widely accepted by the international community that the success of the tourist industry in a particular country or region is directly linked to its ability to offer tourists a safe and pleasant visit. Governments, travel agents, and news media periodically issue warnings about the risks associated with international tourism. Tourists are urged to buy guidebooks and obtain vaccinations as precautions against such risks.

For quite a long time the influence of safety and security for tourism had been ignored in the literature, particularly the issue of safety in the destination country as a determinant of tourism demand. Tourism literature is now turning its attention to matters of safety and security, which were classified among the ten most important world tourism issues for 2004. Previous research has pointed out four major risk factors: crime (de Albuquerque and McElroy, 1999; Alleyne and Boxill, 2003; Barker and Page, 2002; Barker, Page, and Meyer, 2002; Barker, Page, and Meyer, 2003; Brunt, Mawby, and Hambly, 2000; Dimanche and Lepetic, 1999; George, 2003; Lepp and Gibson, 2003; Levantis and Gani, 2000; Lindqvist and Björk, 2000; Mawby, 2000; Roehl and Fesenmaker, 1992); health-related risks (Cartwright, 2000; MacLaurin, 2001; MacLaurin, MacLaurin, and Loi, 2000); terrorism (Coshall, 2003; Kuto and Groves, 2004; Leslie, 1999; Pizam and Fleischer, 2002; Pizam and Smith, 2000; Sönmez, 1998; Sönmez, Apostolopoulos, and Tarlow, 1999; Sönmez and Graefe, 1998; Tarlow, 2003); and war and political instability (Ioannides and Apostolopoulos, 1999; Neumayer, 2004; Richter, 1999; Weaver, 2000). Concern for crime and safety, whether real or perceived, has been clearly identified as adversely affecting tourism behavior, influencing destination choice and experience satisfaction. Political instability and war can increase the perception of risk at a destination. Similarly, terrorism can cause a profound impact on destination image. Health hazards are also regarded as potential issues that can undermine tourism development. All disasters can divert tourism flows away from affected destinations, but war, terrorism, or political instability have a much greater negative psychological effect on potential tourists when planning their vacations (Cavlek, 2002). This applies not only to the time of crisis, but also to the period following it.

Although there is no evidence of a threat from global terrorism in the People's Republic of China (PRC) and the country's image is seen as generally safe, there are some issues related to political instability, health, safety and security concerns that have caused disruptions in growth rates. Since safety and security directly influence decisions in international travel, this chapter researches the consequences of relevant events that have caused major disturbances in inbound tourism in the PRC, namely the Tiananmen Square incident and the outbreak of Severe Acute Respiratory Syndrome (SARS). In addition to these two main issues, which until now were the only ones capable of deterring China's booming domestic and international tourism industry, other important issues will be discussed. Other epidemic diseases (such as HIV/AIDS, avian influenza, and mosquito-borne diseases); crime directed at foreigners in major cities and tourist areas; road and air safety; natural disasters (earthquakes, flooding, and typhoons); cross-strait relations; and restrictions on public demonstrations, political, and religious activities constitute important considerations that will be also addressed throughout the chapter.

Tourism Safety and Security in China

Safety and Security as Seen by Guidebooks

A useful way to study the evolution of the traveler safety situation in China is through guidebooks. Although these do not constitute scholarly works, they convey the impressions of professional travel observers and are widely disseminated among prospective travelers, playing a large part in the creation of a destination image. Although nowadays there is no lack of travel guides, the present work has focused on *Lonely Planet* because it is one of the best-selling English-language guidebooks, and it was the first to be published on China (the next guide to be published on China—*The Rough Guide*—wasn't published until 13 years later). *Lonely Planet's* first edition was issued in October 1984, six years after the country's opening up to international tourism, targeting primarily budget and independent English-speaking travelers, particularly young people. The comparison between *Lonely Planet's* first edition (hereafter referred to interchangeably as the first edition or the 1984 edition) and the latest edition (the 8th, published in August 2002) is very useful to identify the consumer image of China's safety and security evolution over the past 18 years.

The 1984 edition's section on health sounded a bit alarming, starting with the notification that cholera and yellow fever vaccinations were required for travelers going to certain areas. Malaria and hepatitis were identified as serious infectious diseases in China. Tetanus, diarrhea, and drinking water problems also received special attention. The 8th edition expanded on the health risks section, but was more reassuring. It noted that although China had particular health hazards and that some problems can be encountered in isolated areas, it is a healthier place to travel to compared to other parts of the world. Sexually transmitted diseases, with special attention given to HIV/AIDS, were pointed out as something that foreigners should be cautious about, due to the fact that they are becoming more widespread in China.

Regarding physical safety, the first edition presented China as not exactly a crime-free country, but not especially dangerous. However, the authors devoted several paragraphs to the unsettling nature of the Chinese justice. The edition of 2002 identified economic crimes as the most common offenses committed against international travelers. Foreigners were pointed out as natural targets for pickpockets and thieves, with certain cities, like Guangzhou, Guiyang, and Xi'an, as the most notorious examples of this type of crime. High-risk places were mainly train and bus stations. Nevertheless, some more violent crimes, with foreigners being attacked or even killed for their valuables, were reported in more rural locations, thereby stressing that individual traveling to those areas should be regarded as at high risk. Terrorism activities were also reported, although it was highlighted that foreign travelers were not specific targets.

Racism in China is not a real problem. Its existence is not recognized by the Chinese people; however, racial (ethnic) purity is still the desired norm. The isolationist position imposed by the Communist leaders over more than three decades, coupled with a millenary self-centered vision of the world, did really have a lasting effect on Chinese people (Huyton and Ingold, 1997). Although it is unusual to encounter direct racism in the form of insults or to be refused services in China, especially directed at white people coming from prosperous nations, Africans or people of African ancestry and travelers from other Asian nations can face discrimination. The old dual-pricing system for foreigners was identified in the 2002 edition as fundamentally racist. This discriminatory pricing was exemplified in the 1984 edition, referring several times to the higher costs charged to foreigners.

> The cost of hotel rooms depends on what you are. If you have a white face and a big nose then you pay the most. The Chinese also attempt to plug you into the most expensive of the tourist hotels, and to give you the most expensive rooms. They do this for two reasons; they want the money, but also they think you're spectacularly wealthy, and that you'll want to do things in spectacular style . . . they're not trying to rip you off, they're just trying to please you. (Samalgaski and Buckley, 1984, p. 186)

Prices and services showed racial disparities, regardless of the person's willingness to pay. Overseas Chinese (holders of a Chinese passport who reside outside China in countries or regions other than Taiwan, Macao, and Hong Kong) or compatriots (visitors from Taiwan, Macao, and Hong Kong) were frequently refused service, or given poor service (anyway the quality was generally low, as employees had very little knowledge of international standards), just because they paid less than foreign visitors. Foreign visitors, on the other hand, often felt embarrassed and annoyed by their preferential treatment (Zhang, 1995).

This special treatment took place not long after the end of the Cultural Revolution. The hard-line communist leaders' way of thinking that characterized the Cultural Revolution period had fostered anti-foreign sentiments, resulting in foreigners in China being insulted and badly treated. Under the new government's kowtowing policy foreigners received special treatment, while the government relegated its citizens to an inferior condition (Richter, 1983). The campaign against "spiritual pollution" from the West was launched in China in the mid-1980s, but it did not affect tourism, as the attack on spiritual pollution was deliberately kept as a low key internal affair, and most tourists were quite unaware of it (Lynn, 1993). Nonetheless, the ambivalent Chinese attitude toward foreigners has naturally affected how they handle tourists. As a 1930s writer once said, "throughout the ages, Chinese have had only two ways of looking at foreigners, up to them as superior beings or down on them as wild animals. They have never been able to treat them as friends, to consider them as people like themselves" (quoted in Richter, 1989, p. 32).

State-of-the-Art of Safety and Security

Travel warnings and advice issued by governments of the main outbound tourist markets to China were analyzed in an attempt to verify major concerns regarding the safety and security of their citizens while traveling in China. With the aim of acquiring a broad picture of the nature of those concerns, research was conducted in order to identify what type of information had been released to travelers to China in each world region. Government organizations, as well as some international agencies, were the main source of information regarding potential disruptions to tourism in China. Although nine countries were firstly considered as significant to analyze (Japan, USA, UK, Canada, Malaysia, Korea, Germany, Russia, and Australia), it was soon realized that only English-speaking countries, with the exception of Japan, had such information available to their citizens on the

Internet, which is a useful and rapid way to disseminate information. As tourists tend to be better informed about destinations prior to their trip, travel advisories issued by competent entities are of crucial significance. The facts presented below are thus based on the information collected from travel advisories issued by governmental agencies (the content of those warnings was found to be very similar), thus permitting the construction of an image of China's safety and security situation, and how it is regarded by its main tourist-generating markets. (See Table 1.)

Terrorism

Because of the economic damage that can be inflicted on a country's tourism industry, its visibility, and the leverage it may have on governments, some terrorist and organized crime groups have targeted tourism directly. The fact that in the September 11 terrorist attacks, passenger airplanes, which are a key part of the tourism system, were used as weapons, has had a damaging psychological effect. These attacks strikingly impacted the tourism sector worldwide, being more dramatic than any other crisis in recent years (WTO, 2001). In Asia, the situation has deteriorated as a result of regional terrorism, especially the October 2002 Bali bombings, which exacerbated people's reactions regarding Asia as a tourist destination.

There is no evidence of global terrorism in China, although a small number of bomb-related actions and incidents of unrest do occur. Over the past 10 years there has been an increase in bombing events throughout the country. However, this does not constitute a serious threat to tourists, since foreigners are not specific targets. Nonetheless, there is always the risk of indiscriminate attacks against civilian targets in public places, including tourist sites. These bombings are often the result of commercial disputes among Chinese. It is true that terrorist attacks are also common, many of which have been linked to the Eastern Turkistan Islamic Movement (ETIM), also known as the Xinjiang-Uyghur separatist movement. ETIM was designated a terrorist organization by the United Nations in 2002 and is currently active in the Xinjiang Autonomous Region. Xinjiang is largely constituted by Muslim Turkic-speaking minorities (Uygurs, Kazakhas, Kirghizs, and Uzbeks) and there has been ethnic tension between these four groups and the Han people for quite a long time. These conflicts have promoted the upsurge of a movement calling for the Turkic-speaking people to unite and form an East Turkistan state

Table 1 Major Inbound Markets to China That Have Issued Travel Information about Safety and Security

Position in the Rank of China's Inbound Markets	*Country*	*Issued by*
1	Japan	Ministry of Foreign Affairs http://anzen.mofa.go.jp
4	USA	Bureau of Consular Affairs http://travel.state.gov
9	UK	Foreign & Commonwealth Office http://www.fco.goc.uk
10	Australia	Department of Foreign Affairs and Trade http://www.smartraveler.go.au
11	Canada	Consular Affairs Bureau http://www.voyage.gc.ca

under Islam. Since the 1990s, various factions of the ETIM have engaged in a series of violent incidents (supported and funded by Al Qaeda), both inside and outside China, which were responsible for a total of 166 deaths and more than 440 injuries, and for a serious negative impact on social stability in China and in neighboring countries (Wang, 2003).

Crime

Crimes against tourists result in bad publicity for destinations and create a negative image in the minds of prospective visitors. Tour operators tend to avoid destinations that have the reputation for crimes against tourists (Goeldner and Ritchie, 2002). They also play a very important role in creating the image of a destination and can significantly influence international tourism flow toward a country hit by safety and security risks (Cavlek, 2002).

Overall, China is a safe country, with a low but increasing crime rate. Serious crimes against foreigners are rare. Nevertheless, crime does occur both in Chinese cities and in the countryside. Crime directed at foreigners is becoming more frequent in major cities and at tourist sites, which attract thieves and pickpockets. Robberies and attacks on foreigners in popular expatriate bar and nightclub areas in Beijing and Shanghai, and in the shopping district of Shenzhen, are common. Minor thefts and sexual harassment on overnight trains and buses tend also to occur. The most remote areas of China are poorly policed and there is the risk of attack from armed bandits. In Yunnan, drug smuggling and related crimes are increasing. Money exchange on the black market at better rates is frequent in China. Foreigners tempted to exchange money this way, besides breaking the law and possibly having to incur charges, face the risk of shortchanging, rip-offs, and receiving counterfeit currency, which is a problem in China.

Since China started its economic reforms, social institutions (which molded thought and behavior, rewarded compliance, and punished deviance) have been seriously weakened. The loosening of formal and informal controls as a result of the changes in social structure that have accompanied economic reform, alongside the unequal distribution of wealth, has led to a significant increase in crime (Deng and Cordilia, 1999; Xiang, 1999). One of the most notable trends is the dramatic rise in serious economic crimes; it seems that getting rich is becoming an obsession. Indeed, since the official slogan proclaimed that "to get rich is glorious," materialism became the dominant ethos of the reform era. Juvenile delinquency has also drastically increased, becoming more serious and violent in nature; the criminal motivation is mainly money (Xiang, 1999).

Some of the responses adopted by the Chinese government aiming to maintain social order and to reduce crime consist of intensification of programs of legal education that teach people about the law and its requirements. The revival of traditional Confucian values to increase people's awareness of the appropriate balance between individualism and collective responsibilities, and the revitalization of informal social controls programs, can also be felt (Deng and Cordilia, 1999; Xiang, 1999). Indeed, social control has always been successfully attained through informal organizations and indigenous institutions, which regulate much of social life. The empowerment of the masses to take control of their community's welfare (mass-policing) is one of the best ways to engage people in fighting crime.

Health Risks

People are normally more susceptible to health hazards while traveling. These can range from minor upsets to infections caused by serious diseases. The World Health Organization (WHO) reported that the following diseases can occur in China: cholera, hepatitis A, malaria, tuberculosis, and typhoid fever. Mosquito-borne diseases, such as dengue fever and encephalitis B (endemic in rural areas of Southern China from June to August) can also be encountered, although they do not pose a serious risk to travelers. Rabies infection is also frequent; China has 1,000 human rabies cases every year. Travelers planning to visit regions where these diseases are common are advised to take medication against them. The use of mosquito repellent is also recommended.

Western-style medical facilities with international staff are available in large cities in China. However, in rural areas, medical personnel are often poorly trained and have little medical equipment or availability of medications. Air pollution is also a problem throughout China; seasonal smog and heavy particulate pollution are an issue for travelers, especially for those with respiratory problems.

Avian Influenza Epidemics of avian influenza (bird flu) were reported in the beginning of 2004 in parts of Asia and 34 human cases were confirmed in Vietnam and Thailand, with a total of 23 deaths. Although an outbreak of bird flu was confirmed in China, no human cases were reported, but even if travelers were unlikely to be affected, they were warned to avoid bird markets, farms, and places where they might come in contact with live poultry. A second wave of avian influenza infection was reported in late June 2004, when new fatal cases among poultry were communicated to the WHO, and subsequently there were more fatal human cases in Vietnam and Thailand. In July 2004, China was affected by this new outbreak, and one month later it was discovered that pigs had been infected with the strain of avian influenza. Although findings on the possible spread of the infection among pigs (and its transmission to people) are still preliminary, human infection with avian influenza viruses still remains a public health hazard. Travel precautions are being issued in order to provide information to travelers, but no recommendation to avoid the affected areas has been made.

Sexually Transmitted Diseases Although two thirds of the world's population infected with HIV is located in sub-Saharan Africa, the preponderance of new infections is likely to shift to Asia in the coming decades (Burgess, Watkins, and Williams, 2001), being already well established in the region. Recent social and economic changes in China greatly increased the potential for a substantial HIV/AIDS epidemic, which is already causing great concern, given the growth rates observed in the past decade. Ignorance about the disease, poor sterilization practices, and unsafe blood transfusions contribute to HIV/AIDS transmission, as well as transmission of hepatitis. China is one of the world's great reservoirs of hepatitis B infection.

In 2003, China ranked thirteenth in the world, with 840,000 people infected with HIV/AIDS; the number of deaths reached 44,000. It presented an adult prevalence rate (estimated number of adults living with HIV/AIDS) of 0.1%. HIV is currently concentrated in the southwestern province of Yunnan, near the Golden Triangle. The increasing use of drugs, the rapid expansion of open commercial sex activity (prostitution has become a massive industry in China over the past decade, most noticeable in Zhuhai, Shenzhen, and Macao), and a more liberal sexual climate (the emergence of the homosexual "underground") may support an expanding epidemic throughout the country. Some experts speculate that China will have 10 million cases of HIV by 2010 (Harper et al., 2002).

In August 2004, China revised its law on the prevention of infectious diseases to include the first reference in the legal code to AIDS, which reflects a shift in the government's AIDS policy. The new law, which contains specific clauses on blood donation, stipulates that governments of various levels should strengthen prevention and control of AIDS and take measures to prevent the spread of the disease. It also specifies punishments for anyone concealing the spread of a disease, clearly showing a reaction to failed attempts to cover up the extent of the SARS problem.

Severe Acute Respiratory Syndrome (SARS) SARS was first recognized as a new disease in Asia in mid-February 2003. However, it had already started to spread to other parts of the country and to the world since the first case was reported in November 2002 in Guangdong Province. According to the WHO, between November 2002 and July 2003, more than 8,000 cases were reported, causing 774 deaths, from 29 countries and regions on the five continents. The most affected country by this new epidemic was China, with more than 75% of the cases.

In face of this unknown disease, and as a measure of precaution, the WHO decided to issue travel advisories to areas that reported the most SARS cases. Travel advisories are intended to limit further international spread of SARS by restricting and reducing travel to high-risk areas. It was the first time in more than a decade that the WHO had advised travelers to avoid a particular area. Figure 1 shows a chronology of travel recommendations to China issued by the WHO.

In April 2004, the Chinese Ministry of Health reported a total of nine new cases of SARS (including one death) in China. These were the first cases of severe illness and secondary spread of SARS after the 2003 outbreak. However, no further cases in China or anywhere else in the world have been reported since April 29, 2004. On May 18, the WHO reported that the outbreak in China appeared to have been contained with relatively limited secondary transmission. Nonetheless,

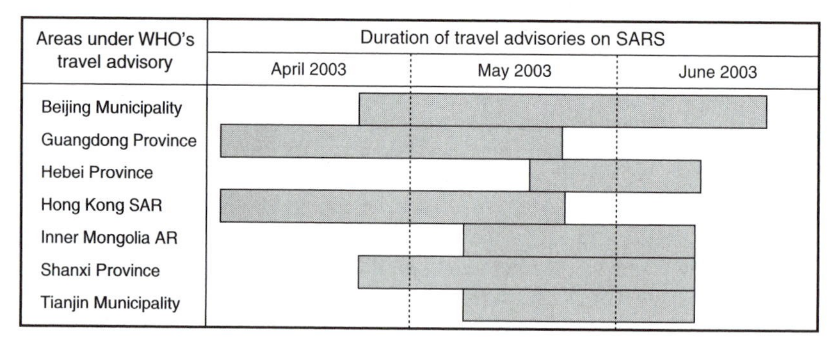

Figure 1 Chronology of travel recommendations to China issued by the WHO.
Source: Breda, 2004.

within a relatively short period of time the SARS epidemic had already caused major damage to China's economy, particularly affecting its tourism industry. The impact of the SARS outbreak on tourism will be expanded upon in the next section.

Local Travel Safety

While in China, foreigners may encounter substantial differences in traveling conditions compared to those in their home countries. In general, many accidents occur, some of them serious, resulting from the poor quality of roads, the often chaotic traffic, and the generally low driving standards (although driving etiquette in China is progressing). Safety standards in public transportation differ from those in the West as well; child safety seats and seat belts are not widely available. Pedestrians and cyclists, if not cautious, are also at risk while near traffic. They are frequently involved in collisions or encounter unexpected road hazards. In fact, it is not unusual to see a pedestrian or a cyclist on a sidewalk being hit by a car or bus driving in the wrong lane.

Air accidents have been reported on internal flights, mainly in routes to the north and east of Beijing. Nonetheless, the US Federal Aviation Administration (FAA) has assessed the Chinese civil aviation authority as Category 1, which means that it complies with international aviation safety standards for overseeing China's air carrier operations. Similarly, there have been several incidents of overcrowded ferries sinking, resulting in the loss of lives. Attacks of piracy in the South China Sea should also be regarded as a threat to yachting safety.

Natural Disasters

China has been greatly affected by natural disasters. The country is located in an active seismic zone and is subject to earthquakes, notably in Inner Mongolia, Yunnan, and Xinjiang. The most recent earthquakes occurred in October 2003 in Gansu Province, measuring 6.1 and 5.8 on the Richter scale. Typhoons can occur along the southern and eastern coasts, affecting Hainan, Guangdong, Fujian, and Zhejiang Provinces during the summer rainy season. Travelers are advised that prior to departing to affected areas they should monitor weather reports. From April to October there are also many severe rainstorms that can cause flooding and landslides.

In 1998, floods along the Yangtze River devastated parts of Central China, killing more than 3,600 people, destroying 5.6 million houses, and swamping 64 million acres of land (Lang, 2002). This situation is not new, and is part of the list of environmental problems that China is facing as a consequence of its rapid economic growth. Loss of forest cover as a result of massive tree clear-cutting over the years (especially during the Great Leap Forward, when huge areas were logged to provide fuel for backyard furnaces in a disastrous campaign to make steel) led to an increasing severity in the flooding. Reforestation and a ban on logging China's natural forests in the upper reaches of the Yellow and Yangtze Rivers, along with the completion of the Three Gorges Dam project, are some of the measures of the Chinese government to help control flooding.

A study using the data envelopment analysis (DEA)-based model for the analysis of vulnerability to natural disasters in China from 1989 to 2000 (Wei, Fan, Lu, and Tsai, 2004) concluded that, in general, the western region was affected more severely. Some provinces in Central China were also badly affected, with Hunan, Guizhou, and Jianxi Provinces being the worst-hit areas.

Political Situation

There are restrictions on undertaking certain religious activities, including preaching and distributing religious materials. Foreigners are also under strictly enforced regulations against any public demonstrations that do not have prior approval from the authorities. Travelers from Australia have been specifically advised to avoid large public gatherings or demonstrations, particularly of a political nature.

The most well-known case of severe measures taken by the Chinese authorities upon religious activities is the ban of the Falungong movement and the imprisonment of some of its followers. Nonetheless, it was not the Falungong movement, but the spread of a Christian-inspired group called the "Shouters" that initiated the "fight for investigation and the banning of heretical teachings campaign, launched by the Chinese leadership" (Kupfer, 2004). The elimination of groups that are perceived as a potential danger to underpinning political unrest, posing an ideological and organizational threat to the Chinese State, is still the guiding principle of the Communist regime.

Tiananmen Square Incident The June 4, 1989, incident in Tiananmen Square showed that the way in which China approaches and solves its domestic economic and political problems will no doubt be reflected in the extent to which foreign tourism is encouraged or constrained. The government's declaration of martial law and the subsequent crackdown on the student democracy movement by the People's Liberation Army (PLA) led to the death of hundreds of protesters. The reaction of the international community to those events and how this new political environment in China has affected tourism will be further developed later in this chapter.

Cross-Strait Relations Since the 1949 military confrontations, political relations between Taiwan and the Mainland did not begin to improve until the 1980s. After almost 40 years of strict restrictions on travel between the two divided states, the ban on travel via a third country was finally lifted in 1987, allowing Taiwan residents to enter mainland China for the purpose of visiting families. Leisure and recreational travel was, however, still prohibited. Notwithstanding the 1987 change in policy, many obstacles still remain in the development of tourism and travel between Taiwan and the Mainland, being highly dependent on the political relations between the two governments.

Nevertheless, as a result of the policy change, the flow of Taiwan visitors to China increased rapidly. Yet, the increased travel activities did suffer a severe setback in 1994, as a result of the global economic recession in the first half of the year. The China Airlines plane crash at Nagoya Airport in central Japan, which resulted in the deaths of 87 Taiwanese passengers; and the Qiandao Lake incident, where 24 Taiwanese visitors were murdered while on a sightseeing tour on a boat on Qiandao Lake in Zhejiang Province, were also instrumental in exacerbating this setback (Huang, Yung, and Huang, 1996). The Chinese authorities' initial dismissal of the Lake tragedy, and its attempt to cover up the case, renewed the political tension between the two governments. It also led a large number of people in Taiwan to reassess their position and shift in favor of Taiwanese independence.

Following the incident, the Taiwanese government temporarily halted group travel to China, as well as other types of cultural exchanges and business activities with the Mainland. As a result, tourist arrivals from Taiwan decreased by 9% in 1994. However, this political tension did not last long, and travel activities were soon reinstated, restoring the normal development of tourist arrivals from Taiwan. In 1995, Taiwan tourists accounted for 3.3% of the country's total arrivals and contributed 19% of China's total tourist receipts. Taiwan tourism has since then become a major component of China's tourism industry. Despite this rapid recovery, the Qiandao Lake tragedy might have caused a long-term impact, similar to the one Tiananmen Square massacre had on the people of Hong Kong, which deeply affected the perception of mainland China, thus strengthening general feelings for independence.

The independence of Taiwan is still in debate. However, it is something that mainland China will never accept, thus posing the question as to whether a clash between the two states will be unavoidable. According to Sheng (2002) a war across the Taiwan Strait is neither inevitable nor imminent, and is less likely in the future since China is rather confident in the face of Taiwan's current political and economic deterioration. This situation gives Beijing the opportunity to exploit, weaken, and paralyze any demand of independence. The PLA has boosted its military pressure over Taiwan by modernizing its warfare capacity. By doing so, China also intends to deter US intervention in the Taiwan Strait, placing emphasis on its strike capability, rather than on its power-projection capability (Sheng, 2002).

However, China's claim of sovereignty over Taiwan and its threat to attack the island if it formally declares independence have led security analysts to see the Taiwan Strait as the most dangerous flashpoint in Asia. Tensions between China and Taiwan have been escalating since the March reelection of President Chen Shuibian, who is a keen independence supporter. The strain was aggravated by the recent announcement of Taiwan's intention to buy weapons in order to help to maintain a balance of power with China, thus permitting it to make a counterstrike to hit Shanghai (China's financial center) if the PLA attacks Taipei.

Observations on Tourism Growth in China

International tourism in China started to develop after 1978 as a result of the "open-door" policy. Since the Chinese government's decision to open the country to the outside world and to promote tourism as a vital economic force to earn foreign exchange earnings to help finance its modernization program, there has been a dramatic increase in tourist arrivals. Tourist arrivals rose from 1.8 million in 1978 up to 91.7 million in 2003, representing a 50-fold increase, with an average annual growth rate of 17%. Although the growth trend of China's international tourism industry has been quite consistent over the last decade (excluding the year 2003), growth rates were not stable during the initial development period, with fluctuations occurring, and even experiencing a major decline in 1989 (the students' demonstration in Tiananmen Square was the cause of a decline of 22.7%). Figure 2 shows the annual percentage variation of international tourist arrivals in China for the period 1978–2003.

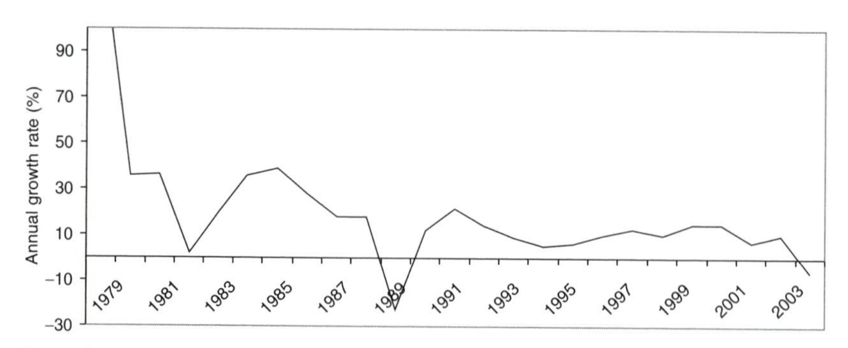

Figure 2 Annual percentage variation of international tourist arrivals in China, 1978–2003. *Source:* China National Tourism Administration.

Over the period as a whole, the average annual growth rate of tourist arrivals showed a downward trend. Problems with accommodation, service, and transport are indicated as possible reasons for this slowdown in the growth of arrivals of overseas visitors. However, Richter (1983) contends that Deng Xiaoping's theories about China's socialist economy might have been considered too controversial, or may have been only temporary policies designed to get Deng Xiaoping into power, and were not necessarily created to be an ongoing program.

The worldwide recession of the early 1980s apparently had an impact on China's tourism industry, as tour cancellations increased from 30% in previous years to 50% in 1982 (Lew, 1987). Overseas visitor arrivals grew only by 2%. To offset the slowdown in tourism, China instituted a number of new policies. Foreign tour operators were allowed to open offices in China, and the "open city" program was introduced, giving more freedom of movement to foreign travelers. This program achieved great popularity, and the number of cities and regions opened to tourists has grown ever since.

The 1989 incident in Tiananmen Square resulted in a severe decline in arrivals from all market segments, with the exception of Taiwanese. However, even before the events of 1989, there was a slowdown of the growth rate in international visitor numbers. Declining interest in China as a destination, as a consequence of overseas perception of poor management, service problems, and congested transport infrastructure, could also have contributed to the low growth rates (Choy, Dong, and Wen, 1986). The figures of total annual visitor arrivals indicate that tourism in China began to pick up shortly after 1989.

In the mid-1990s, China again experienced sluggish growth rates, motivated by a slowdown in the overseas Chinese and the compatriots' segments, partly due to the 1994 incident in Zhejiang Province. Again in the 1990s, China's tourism withstood severe tests—the impact of the Asian financial crisis in 1997 (which was felt mainly in the overseas Chinese market segment) and the devastating floods occurring in the tourist season along the Yangtze River. In May 1999, demonstrations mostly aimed at the United States were held in the major Chinese cities, due to the North Atlantic Treaty Organization's bombing of the Chinese Embassy in Belgrade. As the demonstrations turned violent, the governments of the United States and the United Kingdom issued travel advisories, causing thousands of cancellations from potential visitors (Breda, 2002). But the explosion of xenophobia that hit China during this period had a limited impact on tourism. After a sharp drop in bookings, airlines and hotels reported a return to normality. The lifting of travel advisories from foreign governments also helped the industry bounce back. However, these events did not have a long-lasting negative impact in the Chinese tourism industry; to prove it, in that year, China ranked fifth in the world.

The September 11 terrorist attacks had a severe impact on long-haul tourism, leading to a shift towards intraregional travel, partly compensating the loss of American and European inbound traffic. Intraregional travel is the major kind of travel in the East Asia and the Pacific Region, accounting for nearly 80% of total arrivals, and was a major factor in offsetting the impact felt in the travel and tourism industry. China was the best performing destination within this region (with a more than 6% increase compared to the year 2000), partly because of the close proximity to its main generating markets. For example, the Japanese outbound market, one of the world leaders in the field, replaced destinations in America by China, Thailand, and Australia (WTO, 2001). Despite the fact that China showed a robust increase in tourist arrivals over those in 2000, these events contributed to slow down China's growth rates.

In 2003, China again faced a severe test. Until the SARS epidemic became public, China was one of the few countries that did not experience a decline in tourism, even during the recent war in Iraq. Despite the good results in the beginning of the year, figures had been significantly impacted as a result of the WHO's travel advisory for SARS affected areas in China (Ap, 2003), resulting in a 7.1 decline in total arrivals over 2002. This was the most damaging event for the Chinese tourism industry since the Tiananmen Square incident in 1989.

Implications of the Tiananmen Square Incident

Since the traveler must physically be in the destination country to consume the tourist product that it has to offer, any event that persuades the potential traveler to either stay at home or travel elsewhere directly impacts that destination's exports earnings (Roehl, 1990). There is no doubt the Tiananmen Square conflict severely damaged the international tourism industry of the PRC, at least in the short term, mostly because of the economic sanctions imposed by the world community. "Tour cancellations and a drop in foreign business activity sent hotel occupancy rates in the PRC to the lowest point since the country opened its doors to tourists in 1978" (Gartner and Shen, 1992, p. 47).

Political events of this nature clearly influence tourist demand. The low occupancy rates reflected many hotels' reliance on business travelers. Business visits were affected by both perceptions of stability, which influenced business confidence, and also by the formal and informal sanctions that were imposed on corporations conducting business in China (Hall, 1994). The political unrest of 1989 has led to considerable difficulties for planning and investing within the Chinese tourism industry and posed substantial problems for improving the image of China as a tourist destination. "The conflict in Tiananmen Square was carried out by major news networks throughout the world and, owing to the nature of the conflict, did not portray the PRC in a light favorable to improve its tourist image" (Gartner and Shen, 1992, p. 49). "Many people in western nations demonstrated moral support for the democracy-loving Chinese students by not traveling to China" (Yu, 1992, p. 10).

In a study targeting the mature travel market in the United States, before and after the conflict (Gartner and Shen, 1992), the extent of the damage to China's tourism image was analyzed. It was concluded that its overall image was still favorable and positive. The hospitality component appeared to be directly affected by the conflict, much more than the image of the attractions of its tourist sites. "Safety and security, pleasant attitudes of service personnel, receptiveness of local people to tourists, and cleanliness of environment were all down significantly, indicating that respondents felt the PRC was less likely, after Tiananmen Square, to provide the hospitality needed for an enjoyable visit" (Gartner and Shen, 1992, p. 51).

Not all countries of origin responded to the Tiananmen Square incident in the same way. While almost all tourist-generating markets for China registered recessions in 1989, Taiwan and the Soviet Union became the top generating markets for China's international tourism industry, at the time when tourists from Western democratic countries declined.

> The decline of tourists from the Western democratic countries immediately after the Tiananmen Incident is logical and understandable. The perception of China as an international destination held both by the tourists and the travel industry in the West was dramatically altered by the anti-democratic actions of the Chinese government in 1989. As moral support for the democratic demonstrators in China, tourists cancelled their already scheduled trips or put off their travel plans to a later date. (Yu, 1992, p. 11)

The drastic decline in the number of tourists was evident at Beijing's joint-venture hotels, where all hotels reduced both their Chinese and expatriate staffs, and most remaining employees were working at some 65% of their normal wage package (Breda, 2002). It was also estimated that 620,000 tourism workers underwent compulsory political indoctrination aiming "to cleanse their socialist minds, deepen their love of the Communist Party, and, alarmingly, to cultivate their suspicions of foreigners" (Hall, 1994, p. 123).

The crackdown on the students' demonstration in Beijing definitely created a new environment for Chinese tourism, which affected both the Chinese and global travel industries. There was an immediate drop in the number of incoming visitors, a total decrease of 23%, and a 17% decrease in terms of international tourism receipts. Visitation from Japan and the United States, China's two largest markets and sources of high-expenditure visitors, showed even larger declines.

Roehl (1995) estimated that the impact of the Tiananmen Square incident on arrivals was greater than previously estimated. Overall, his study suggests that the impact of the events led to a decrease of 11 million compatriot arrivals than might otherwise have occurred. Likewise, foreign visitors registered more than 560,000 fewer arrivals. The incident also affected foreign investment in China, particularly in the hotel industry, which had serious consequences for both investors and lending institutions. A breakdown of the quarter-by-quarter tourist arrivals in 1989 is shown in Figure 3.

Although political events greatly affect the tourism industry, it seemed that the situation in China in 1989 only had a short-term impact. The crisis was between the government and its internal critics; there was no violence directed towards international visitors. China's tourism industry responded to this event rather well as it only experienced a 17% decline in receipts in 1989 and was fully recovered 2 years

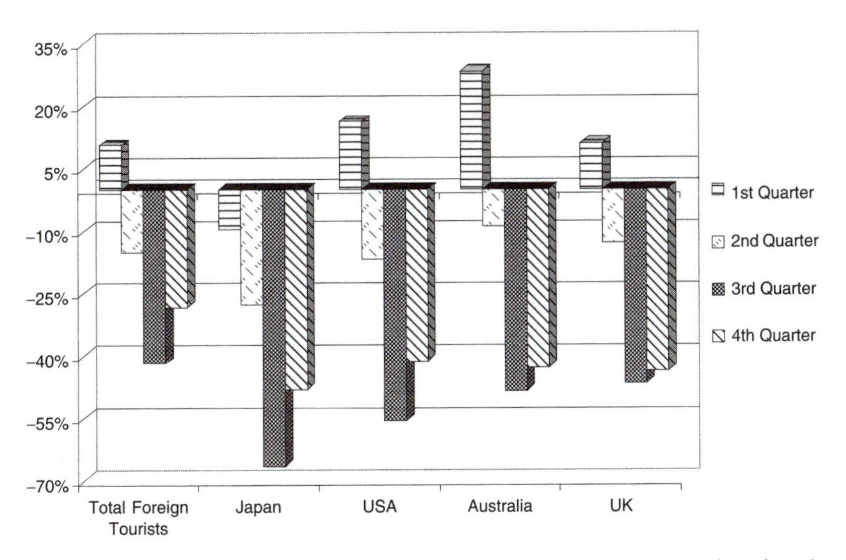

Figure 3 Percentage variation in China's arrivals, by quarter, from its major inbound tourist markets in 1989.
Source: China National Bureau of Statistics.

later. This quick recovery can be attributed especially to the "continued government's commitment to reform and open-door policies; the rapid growth of the Chinese economy; and the industry's successful responding strategies in terms of product development, market positioning, and overseas promotion" (Jenkins and Liu, 1997, p. 105).

After the Tiananmen incident, the government attempted to rebuild its international tourism industry by improving the country's tourist image. A press release from the CNTA in June 1989 stated:

> The CNTA solemnly proclaims that the safety of overseas tourists who come to China has never been affected and can be guaranteed. Tourists may carry on their visits and tours as planned. They are welcome to visit China and do not need to change their scheduled travel plans. (Quoted in Wei, Crompton, and Reid, 1989, p. 322)

Impact of the SARS Outbreak on China's Tourism Industry

China's booming domestic and international tourism industry has recently suffered losses in tourism and related service industries as a result of the SARS epidemic. In light of the events concerning the disease, the WHO advised international travelers to avoid visiting some areas in China that had the most SARS cases. China was the worst affected country.

The outbreak of SARS in China led to a sharp decline in inbound and domestic travel, with social and economic impact, but also had a disproportionately large psychological influence on the public, considering the relatively low morbidity and mortality of the disease. The pronounced psychological impact of SARS can be attributed to a combination of two aspects regarding information about the illness

(Breda, 2004). First, there was a rapid transmission of information about the number of people infected by SARS, as a result of modern media and highly developed networks of communication. Second, there was insufficient medical information on SARS and great uncertainty over the nature of the disease. The lack of accurate, timely, and transparent provision of information on the nature and extent of SARS increased the public's fears, caused second-guessing, and naturally led to an exaggerated perception about the danger of the disease. Concealing health information from tourists, as well as not taking adequate measures to prevent the outbreak of communicable diseases, can be almost as lethal for tourism as the disease itself.

Tourism was thus especially affected by the SARS-induced panic (McKercher and Chon, 2004). Even some destinations that had not recorded any cases of infection suffered almost as much as the areas actually affected. The rapid and wide geographical spread of the disease by travelers, cases of transmission during hotel stays, in restaurants, places of entertainment, or even during airplane trips, made SARS a phenomenon that was perceived to be linked with tourism itself.

The intraregional tourism market is an important source of visitors to China. As the SARS epidemic started to spread into other countries within the Asia-Pacific region, China suffered a major decline in tourist arrivals. Some airlines that offer service to China cancelled regularly scheduled flights due to insufficient bookings. Governments from some foreign countries advised their citizens not to visit China, thus causing the cancellation of a significant number of package tours (Chien and Law, 2003; Overby, Rayburn, Hammond, and Wyld, 2004).

China's inbound travel suffered seriously, but had a somewhat lower accumulated loss of 11.7% in the first two quarters of the year, due to the positive results in the first months of 2003. The worst period recorded was during the months of April and May, both for foreign and compatriot arrivals, registering decreases of 61.8 and 25.1, respectively. With the SARS outbreak over in June, decrease rates started to become less and less accentuated, showing that recovery was under way. Foreign arrivals suffered greater losses and took a little longer to recover. At the beginning of 2004, arrivals from both markets already showed positive growth rates. Figure 4 shows the monthly variation of visitor arrivals in China for 2003 and the first 5 months of 2004.

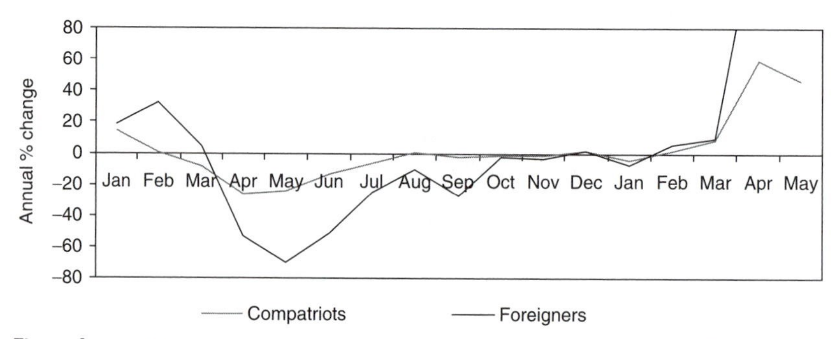

Figure 4 Monthly variation of visitor arrivals in China for 2003 and the first five months of 2004.
Source: China National Bureau of Statistics

Data for the period 1998–2002 regarding compatriot and foreign markets sending the most visitors to China were used to estimate what might have occurred with arrivals in China in 2003 if visitor growth trends had remained unchanged. Particularly, regression analysis was used to develop equations to estimate the number of arrivals in 2003. The dependent variable, the yearly total number of visitors from a particular source, was modeled as a function of time. In all cases the linear trend was statistically significant at the 0.95 confidence level; thus, the derived equation was used to estimate arrivals in 2003. Table 2 shows estimates of the impact of SARS on tourist arrivals in China from selected countries in 2003.

If the growth trend in foreign arrivals had continued through 2003, the total number of visitors in China would have been nearly 15 million (as estimated by the regression of total foreign visitors against time, $R^2 = 0.99$), which means that foreign visitors registered 3 million fewer arrivals. Overall, in 11 of the 12 selected countries, when estimated arrivals for 2003 are compared to actual arrivals, the decline is greater than when arrivals in 2002 are used to estimate what should have happened in 2003. Similarly, regression results suggest that in 2003 China received 9 million fewer compatriot visitors than would have otherwise been expected. When applying the same model to compare estimated international tourism receipts (21,998.1 million US$) with actual receipts (17,396.0 million US$) in 2003, it reveals that the impact of SARS was greater than just comparing it to 2002 values (4,602.1 against 1,613.1 million US$).

Table 2 Estimates of the Impact of the SARS Outbreak on Arrivals in China from Selected Countries in 2003 (in Thousands)

Country of Origin	*2003 Arrivals*	*Forecasted 2003 Arrivals*	*Forecast Method*[a]	*Naive Impact Method*[b]	*Forecasted Impact Model*[c]
Japan	2,160.3	3,159.3	R^2 .97	−765.2	−999.0
Korea	1,945.1	2,455.5	R^2 .99	−179.2	−510.4
Philippines	457.5	551.2	R^2 .97	−51.1	−93.7
Singapore	398.3	523.2	R^2 .94	−98.8	−124.9
Malaysia	429.9	639.1	R^2 .95	−162.5	−209.2
USA	821.2	1,204.7	R^2 .96	−300.0	−383.5
Canada	230.0	307.6	R^2 .97	−61.3	−77.6
UK	287.8	359.4	R^2 .96	−55.2	−71.6
Germany	221.8	301.4	R^2 .99	−60.0	−79.6
France	155.9	243.7	R^2 .99	−66.2	−87.8
Russia	1,380.7	1,471.3	R^2 .96	−109.1	−90.6
Australia	245.2	312.5	R^2 .98	−46.1	−67.3
Other foreign	2,669.2	3,180.2	R^2 .97	−300.0	−511.0
Total foreign	11,394.0	14,710.6	R^2 .99	−2,045.5	−3,316.6
Compatriots	77,527.4	86,908.2	R^2 .99	−3,280.8	−9,380.8

[a]A linear trend model with time as the independent variable was used to estimate expected 2003 arrivals. The linear trend was statistically significant at the 0.95 confidence level. The model's adjusted R^2 is presented.
[b]Naive impact model measures the impact of SARS as arrivals in 2003 minus arrivals in 2002.
[c]Forecasted impact model measures the impact of SARS as actual arrivals in 2003 minus forecasted arrivals for 2003.
Source: China National Bureau of Statistics.

According to the International Labour Organization (ILO), countries or areas directly affected by SARS were estimated to lose more than 30% of their travel and tourism employment. The World Travel and Tourism Council (WTTC) estimated that China was expected to lose more than 2.8 million jobs. However, if also taking into consideration the indirect impact of SARS, its real impact would be even greater. China was expected to suffer, directly and indirectly, a SARS related loss of 6.8 million jobs and 20.4 billion US$ worth of GDP (WTTC, 2003).

Although the SARS outbreak significantly slowed the development of Chinese tourism, it also facilitated the reorganization of Chinese tourist agencies. Many hotels, restaurants, and other attractions remained closed while the public continued to avoid such frequented locations; however they seized the opportunity to undertake renovation projects and to introduce unprecedented hygiene measures, in an attempt to build consumer confidence (Breda, 2004). This slow growth period was thus used to perform renovations and employee training, representing a means of improving China's service industry. During the post-SARS period, the tourist industry of various parts of China, driven by new changes and new market demands, has been absorbing new ideas and approaches to future strategies. These positive developments and improved public health measures have added weight to arguments that SARS, although serious, contributed to the improvement of sanitary conditions and caused only a temporary shock to economic growth.

The industry players designed aggressive revitalization campaigns in an attempt to accelerate recovery; and a series of high-profile special events have been organized, beginning with the ASEAN tourism ministers meeting in August 2003, the WTO General Assembly in October, and the travel fair in Kunming in November, all part of the "seeing is believing" campaign (WTO, 2003). The Women's World Cup in 2007, combined with the Olympics in Beijing in 2008 and the International World's Fair and Exposition to be held in Shanghai in 2010, which is considered the world's third largest event after the Olympics and the World Cup, will contribute to boosting the image of the country.

Conclusion

There was hardly any international tourism in China before the Chinese government's decision to open the country to the outside world, and to promote tourism as a vital economic force to earn foreign exchange to help finance its modernization program. With the introduction of reform and the open-door policy since 1978, China entered the international tourist market. In recent decades, tourism has been a boom industry in China, and has come to play an extremely significant role in economic and regional development, as well as in international relations. It has become one of the most important tourist destinations in the world. In 2002, it ranked fifth in the world, showing a substantial increase over the total international tourist arrivals achieved during the initial stage of tourism development.

The spectacular boom experienced during this period was partly due to the fact that initial policies affecting tourism were directed toward maximizing growth rates of visitor arrivals, especially foreign travelers. However, unlike the domestic and compatriot tourism industry, which is conditioned largely by location and access, the foreign tourism market is competitive and risky; it is highly dependent upon fashion trends and political, economic, and social stability. Experience and studies have shown that the special character of the tourism industry makes it more liable to independent events than other sectors. It is vulnerable to variations in politics and economics, as well as any major change in policy or ideology, and these changes can significantly modify its development process. However, until now there is no evidence that China might want to change its reform and open-door policy; on the contrary, major policies within the tourism sector show a greater openness to the outside world. The increased dependence on tourism as a source of economic growth also shows that it would be extremely difficult for China to halt tourism development. The WTO estimates that China will receive 130 million foreign tourists by 2020, making China the world's number one tourist destination.

Concept Definitions

Terrorism Use of premeditated violence (or threat of violence) by organized groups, against civilians or unarmed military personnel, in order to attain political, religious, or ideological goals.
Crime Any act punishable by law, motivated by economic, political, racial, or religious reasons. It can range from petty offenses to violent crimes.
Health hazard Any source of danger that can be harmful to people's physical condition, ranging from minor upsets to infections caused by serious diseases.
Natural disaster A phenomenon not caused by humans, involving the structure or composition of the earth, ranging from eruptions, avalanches, or earthquakes to landslides, floods, hurricanes, or typhoons.
Political instability Disturbances motivated by political, racial, ethnic, or religious conflicts, which can lead to social disorder and instability.

Review Questions

1. What are the main factors regarding safety and security affecting the tourism industry? What might be the implications of those issues for destinations?
2. How do the main tourist markets see the safety and security situation in China, and how do they perceive it as a tourist destination?
3. The growth trend of China's international tourism industry has not been consistent over the years. What factors have caused disruptions in tourism growth?
4. Explain the major implications of the Tiananmen Square incident to China's tourism industry.
5. Why did the SARS outbreak have such a repercussion on China's tourism sector?

References

Alleyne, D., and Boxill, I. (2003). The impact of crime on tourist arrivals in Jamaica. *International Journal of Tourism Research,* 5, 381–391.

Ap, J. (2003). Encountering SARS: A perspective from an infected area. *e-Review of Tourism Research (eRTR),* 1(1).

Barker, M., and Page, S. J. (2002). Visitor safety in urban tourism environments: The case of Auckland, New Zealand. *Cities,* 19(4), 273–282.

Barker, M., Page, S. J., and Meyer, D. (2002). Modeling tourism crime: The 2000 America's Cup. *Annals of Tourism Research,* 29(3), 762–782.

Barker, M., Page, S. J., and Meyer, D. (2003). Urban visitor perceptions of safety during a special event. *Journal of Travel Research,* 41, 355–361.

Breda, Z. (2002). *Tourism in the People's Republic of China: Policies and Economic Development.* Unpublished MA Thesis, Universidade de Aveiro, Aveiro, Portugal.

Breda, Z. (2004). The impact of severe acute respiratory syndrome (SARS) on China's tourism sector. *Tourism Research Journal,* 1(2), 5–14.

Brunt, P., Mawby, R., and Hambly, Z. (2000). Tourist victimisation and the fear of crime on holiday. *Tourism Management,* 21, 417–424.

Burgess, J., Watkins, C. W., and Williams, A. B. (2001). HIV in China. *Journal of the Association of Nurses in AIDS Care,* 12(5), 39–47.

Cartwright, R. (2000). Reducing the health risks associated with travel. *Tourism Economics,* 6(2), 159–167.

Cavlek, N. (2002). Tour operators and destination safety. *Annals of Tourism Research,* 29(2), 478–496.

Chien, G. C. L., and Law, R. (2003). The impact of the severe acute respiratory syndrome on hotels: A case study of Hong Kong. *International Journal of Hospitality Management,* 22(3), 327–332.

Choy, D. J. L., Dong, G. L., and Wen, Z. (1986). Tourism in PR China: Market trends and changing policies. *Tourism Management,* 7(3), 197–201.

Coshall, J. T. (2003). The threat of terrorism as an intervention on international travel flows. *Journal of Travel Research,* 42, 4–12.

de Albuquerque, K., and McElroy, J. (1999). Tourism and crime in the Caribbean. *Annals of Tourism Research,* 26(4), 968–984.

Deng, X., and Cordilia, A. (1999). To get rich is glorious: Rising expectations, declining control, and escalating crime in contemporary China. *International Journal of Offender Therapy and Comparative Criminology,* 43(2), 211–229.

Dimanche, F., and Lepetic, A. (1999). New Orleans tourism and crime: A case study. *Journal of Travel Research,* 38(1), 19–23.

Gartner, W. C., and Shen, J. Q. (1992). The impact of Tiananmen Square on China's tourism image. *Journal of Travel Research,* 30(4), 47–52.

George, R. (2003). Tourist's perceptions of safety and security while visiting Cape Town. *Tourism Management,* 24, 575–585.

Goeldner, C. R., and Ritchie, J. R. B. (2002). *Tourism: Principles, Practices, Philosophies,* 9th ed. New York: John Wiley & Sons.

Hall, C. M. (1994). *Tourism in the Pacific Rim: Development, Impacts and Markets.* Melbourne, Australia: Longman Cheshire.

Harper, D., Cambon, M., Mayhew, B., Gaskell, K., Miller, K., Huhti, T., et al. (2002). *Lonely Planet: China,* 8th ed. Victoria, BC Canada: Lonely Planet Publications.

Huang, C.-T., Yung, C.-Y., and Huang, J.-H. (1996). Trends in outbound tourism from Taiwan. *Tourism Management,* 17(3), 223–228.

Huyton, J. R., and Ingold, A. (1997). Some considerations of impacts of attitude to foreigners by hotel workers in the People's Republic of China on hospitality service. *Progress in Tourism and Hospitality Research,* 3, 107–117.

Ioannides, D., and Apostolopoulos, Y. (1999). Political instability, war, and tourism in Cyprus: Effects, management, and prospects for recovery. *Journal of Travel Research,* 38(1), 51–56.

Jenkins, C. L., and Liu, Z. H. (1997). Economic liberalization and tourism development: The case of the People's Republic of China, in F. M. Go and C. L. Jenkins (eds.), *Tourism and Economic Development in Asia and Autralasia*. London: Pinter, pp. 103–121.

Kupfer, K. (2004). Christian-inspired groups in the People's Republic of China after 1978: Reaction of State and Party authorities. *Social Compass,* 51(2), 273–286.

Kuto, B. K., and Groves, J. L. (2004). The effect of terrorism: Evaluating Kenya's tourism crisis. *e-Review of Tourism Research (eRTR),* 2(4).

Lang, G. (2002). Forests, floods and the environmental state in China. *Organization and Environment,* 15(2), 109–130.

Lepp, A., and Gibson, H. (2003). Tourist roles, perceived risk and international tourism. *Annals of Tourism Research,* 30(3), 606–624.

Leslie, D. (1999). Terrorism and tourism: The Northern Ireland situation—A look behind the veil of certainty. *Journal of Travel Research,* 38(1), 37–40.

Levantis, T., and Gani, A. (2000). Tourism demand and the nuisance of crime. *International Journal of Social Economics,* 27(7/8/9/10), 959–967.

Lew, A. (1987). The history, policies and social impact of international tourism in the People's Republic of China. *Asian Profile,* 15(2), 117–128.

Lindqvist, L.-J., and Björk, P. (2000). Perceived safety as an important quality dimension among senior tourists. *Tourism Economics,* 6(2), 151–158.

Lynn, M. F. (1993). *The Development and Impact of Foreign Tourism in China and Thailand*. Unpublished MA Thesis, University of Hong Kong, Hong Kong.

MacLaurin, T. L. (2001). Food safety in travel and tourism. *Journal of Travel Research,* 39, 332–333.

MacLaurin, T. L., MacLaurin, D. J., and Loi, L. S. (2000). Impact of food-borne illness on food safety concerns of international air travelers. *Tourism Economics,* 6(2), 169–185.

Mawby, R. I. (2000). Tourists' perception of security: The risk-fear paradox. *Tourism Economics,* 6(2), 109–121.

McKercher, B., and Chon, K. (2004). The over-reaction to SARS and the collapse of Asian tourism. *Annals of Tourism Research,* 31(3), 716–719.

Neumayer, E. (2004). The impact of political violence on tourism: Dynamic cross-national estimation. *Journal of Conflict Resolution,* 48(2), 259–281.

Overby, J., Rayburn, M., Hammond, K., and Wyld, D. C. (2004). The China syndrome: The impact of the SARS epidemic in Southeast Asia. *Asia Pacific Journal of Marketing and Logistics,* 16(1), 69–94.

Pizam, A., and Fleischer, A. (2002). Severity versus frequency of acts of terrorism: Which has a larger impact on tourism demand? *Journal of Travel Research,* 40, 337–339.

Pizam, A., and Smith, G. (2000). Tourism and terrorism: A quantitative analysis of major terrorist acts and their impact on tourism destinations. *Tourism Economics,* 6(2), 123–138.

Richter, L. K. (1983). The political implications of Chinese tourism policy. *Annals of Tourism Research,* 10(4), 395–413.

Richter, L. K. (1989). *The Politics of Tourism in Asia*. Honolulu, HI: University of Hawaii Press.

Richter, L. K. (1999). After political turmoil: The lessons of rebuilding tourism in three Asian countries. *Journal of Travel Research,* 38(1), 41–45.

Roehl, W. S. (1990). Travel agent attitudes toward China after Tiananmen Square. *Journal of Travel Research,* 29(2), 16–22.

Roehl, W. S. (1995). The June 4, 1989, Tiananmen Square incident and Chinese tourism, in A. Lew and L. Yu (eds.), *Tourism in China: Geographical, Political and Economic Perspectives*. Boulder, CO: Westview Press.

Roehl, W. S., and Fesenmaker, D. R. (1992). Risk perceptions and pleasure travel: An explanatory analysis. *Journal of Travel Research*, 30(4), 17–26.

Samalgaski, A. S., and Buckley, M. B. (1984). *Lonely Planet: China,* 8th ed. Victoria, BC Canada: Lonely Planet Publications.

Sheng, L. (2002). Peace over the Taiwan Strait? *Security Dialogue,* 33(1), 93–106.

Sönmez, S. F. (1998). Tourism, terrorism and political instability. *Annals of Tourism Research,* 25(2), 416–456.

Sönmez, S. F., Apostolopoulos, Y., and Tarlow, P. (1999). Tourism in crisis: Managing the effects of terrorism. *Journal of Travel Research,* 38(1), 13–18.

Sönmez, S. F., and Graefe, A. R. (1998). Influence of terrorism risk on foreign tourism decisions. *Annals of Tourism Research,* 25(1), 112–144.

Tarlow, P. (2003). Ideas on how tourism can confront the terrorism menace. *e-Review of Tourism Research (eRTR),* 1(1).

Wang, J. Z. (2003). Eastern Turkistan Islamic movement: A case study of a new terrorist organization in China. *International Journal of Offender Therapy and Comparative Criminology,* 47(5), 568–584.

Weaver, D. B. (2000). The exploratory war-distorted destination life cycle. *International Journal of Tourism Research,* 2, 151–161.

Wei, L., Crompton, J. L., and Reid, L. M. (1989). Cultural conflicts: Experiences of US visitors to China. *Tourism Management,* 10(4), 322–332.

Wei, Y., Fan, Y., Lu, C., and Tsai, H.-T. (2004). The assessment of vulnerability to natural disasters in China by using the DEA method. *Environmental Impact Assessment Review,* 24, 427–439.

World Tourism Organization (WTO). (2001). *Special Report Number 18—Tourism After 11 September 2001: Analysis, Remedial Actions and Prospects*. Madrid: WTO.

WTO. (2003). *Special Report Number 25—Fifth Meeting of the Tourism Recovery Committee*. Madrid: WTO.

World Travel and Tourism Council (WTTC). (2003). *China: Special SARS Analysis: Impact on Travel and Tourism*. London: WTTC.

Xiang, G. (1999). Delinquency and its prevention in China. *International Journal of Offender Therapy and Comparative Criminology,* 43(1), 61–70.

Yu, L. (1992). Emerging markets for China's tourism industry. *Journal of Travel Research,* 31(1), 10–13.

Zhang, G. R. (1995). China's tourism development since 1978: Policies, experiences and lessons learned, in A. Lew and L. Yu (eds.), *Tourism in China: Geographical, Political and Economic Perspectives*. Boulder, CO: Westview Press, pp. 3–17.

12

When Wildlife Encounters Go Wrong: Tourist Safety Issues Associated with Threatening Wildlife

Gianna Moscardo, Matthew Taverner, and Barbara Woods

Learning Objectives

- To have an understanding of the different types of risk to tourist safety associated with wildlife encounters.
- To be able to list the core factors associated with wildlife attacks on tourists.
- To understand the similarities with, and differences between, negative wildlife encounters, and other tourist safety issues.
- To be able to describe the variables that influence tourists' risk perceptions of wildlife encounters.
- To be able to use the five main factors that contribute to the risks associated with wildlife encounters to develop management options.

Introduction

While much has been written about the attractiveness of wildlife encounters for tourists to nature-based destinations and the need to manage the negative impacts of the tourists on the wildlife, very little has been published in the academic literature on how to manage the negative impacts of the wildlife on the tourists. This chapter will explore the potential impacts of wildlife on tourist safety, as well as on tourists' perceptions of destination attractiveness and travel choice behaviors.

The chapter will describe key results from two recent studies conducted by the authors. The first was a critical incidents study of worst wildlife tourism experiences that generated a substantial number of incidents in which tourists reported being threatened or physically harmed by wildlife. These incidents included encounters from many different places and the results will be placed in an international context in order to understand the extent and nature of this type of threat to tourist safety. The second study is a survey of visitors to the North Queensland region of Australia that explored tourists' perceptions of the risks associated with different wildlife in the region and the impact these risk perceptions had on their travel behaviors within the region. The chapter will then conclude with implications for the management of destinations with potentially threatening wildlife populations.

What Are the Types of Negative Wildlife Encounters Tourists Can Have?

Negative tourist wildlife encounters can be organized into three main types: transport accidents related to the presence of wildlife on or near transport corridors, the transfer of diseases to tourists from wildlife, and encounters where tourists are attacked or harassed by wildlife. As with many others areas of tourist safety (see Bentley et al., 2001, for a review and discussion of unintentional injuries to tourists, and Schiebler et al., 1996, for a discussion of crime against tourists) very little reliable, systematic, or consistent data is available for analysis. Thus the following sections provide only a broad overview of the extent and nature of the threats to tourist safety associated with negative encounters with wildlife. Further, although there is insufficient data to be able to quantify these threats, the review does provide evidence that these risks exist and suggests some areas for further consideration.

Wildlife and Transport Accidents

Armour and Macdonald (1998) provide a detailed published analysis of animal-caused fatalities in British Columbia, Canada. In this report they describe factors associated with 133 human fatalities attributed to encounters with animals in the state between 1969 and 1997. They found that 38% of these fatalities could be attributed to motor vehicle accidents resulting from collisions with animals or from swerving to avoid animals on roads. They concluded that "while there is a tendency to focus on and sensationalize fatalities caused by wild animal attacks, . . . the number of deaths due to bear attacks was equal to the number caused by motor vehicles hitting moose alone" (1998, p. 26).

According to the Florida Museum of Natural History, on average 130 people die each year in the United States as a result of collisions with deer (University of Florida and Florida Museum of Natural History (UFl/FMNH), 2003). In an analysis of the causes of road traffic accidents among international travelers in Australia the problem of collisions and avoidance of animals on roads is also noted but not quantified (Wilks et al., 1999).

Wildlife and Disease

According to Re and Gluckman (2003) between 15% and 37% of short-term international travelers experience some sort of health problem while on holiday and many of these are related to illnesses from animal bites. In an Australian study nearly 40% of all returning international travelers presenting to one hospital with a fever had a disease resulting from a mosquito bite or transmitted by insect bites from wildlife in the destination (O'Brien et al., 2001). In a similar study in Norway, 4% of a sample of 940 travelers to Africa returned to Norway with African tick bite fever, while in one year in Sweden 107 cases of dengue fever (transmitted by mosquitoes) were diagnosed in returning international travelers (Lindback et al., 2003).

Table 1 lists some of the more common diseases passed from animals to humans either through direct contact or through insect bites. Comprehensive statistics are not easily accessible as there are few standard systems for recording these diseases and many minor instances that require little medical attention may go unreported. Generally, however, there is agreement that growth in tourism to wildlife areas, particularly in developing countries, is associated with an increasing incidence of these zoonotic diseases in returning international travelers (Choi, 2003; Gill et al., 2000; Jensenius et al., 2003; Lindback et al., 2003; Moore et al., 2002; Spira, 2003; Wilde et al., 2003).

Wildlife Attacks

In addition to the diseases that can be passed through animal bites, many bites themselves can result in extensive injury and death. In one discussion of travel related diseases, for example, allergic and painful reactions to insect and spider bites are included as a separate category of travel disease (Spira, 2003), and although Armour and Macdonald (1998) note that there are multiple causes of animal related fatalities in their Canadian study, they still reported 41 fatal wildlife attacks on humans in the 28-year period under study. These were broken down into 19 bear attacks, 5 cougar attacks, 1 snakebite, and 16 insect bites or stings.

Table 1 A Selection of Diseases Resulting from Wildlife Encounters

Transmission	*Disease*
Passed by mosquitoes	Malaria
	Yellow fever
	Japanese B encephalitis
	Dengue fever
Direct from wildlife bites	B virus, especially from monkeys
	Rabies
Tsetse flies	African trypanosomiasis
Ticks	Typhoid
	Rickettsia

Source: Armitage and Wolfe, 2003; Choi, 2003; Huff and Barry, 2003; Jensenius et al., 2003; Moore et al., 2002; Wilde et al., 2003.

Bears are associated with many problem encounters with tourists and recreationists in North America. Armour and Macdonald (1998) report 19 fatal bear attacks in British Columbia between 1969 and 1997, with other statistics showing a further 100 people injured in bear attacks (Anon, 1996). According to Clark, van Manen, and Pelton (2002), between 1990 and 1998, 1,414 negative encounters were reported between bears and visitors in the Great Smoky Mountains National Park, 18 of which were associated with human injury. Gunther (1998) provides similar statistics for Yellowstone National Park, where 23 people were injured by bears during the period from 1980 to 1997. Although this represents a very small number when compared to the 47 million visitors who came to the park during the same time, Gunther notes that an increase in backcountry recreation seems to be associated with an increase in human injuries from bear encounters.

Cougars, also called mountain lions, are another North American animal associated with attacks on humans. Chester (2000) provides a summary of North American statistics showing a similar pattern to that reported for bears with small but increasing numbers of negative encounters. According to Chester (2000) there have been on average 14 attacks each year in the United States since 1970. Sharks are another predator responsible for human injury and deaths. The University of Florida and Florida Museum of Natural History have created an International Shark Attack File (ISAF) accessible on the Internet where it is reported that 86 people were attacked by sharks in 2002, with most occurring in Florida, Australia, and South Africa (UFl/FMNH, 2003). Crocodiles are responsible for much media attention when they attack humans, with one state in Australia, Queensland, reporting 12 attacks, including four fatal attacks from 1985 to 2001 (M. Read, Coordinator, Crocodile Management Unit, Queensland National Parks and Wildlife Service, Personal Communication, November, 2003). Alligators have a more aggressive record with 365 attacks including 13 fatalities reported for six southern US states (UFl/FMNH, 2003).

Other animals that have been associated with human injuries and deaths include dingoes in Australia (Roberts, 2001), snakes (Casper and Hay, 1998), elephants (Williams et al., 2001), marine jellyfish (Gordon, 1997), as well as scorpions, spiders, and insects (Bowman, 2003). In addition, a number of other species have been associated with aggressive, although non-fatal, attacks on humans including birds (Thomas and Jones, 1999), a variety of Australian native wildlife, most notably kangaroos (Skira and Smith, 1991), elk, moose, bison, and wild pigs (Rusch, 1999), lions, sea lions, fish, bats, raccoons, and squirrels (UFl/FMNH, 2003), and biting insects such as midges (Blackwell and Page, 2003).

Four key sets of points need to be made about these negative human–wildlife encounters. Firstly, while the information reported in this section has provided some insight into the phenomenon of wildlife attacks on tourists, it is difficult to determine the extent of this particular threat to tourist safety as no standard system exists for reporting or recording such incidents; and it is likely that many attacks which result in limited or no actual physical harm to the tourist are not reported at all. Secondly, it is important to note that in many of these cases the incidents are related to people traveling for holiday or recreational purposes. Separate statistics are available for animal related injuries and fatalities that occur with domestic animals or as part of work or everyday residential behavior. There have been some attempts to examine the characteristics of victims of wildlife injuries (see Armour and Macdonald, 1998, and UFl/FMNH, 2003, for examples) but no consistent patterns can be identified because very little analysis is available and where evidence has been examined, the results differ greatly according to the nature of the wildlife and the type of incident involved. The third key point is that although actual attacks appear to be rare, generally it is believed that the incidence of negative wildlife–human encounters is increasing. Typically this is attributed to the rise in tourism in general, but more specifically to the growth in adventure tourism, wildlife based tourism, and tourism into more remote areas (Armour and Macdonald, 1998; Chester, 2000; Gordon, 1997; UFI/FMNH, 2003).

Finally, there has been very little research into the factors that contribute to specific negative encounters. In some cases, such as with spiders and marine jellyfish, it appears that mere entry into the appropriate environment at the relevant time of year or day is all that is needed for a negative encounter to become possible. In others, such as with kangaroos, problems with feeding animals and the presence of food may be critical to the development of negative incidents. In other cases, such as with bears, it may be that humans venture too close to the animal, resulting in a defensive attack. On the whole, however, there have been few systematic attempts to examine the behavior of tourists and/or their responses to, and perceptions of, threatening or dangerous wildlife.

Exploring Negative Wildlife Encounters: A Critical Incidents Approach

The following section provides the results of a critical incidents study of tourist perceptions of negative wildlife encounters that suggests that this risk to tourist safety may be more widespread than official statistics indicate, as it is based on reporting from a sample of visitors rather than an analysis of official incident reports. Further, the details provided in these qualitative descriptions provide some insight into the factors that may be associated with negative wildlife encounters.

This study was originally intended to explore the nature of both positive and negative wildlife experiences reported by tourists, with the aim of developing a more detailed understanding of the service and site management dimensions specific to this type of tourism. A critical incidents approach (see Chell, 1998, for a more detailed discussion of the critical incidents technique) was used in order to elicit detailed descriptions of positive and negative wildlife experiences that could be explored for service quality and management themes. The survey was conducted with a total sample of 790 respondents made up of residents and tourists to the North Queensland region of Australia (see Woods, 2000, for a more detailed description of the total study). The experiences reported ranged from safaris in India and South America to families visiting local wildlife parks and sanctuaries. The reported critical incidents occurred in a variety of environments including coral reefs, mountains, rainforests, and various captive settings such as zoos and wildlife parks. Of particular interest to the present discussion were the negative critical incidents reported by 65% of the sample and these were firstly examined for the existence of major themes. It was this preliminary analysis that identified the prevalence of animal bites or attacks. This was the second most common theme to emerge after the category of poor management of captive animals. Just over 38% of all the negative incidents reported included some aspect of a threatening or dangerous encounter with wildlife while on holidays.

These 195 critical incidents related to wildlife attacks on tourists were then analyzed in more detail. The majority of the incidents reported did not result in an injury to the tourist that required medical attention. The most common type of encounter involved the tourists reporting being chased or harassed by an animal (44% of the incidents reported), with a further 36% of the sample reporting minor injuries (not requiring medical attention) from an attack or bite. Examples of the first category include "The kangaroo stole my young daughter's whole bag of popcorn. It scared us" and "The monkeys in Indonesia kept trying to attack us—jumped all around us in groups—looking for food or us." Examples of the second category include "Our small grandson was hit in the face by an angry wallaby—we won't be going back," and "I was cooking a fish on the campfire and an eagle hit me in the head while trying to get the fish. It hurts when they peck you."

In addition, 15% of the incidents involved the tourists reporting feelings of fear associated with finding themselves close to an animal seen as dangerous. One respondent from New Zealand provided an example of this type of encounter when he stated that "A two-meter bronze whaler [shark] came within 3 meters of me while surfing in Queensland with only three of us at dusk and 300 meters from the shore. Scary!" In another example, "I nearly **** myself when I found a snake in my tent." The remaining 5% of the incidents involved an attack resulting in injury requiring medical attention. For example, "I was swimming and saw a brown snake. The snake bit my foot and I was rushed to hospital."

Table 2 provides a summary of some of the other features of these negative incidents. Given that the study was conducted in Australia it is not surprising to find that the majority of the incidents reported were in Australia. The wide range of animals associated with the attacks however, was unexpected. In total 60 different types of animals were reported in these negative wildlife encounters and they ranged from those which would usually be seen as threatening, such as crocodiles, sharks, lions, snakes, and bears, to those not generally associated with attacks on humans, such as horses, koalas, dolphins, and birds.

Table 2 Key Features of Negative Wildlife Incidents Involving Animal Attacks

Location of the Incident	*Type of Animal (10 most Commonly Reported)*	*Tourist Actions Reported*
Australia (70%)	Kangaroos (15%)	No specific actions recorded (62%)
Americas (8%)	Monkeys (9%)	
Asia (4%)	Snakes (9%)	Feeding wildlife (8%)
Africa (1%)	Mosquitoes (6%)	Wildlife attracted to campsite (8%)
Europe (0%)	Elephants (5%)	
Other (17%)	Jellyfish (5%)	Walking on a trail (6%)
	Crocodiles (4%)	
	Komodo dragons (4%)	Wildlife attracted to picnic (6%)
	Magpies (4%)	Surfing (1%)
	Monitor lizards (4%)	Wildlife photography (1%)
		Driving through sanctuary (1%)
		Swimming (1%)
		Other (8%)

The list also included insects and some of the descriptions did include concerns over the *possibility* of contracting diseases from the animals. In one incident the respondent wrote about an incident with monkeys in Indonesia: "I don't like to be harassed and grabbed by animals—these monkeys could have rabies or other diseases." In addition a number of other factors appeared to be related to the negative wildlife encounters including the connection between food and wildlife feeding noted in other literature. For example, one Australian respondent reported that "Wallabies were around the campsite and vehicle scrounging for food and at times I had to push them away and were concerned they would turn aggressive." Another reported problems with goannas, stating, "The goannas were accustomed to being fed. They were not used to people not feeding them, so they became very aggressive."

Another common theme that emerged from these critical incidents was fear and anxiety associated with not knowing how to react to the wildlife encountered. For example, "A large male roo [kangaroo] attacked and seriously injured a male member of our party. The roo was put down but there was a fear/danger of real threat—a strong attacker—a lack of knowledge about the best way to handle the situation." In a similar fashion, many of the incidents were associated with unexpected encounters with wildlife perceived to be threatening or actually acting in an aggressive manner. For example, in one incident involving a snake—"I didn't expect to come across it and got a fright." In another "The emu approached me from behind startling me and stealing my sandwich from my hands and generally making a nuisance of itself. Personal safety was in jeopardy. Fear and humiliation."

In summary, this exploratory critical incidents study confirmed the prevalence of negative wildlife encounters in tourist experiences. In particular, the analyses of the tourists' accounts of these negative wildlife encounters highlighted the issues of fear, a lack of information and preparedness to deal with threatening wildlife, the association between food and negative wildlife encounters, and the negative impacts of biting insects on holiday satisfaction. In short, negative wildlife encounters do appear to pose a number of risks to tourists' safety and have potentially negative impacts on tourist experiences.

Comparisons with Other Tourist Safety Issues

Negative wildlife encounters share a number of similarities with other tourist safety issues such as crime, terrorism, and natural disasters. Firstly, there is the issue of media coverage of negative wildlife encounters. Animal attacks, in particular, are often given substantial and inaccurate media coverage (Gordon, 1997; Rosenberg, 2001). One media monitoring study, which reviewed coverage of Australia's Great Barrier Reef (GBR) in local, regional, and national newspapers over a 4-month period found that 16% of all reef stories were about wildlife attacks on tourists and 24% were about sightings of dangerous animals. In addition, all but one of the front-page headlines about the GBR were about dangerous animal sightings or attacks. Further analysis of these stories indicated that most were generated from a single instance (CRC Reef, 2000).

Secondly, there has often been reluctance on the part of local and regional tourist associations and relevant government authorities to provide tourist education on these safety issues (Peach and Bath, 1996; Rosenberg, 2001). Attempts to develop tourist safety education programs in this area have been hampered both by concerns over the potentially negative effect on tourist numbers by advertising the existence of a tourist safety threat (Roberts, 2001), and by a lack of evidence about the effectiveness of various education options and approaches. These are also problems encountered in other tourist safety issues such as crime (Dimanche and Lepetic, 1999; Schiebler et al., 1996), transport accidents (Prideaux, 2003), political violence (Hall and O'Sullivan, 1996), and natural disasters (Tilson and Stacks, 1997).

While negative wildlife encounters are similar in some ways to other tourist safety issues, this risk to tourist safety also has some more unique features that need to be considered. Firstly, it is certainly the case that dangerous and threatening wildlife can be a major tourist attraction. For example, in the critical incidents study described in a previous section, many of the animals reported in the best wildlife experiences were those also listed in the negative encounter descriptions including kangaroos (10%), birds (14%), and crocodiles (10%). Ryan and Harvey (2000) also report that despite research suggesting that there is some ambivalence in attitudes toward saltwater crocodiles, they are a major tourist attraction in Northern Australia. Other potentially dangerous animals, which are generally seen as attractive to tourists, are bears, large cats such as lions and tigers, and sharks. In a study by Woods (2000) many of the animals listed as favorite or preferred included those generally seen as dangerous, such as tigers (ranked 7th), lions (ranked 12th), sharks (ranked 14th), crocodiles (ranked 15th), snakes (ranked 17th), and bears (ranked 20th). In addition, many of the other animals listed as favorites were also those involved in the negative critical incidents reported in a previous section including kangaroos (ranked 8th), monkeys (ranked 13th), and birds (ranked 5th). This research is part of a tradition of studies exploring preferences for, and attitudes towards, different wildlife and domestic animal species (Adams et al., 1986; Bixler et al., 1994; Gray, 1993; Kellert, 1989). It appears that a number of factors are associated with these attitudes and preferences but generally it seems that humans prefer animals that are larger, seen as attractive, seen as possessing intelligence, and that have positive symbolic attachments such as representing wilderness and courage (Glickman, 1995; Kellert et al., 1996; Reynolds and Braithwaite, 2001; Woods, 2000). On the other hand, spiders, rodents, and most insects are seen as unattractive and are rarely used in tourist attractions (Bixler et al., 1994; Gray, 1993; Kellert, 1989; Woods, 2000).

In short, it appears that wildlife can be classified into three categories. There are those animals that are recognized as potentially dangerous but which are also attractive to some tourists, such as bears, crocodiles, lions, and tigers. There are those that are attractive and not seen as dangerous but which can be involved in negative incidents, such as birds and kangaroos, and there are those that are potentially dangerous and unattractive, such as spiders and insects.

In the first instance there is some commonality with adventure tourism activities in that some tourists are likely to actively pursue risky situations as the risk is part of the attraction. Almagor (1985), for example, talks about the challenges faced by African safari guides in preventing tourists from getting into dangerous situations with wildlife. This exploration of the motivations of the two groups highlighted a difference in their motivations, with guides concerned about visitor safety and the visitors seeking challenging and life-changing experiences through close contact with the animals. Several authors in the area of adventure or high-risk recreation and tourism have noted that for some individuals a core element of the experience is the risk (Grant et al., 1996; Robertson, 1999; Ryan, 2003). While the authors could find no studies of the motivations of visitors seeking close contact with dangerous wildlife species, anecdotal evidence and the existence of certain forms of tourism, such as swim with shark tours, suggest that there exists a group of tourists who deliberately seek high-risk encounters with wildlife such as bears, sharks, and other large predators. One of the few studies of relevance here (Galloway and Lopez, 1999) found that visitors to national parks who scored highly on a sensation seeking scale and sought high-risk adventure activities were also more likely than others to actively seek interactions with wildlife and less likely to state that they would avoid potentially dangerous wildlife.

Ryan (2003) and Page and Meyer (1996) note that tourists are also likely to engage in risky behaviors because they often find themselves in unfamiliar settings with limited knowledge. In this instance the tourist is not seeking a high-risk encounter. A number of negative wildlife encounters can be explained by this phenomenon. The critical incidents study described in a previous section highlighted the unexpected nature of many negative wildlife encounters and the prevalence of food or wildlife feeding as a factor in negative encounters. So a second type of risk with regard to wildlife and tourists is that of tourists engaging in behaviors that they do not perceive as risky but which can have negative outcomes. In some cases the behaviors may be prompted by a desire to get closer to the wildlife and this is likely to occur in situations where the animal is attractive but not perceived as dangerous. One particular management challenge in this category is that of discouraging tourists and tour operators from feeding wildlife. Such feeding is often associated with negative encounters between visitors and birds and mammals such as monkeys and kangaroos. In other cases the behaviors may simply reflect a lack of knowledge about the presence of particular species and/or how to avoid risky situations. This is most likely to occur with animals that are not attractive.

Understanding Risk Perception and Behavior Related to Negative Wildlife Encounters: Some Empirical Evidence

The previous reviews and the exploratory critical incidents study have suggested a number of factors related to negative wildlife tourist encounters that are worthy of research attention. In order to explore some of these issues one of the authors conducted a survey of visitors to the North Queensland region focusing on their perceptions of threatening and attractive wildlife species, the risks they associated with these species, and the impacts of their awareness of these species on their travel behaviors and decisions. The North Queensland region is a major destination for both domestic and international visitors in Australia. In particular it relies heavily on its natural environments, including the World Heritage listed Wet Tropics Rainforest and the Great Barrier Reef, to attract visitors.

Within these environments, the wildlife is a major component of the destination's attractiveness for tourists. Surveys of visitors to the region have found that more than 60% of tourists report that seeing wildlife is one of the factors in their choice of the region as a holiday destination. These tourists also listed coral, fish, turtles, and sharks as the wildlife they would most like to see on the Great Barrier Reef and birds, kangaroos, crocodiles, butterflies, and cassowaries as the animals they would most like to see in the Wet Tropics region (Galletly and Moscardo,

2001; Moscardo, 2001). Some of these attractive species, such as crocodiles and sharks, also pose a risk to tourist safety. In addition to the wildlife that attract visitors to the region there are a number of other species which pose risks to tourists' safety including marine jellyfish, especially the box jellyfish and Irukandji, venomous snakes, and scrub ticks and the potential to contract dengue and Ross River virus from mosquito bites.

A total of 382 usable questionnaires were completed by domestic and international tourists to the region who were surveyed at a major tourist attraction and transport node in the region. The sample was achieved with a response rate of 91%. Table 3 summarizes the key sociodemographic and travel related characteristics of the sample. As can be seen the majority of the tourists surveyed were international visitors, with the majority (45%) in the 21 to 30 year age group. The majority (72%) were on their first visit to the region and most were independent travelers. This profile fits with available profiles of visitors to the region (Tourism Queensland, 2003).

The first questions sought to determine awareness of dangerous or threatening wildlife in the region and the attractiveness of seeing different wildlife species in different settings. The answers to these questions are summarized in Table 4. These responses confirm a number of features already identified. First, some wildlife perceived as dangerous, particularly large predators, are also very attractive to visitors, although this study shows that for some visitors these animals are attractive only within the confines of a controlled tourist setting such as a zoo or aquarium. Secondly, insects and rodents are seen as both dangerous and unattractive, thus confirming previous studies of wildlife preferences. Thirdly, perceptions of which animals are dangerous are not always accurate. The list of wildlife in the region seen as dangerous includes animals such as the blue ringed octopus that are not found in the region, as well as animals which do not generally pose much of a risk, such as spiders and cassowaries. These inaccuracies are notable given that the

Table 3 Sociodemographic and Travel Related Characteristics of the Sample

Variables	*Percent of Sample*	*Variables*	*Percent of Sample*
Age (years)		Usual place of residence	
• <21	11%	• Australia	39%
• 21–30	45%	• UK/Ireland	28%
• 31–40	9%	• USA/Canada	6%
• 41–50	8%	• Europe	23%
• 51–60	14%	• Asia	1%
• 61–70	10%	• Other	4%
• >70	3%		
Travel party		Previous visits to region	
• Alone	14%	• None	72%
• With spouse/partner	49%	• One	9%
• Family	11%	• More than one	19%
• Family and friends	5%		
• Friends	18%		
• Tour	3%		

Table 4 Wildlife Described as Dangerous, Wildlife Tourists Would Like to See, and Wildlife Tourists Would Not Like to See

Wildlife Described as Dangerous	*Wildlife Visitors do not Want to See*	*Wildlife Visitors Want to See in Nature*	*Wildlife Visitors Want to See in a Controlled Setting*
Marine jellyfish	Snakes	Kangaroos	Crocodiles
Snakes	Spiders	Koalas	Sharks
Crocodiles	Marine jellyfish	Crocodiles	Snakes
Spiders	Crocodiles	Tropical fish	Spiders
Sharks	Sharks	Birds	Marine jellyfish
Stonefish	Insects	Other marsupials	Dingoes
Dingoes	Mosquitoes	Cassowaries	Cassowaries
Cassowaries	Dingoes	Whales	Marine life and corals
Insects	Stonefish	Dolphins	Kangaroos
Sea snakes	Reptiles	Sharks	Koalas
Cone shells	Kangaroos	Snakes	Whales
Venomous sea creatures	Sea snakes	Marine life and corals	Birds
Blue ring octopuses	Rodents	Turtles	Tropical fish
Marsupials	Cassowaries	Dingoes	Other marsupials
	Birds	Spiders	

Note: The responses are listed in order of frequency of reporting by the sample.

majority of the sample described themselves as reasonably prepared in terms of the information they had about dangerous wildlife before they arrived in the region, and that 56% reported seeking further information in the region. An analysis of the sources of information on dangerous wildlife accessed prior to their trip provides some insight into these patterns of the findings. The most common sources were television programs (28%), guidebooks (21%), friends (20%), and print media articles (13%). Very few respondents used sources likely to provide accurate information about threatening wildlife in the region.

The second part of the survey questionnaire asked visitors to answer a series of questions about their perceptions of the risks associated with encounters with eight target wildlife species—saltwater crocodiles, freshwater crocodiles, box jellyfish, Irukandji, venomous snakes, poisonous spiders, sharks, and cassowaries. These species were chosen because of operator and management concerns and media coverage, not on some objective measure of the risk they actually pose to tourists. The surveyed tourists were asked to rate both the likelihood of an encounter with these target species and the seriousness of an encounter with the target species and the results of these questions are summarized in the matrix in Figure 1. The most noteworthy feature of this matrix is that the only species seen as likely to be encountered in the region was the poisonous spider. There are very few poisonous spiders in the region and spider bites requiring medical treatment are very rare (Sutherland and Trinca, 1978).

The third section of the questionnaire asked the tourists to rate the risk of a negative encounter in the region with a crocodile (the two types were combined as visitors made little distinction between them), marine stingers (the two types were

Encounter is unlikely and consequences would be not serious	Encounter is unlikely but consequences would be serious
Cassowary	Saltwater crocodile Freshwater crocodile Venomous snake Shark Box jellyfish Irukandji
Encounter is likely and consequences would be not serious	**Encounter is likely and consequences would be serious**
	Poisonous spider

Figure 1 Risk matrix combining perceptions of likelihood and seriousness of wildlife encounters.

combined again because of few differences in visitors responses), venomous snakes, sharks, cassowaries, and poisonous spiders, as compared to a number of other safety issues. These risk ladders were adapted from those used by Riley and Decker (2000) and the responses are summarized into a single risk ladder presented in Figure 2.

The median risk rankings given by the tourist sample were all greater than riding a horse for pleasure and the median risk associated with marine stingers was greater than the risks involved in driving a motor vehicle or tractor. In all cases the perceived risks associated with the wildlife encounters were higher than the actual risks. Only one death has been confirmed from a cassowary attack in the last 80 years (Kuranda Envirocare, 2003), with no deaths and only one attack recorded for sharks in the North Queensland region in the year 2002 (AAP, 2003), and no deaths recorded for poisonous spiders (Sutherland and Trinca, 1978). According to Bush (2003), the death rate from snakebites in Australia as a whole is less than 0.5 deaths per million people each year. There have been 12 documented crocodile attacks in the region since 1985 with four fatalities (M. Read, Coordinator, Crocodile Management Unit, Queensland National Parks and Wildlife Service, Personal Communication, November, 2003), and there were two confirmed fatalities from marine stingers in the region in the year 2002 (Bailey et al., 2003).

The fourth section of the questionnaire asked tourists to list what, if any, tourist activities would they change or not participate in because of threatening wildlife. More than two thirds of the sample (69%) listed some activity that they believed was influenced by their perceptions of risk associated with wildlife in the region. The most commonly mentioned activity was swimming with 42% of the respondents stating that they would not go swimming or would limit their swimming to certain locations and times to avoid negative wildlife encounters. Other affected activities included bush walking, diving and snorkeling, and visiting specific locations associated with threatening wildlife.

Finally, the questionnaire asked the tourists to offer recommendations to regional management organizations for improving the provision of information about wildlife safety issues to visitors. The open-ended answers to this question revealed a number of themes covering both the need for more information, specifics about how to

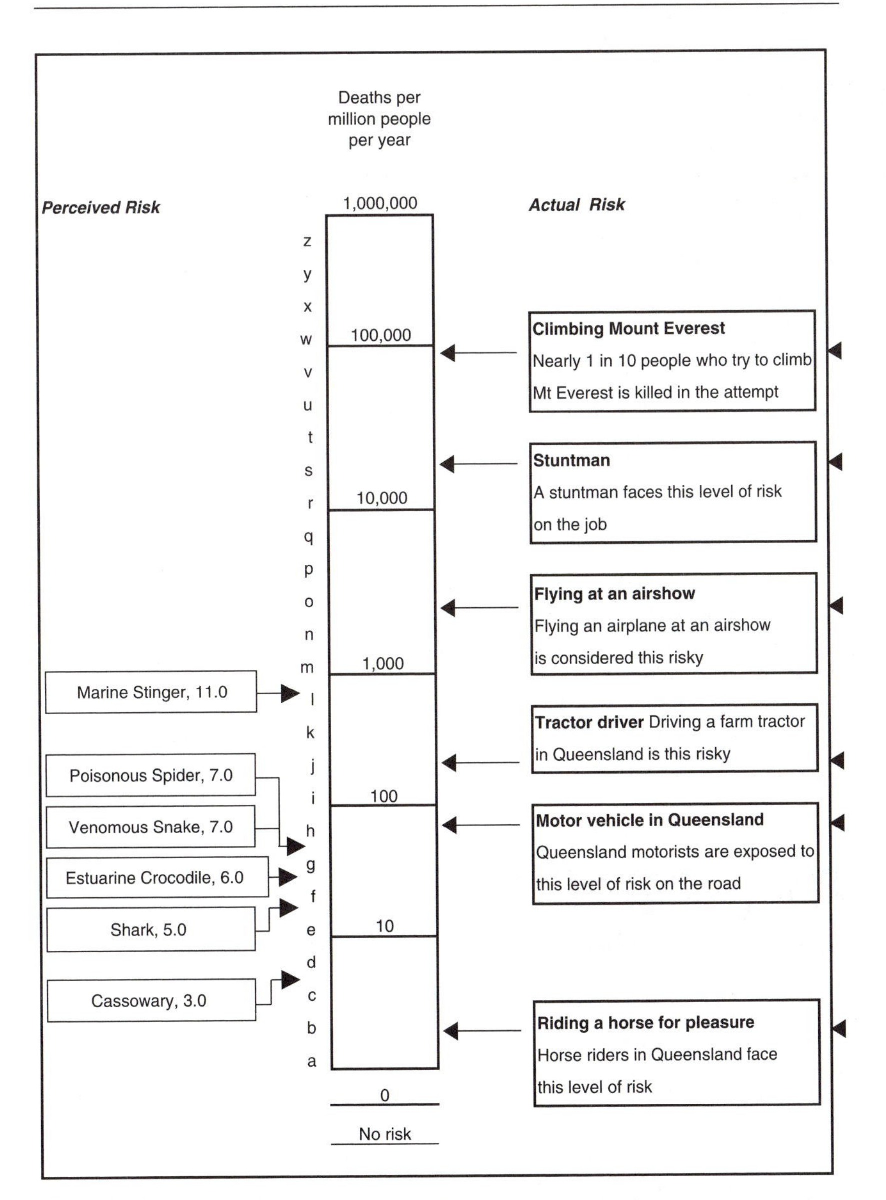

Figure 2 Risk ladder of perceived species risks vs actual activity risks.

deliver the information, concerns over creating unnecessary fear, and discussions of responsibility for managing the risks associated with wildlife encounters. Nearly one tenth of the responses given to this question (9%) suggested that further information was not necessary either because what existed was sufficient ("I don't think there is much to warn people about; I've seen signs everywhere there is a potential risk") or because the core responsibility for managing risk in wildlife encounters lay with the individual visitor ("They do a very good job anyway—the ignorance and carelessness of visitors is to blame for the majority of incidents involving dangerous wildlife"). However, many of the respondents offered various options for better informing visitors, suggesting that they felt that more could be done and that the responsibility for providing information lay with the local authorities. Comments in this vein included "We only found out whilst here about the mosquito—Ross River and dengue—two of our party have fallen ill with mosquito bites. Doctor reckons it is Ross River. We were NOT prepared for this and feel it should be more publicized esp. to tourists." The other comments of particular interest to the present discussion were those that highlighted the need for tourist education campaigns that informed rather than frightened visitors. Some of these respondents seemed to be suggesting that too many warnings could create unnecessary fear, for example, "Show people how to avoid risk rather than tell people how dangerous and make people feel insecure." This was supported by several comments that suggested that some visitors had become frightened by existing information, for example, "I have been put off swimming in the ocean even after stinger season was over because jellyfish freak me out."

In summary, this study highlighted the contrasts in visitor responses to different wildlife species. For some visitors a species is seen as dangerous but still attractive, for others the same species can be both dangerous and unattractive, while for still others it can be attractive but not seen as dangerous. Visitors were also generally not well informed and the majority of the present sample felt that more could be done to inform them about the risks associated with wildlife in the region. Although visitors generally overestimated the level of risk associated with wildlife, nearly one third saw no reason to change their behaviors in response to potential risks from wildlife encounters. Finally, the study also found evidence of differences in visitor perceptions of responsibility for managing the risks associated with wildlife encounters.

Toward a Framework for Understanding and Managing Negative Encounters between Wildlife and Tourists

The literature and research reviewed in the preceding sections suggests that there are three main dimensions or characteristics of wildlife tourist encounters that contribute to actual and perceived risk and perceived responsibility for risk management. These are:

1. The type of negative encounter—traffic accidents, disease and attacks, or harassment.
2. The type of wildlife perceptions held by visitors—dangerous but attractive, attractive but not recognized as potentially dangerous, and unattractive.
3. The type of visitor motivation—encounters where tourists deliberately seek close encounters with dangerous species, situations where tourists seek close contact with wildlife but are unaware of potentially dangerous outcomes, and completely accidental encounters where contact with wildlife was not sought at all.

To these we can also add whether or not the encounter occurs as part of a structured tourist activity, such as on a tour or within a wildlife life park, or as an activity independent of any tour operation. Ryan (2003) notes that visitors may pay less attention to risks when they are part of a tour as they assume the tour operator has taken responsibility for risk management. Finally, we can also add the dimension of differences between particular wildlife species. Animals differ in variables such as their prevalence, behavior patterns, size, and the toxicity of their venom, if they are venomous.

These five variables can combine in different ways to create different risk situations with different implications for management. For example, tourists who deliberately seek close contact with bears, even though they are aware of the risks associated with that contact, create a different management challenge than do visitors who feed kangaroos in a wildlife park and are unaware of the potentially negative consequences. In the latter case the visitor is likely to see the park managers as responsible for any negative outcomes, whereas in the former case the individuals may be prepared to take responsibility for any negative outcomes. In the former case education campaigns and fines may be the only management options available, while in the latter case a wider range of options is available including education campaigns, control of which specific animals are fed, and the provision of alternative activities that allow for close contact.

Both the available evidence and common sense suggest that information and risk management programs are very important in all the risk situations that have been outlined. The challenge is to develop effective programs. This requires first a recognition that these threats to tourist well-being exist and that visitors want and have the right to be informed of these risks. The risk perceptions study reported in this chapter did find some evidence that the presentation of information about potentially threatening wildlife can have negative impacts on tourists' behavior and destination images. The study also found, however, that many of the threatening species were also of interest to many visitors, especially if presented in a controlled setting. For regional planning, then, it is important to understand the range of visitor perceptions of dangerous wildlife. It is possible that the presence of threatening wildlife species may be an attraction in itself. A core research need here is to determine the range of perceptions of wildlife and the size and nature of those markets seeking contact with these threatening species. Preliminary work using the concept of mindfulness/mindlessness from social psychology may be useful in understanding these dynamics of wildlife perceptions (see Moscardo, 2004). In addition, a major factor associated with negative wildlife encounters is close contact, and in many cases visitors actively seek that contact. It may also be that much more can be done on the development of ways to provide visitors with opportunities to have close, but safe, contact with the wildlife (see Matt and Aumiller, 2003, for an example). Finally there is a clear need to have systematic evaluations of successful education campaigns in this area.

In summary more research is needed to determine:

- The frequency of different types of risk situations associated with wildlife;
- The factors that contribute to negative outcomes in these situations;
- Visitor perceptions of the wildlife; and
- Visitor perceptions of the risks associated with wildlife in a region.

Concluding Remarks

The issue of negative encounters between tourists and wildlife also has implications for the ecological sustainability of tourism in a region. A common response to an animal attack on a visitor is the destruction of the animals and in some cases reductions in the population of the species. Arguably this is a rarely recognized negative environmental impact of tourism. As visitors move into more and more remote areas this could become an increasingly important management issue. To be sustainable, managers in the tourism sector in areas with potentially threatening wildlife species need to find ways to manage the encounters between tourists and wildlife for positive outcomes for both sets of participants. In the words of one study participant, "I think the tourists are more of a danger to the wildlife."

Concept Definitions

Wildlife encounter A wildlife encounter refers to any type of contact between humans and other animals (including insects, birds, and marine creatures). These encounters can take place in both natural settings and in places where animals are captive, such as zoos and wildlife parks.

Zoonotic disease Any disease that can be contracted by humans from contact with other animals.

Critical incident A critical incident can be defined as a finite episode of human behavior that is usually easily remembered and described by the participant, and that can be examined to provide an understanding of the causes and outcomes of the incident.

Review Questions

1. Provide an example of each of the main types of risks to tourist safety associated with wildlife encounters.
2. Describe three of the core factors that appear to be associated with wildlife attacks on tourists.
3. What features do negative wildlife encounters share with other tourist safety issues?
4. What are the five main factors that contribute to the risks associated with wildlife encounters?
5. Outline the gaps in our knowledge of negative tourist wildlife encounters and suggest a research program to address these.

References

AAP. (2003). Sharks caught after swimmer killed in attack. *The Sydney Morning Herald*. http://www.smh.com.au/articles/2003/02/09/1044725665481.html. Accessed October 2003.

Adams, C. E., Newgard, L., and Thomas, J. K. (1986). How high school and college students feel about wildlife. *American Biology Teacher,* 48(5), 263–267.

Almagor, V. (1985). A tourist's "vision quest" in an African game reserve. *Annals of Tourism Research,* 12(1), 31–48.

Anonymous. (1996). Bear-human conflicts in British Columbia. www.mala.bc.ca/www/discover/rmot/project.htm. Accessed November 2003.

Armitage, K. B., and Wolfe, M. S. (2003). Update on travel medicine. *Patient Care,* 37(6), 46–54.

Armour, R., and Macdonald, J. (1998). Animal caused fatalities in British Columbia, 1969 to 1997. *Quarterly Digest,* 8(1/2), www.VS.gov.bc.ca/stats/quarter/ql_2_98/. Accessed November 2003.

Bailey, P. M., Little, M., Jelinek, G. A., and Wilce, J. A. (2003). Jellyfish envenoming syndromes: Unknown toxic mechanisms and unproven therapies. *Medical Journal of Australia,* 178(1), 34–37.

Bentley, T., Meyer, D., Page, S., and Chalmers, D. (2001). Recreational tourism injuries among visitors to New Zealand: An exploratory analysis using hospital discharge data. *Tourism Management,* 22, 373–381.

Bixler, R. D., Carlisle, C. L., Hammitt, W. E., and Floyd, M. F. (1994). Observed fears and discomforts among urban students on field trips to wildland areas. *Journal of Environmental Education,* 26(1), 24–33.

Blackwell, A., and Page, S. J. (2003). Biting midges and tourism in Scotland, in J. Wilks & S. J. Page (eds.), *Managing Tourist Health and Safety in the New Millennium.* Amsterdam: Pergamon, pp. 177–196.

Bowman, M. J. (2003). From stingers to fangs: Evaluating and managing bites and envenomations. *Trauma Reports,* 4(3), 1–12.

Bush, B. (2003). Snakes harmful and harmless. http://members.iinet.net.au/~bush/myth.html. Accessed October 2003.

Casper, G., and Hay, R. (1998). Timber rattlesnake bites. www.mpm.edu/collect/vertzo/herp/timber/bites.htm. Accessed November 2003.

Chell, E. (1998). Critical incident technique, in G. Symon and C. Cassell (eds.), *Qualitative Methods and Analysis in Organisational Research.* London: Sage, pp. 51–72.

Chester, T. (2000). Mountain lion attacks on people in the U.S. and Canada. www.topangaonline.com/nature/lionatk.html. Accessed November 2003.

Choi, C. (2003). Travel-related infections. *Topics in Emergency Medicine,* 25(2), 182–194.

Clark, J. E., van Manen, F. T., and Pelton, M. R. (2002). Correlates of success for on-site releases of nuisance black bears in Great Smoky Mountains National Park. *Wildlife Society Bulletin,* 30(1), 104–111.

CRC Reef. (2000). *Media Monitoring Report 1.* Townsville, Australia: CRC Reef Research Centre.

Dimanche, F., and Lepetic, A. (1999). New Orleans tourism and crime. *Journal of Travel Research,* 38(1), 19–23.

Galletly, A., and Moscardo, G. (2001). *Understanding Visitor-Wildlife Interactions.* Townsville, Australia: CRC Reef Research Centre.

Galloway, G., and Lopez, K. (1999). Sensation seeking and attitudes to aspects of national parks. *Tourism Management,* 20, 665–671.

Gill, J., Stark, L. M., and Clark, G. G. (2000). Dengue surveillance in Florida, 1997–98. *Emerging Infectious Diseases,* 6(1), 30–36.

Glickman, S. E. (1995). The spotted hyena from Aristotle to the Lion King: Reputation is everything. *Social Research,* 62(3), 501–538.

Gordon, D. G. (1997). When do animals really attack? *Current Science,* 82(15), 8–9.

Grant, B. C., Thompson, S. M., and Boyes, M. (1996). Risk and responsibility in outdoor recreation. *Journal of Physical Education, Recreation and Dance,* 67(7), 34–36.

Gray, G. G. (1993). *Wildlife and People.* Chicago: University of Illinois Press.

Gunther, K. A. (1998). Yellowstone National Park bear-related injuries and fatalities 1975–1997. *Information Paper No. BMO-1*. www.griztrax,net/Yellowstone/Griz_injury_statistics.html. Accessed November 2003.

Hall, C. M., and O'Sullivan, V. (1996). Tourism, political stability and violence, in A. Pizam and Y. Mansfeld (eds.), *Tourism, Crime and International Security Issues*. Chichester, UK: John Wiley & Sons, pp. 105–121.

Huff, J. L., and Barry, P. A. (2003). B-virus (*Cercopithecine herpesvirus* 1) infection in humans and macaques—Potential for zoonotic diseases. *Emerging Infectious Diseases,* 9(2), 246–251.

Jensenius, M., Fournier, P., Vene, S., Hoel, T., Hasle, G., Henriksen, A. Z., Hellum, K. B., Raoult, D., and Myrvang, B. (2003). African tick bite fever in travelers to rural sub-equatorial Africa. *Clinical Infectious Diseases,* 36(11), 1411–1418.

Kellert, S. R. (1989). Perceptions of animals in America, in R. J. Hoage (ed.), *Perceptions of Animals in American Culture*. Washington, DC: Smithsonian Press, pp. 5–24.

Kellert, S. R., Black, M., Rush, C. R., and Bath, A. J. (1996). Human culture and large carnivore conservation in North America. *Conservation Biology,* 10(4), 977–990.

Kuranda Envirocare (2003). The Cassowary. http://envirocare.org.au/Endangered/Cassowaries/dangers.htm. Accessed October 2003.

Lindback, H., Lindback, J., Tegnell, A., Janzon, R., Vene, S., and Ekdahl, K. (2003). Dengue fever in travelers to the tropics, 1998 and 1999. *Emerging Infectious Diseases,* 9(4), 438–443.

Matt, C., and Aumiller, L. (2002). A win-win situation: Managing to protect brown bears yields high wildlife viewer satisfaction at McNeil River State Game Sanctuary, in M. J. Manfredo (ed.), *Wildlife Viewing*. Corvallis: Oregon State University Press, pp. 351–364.

Moore, D. A. J., Edwards, M., Escombe, R., Agranoff, D., Bailey, J. W., Squire, S. B., and Chiodini, P. L. (2002). African trypanosomiasis in travelers returning to the United Kingdom. *Emerging Infectious Diseases,* 8(1), 74–77.

Moscardo, G. (2001). *Visitor-Rainforest Wildlife Interactions*. Cairns, Australia: Rainforest CRC.

Moscardo, G., Woods, B. and Saltzer, R. (2004). The role of interpretation in wildlife tourism in K. Higgerbottan (ed) , *Wildlife tourism impacts, planning and management*. Altona, Victoria Common Ground Publishing. 231–252

O'Brien, D., Tobin, S., Brown, G. V., and Torresi, J. (2001). Fever in returned travelers: Review of hospital admissions for a 3-year period. *Clinical Infectious Diseases,* 33(5), 603–616.

Page, S. J., and Meyer, D. (1996). Tourist accidents. *Annals of Tourism Research,* 23(3), 666–690.

Peach, H., and Bath, N. (1996). Health and safety information needs of tourists and the tourist industry in northern Queensland. Paper presented at the *Asia Pacific Tourist Association Conference,* Townsville, Australia.

Prideaux, B. (2003). International tourists and transport safety, in J. Wilks and S. J. Page (eds.), *Managing Tourist Health and Safety in the New Millennium*. Amsterdam: Pergamon, pp. 142–154.

Re, V. L., and Gluckman, S. J. (2003). Fever in the returned traveler. *American Family Physician,* 68(7), 1343–1357.

Reynolds, P. C., and Braithwaite, D. (2001). Towards a conceptual framework for wildlife tourism. *Tourism Management,* 22, 31–42.

Riley, S. J., and Decker, D. J. (2000). Risk perception as a factor in wildlife stakeholder acceptance capacity for cougars in Montana. *Human Dimensions of Wildlife,* 5(3), 50–62.

Roberts, M. (2001, May 14). Very dangerous liaisons: Wildlife in Australia. *Newsweek International,* p. 33.

Robertson, D. (1999). Beyond twister: A geography of recreational storm chasing on the southern plane. *Geographical Review,* 89(4), 533–553.

Rosenberg, H. (2001). Reality bites. *The Animals'Agenda,* 21(5), 35.

Rusch, L. (1999). Unexpected threats. *Backpacker,* June, 32–24.

Ryan, C. (2003). Risk acceptance in adventure tourism—Paradox and context, in J. Wilks and S. J. Page (eds.), *Managing Tourist Health and Safety in the New Millennium.* Amsterdam: Pergamon, pp. 55–66.

Ryan, C., and Harvey, K. (2000). Who likes saltwater crocodiles? *Journal of Sustainable Tourism,* 8(5), 426–433.

Schiebler, S. A., Crotts, J. C., and Hollinger, R. C. (1996). Florida tourists' vulnerability to crime, in A. Pizam and Y. Mansfeld (eds.), *Tourism, Crime and International Security Issues.* Chichester, UK: John Wiley & Sons, pp. 37–50.

Skira, I., and Smith, S. (1991). Feeding wildlife in national parks, in *Proceedings of the Fifth Australasian Regional Seminar on National Parks and Wildlife Management,* Tasmania, publisher pp. 182–187.

Spira, A. M. (2003). Assessment of travelers who return home ill. *The Lancet,* 361(9367), 1459–1476.

Sutherland, S. K., and Trinca, J. C. (1978). A survey of 2144 cases of bites by the red-back spider (*Latrodectus mactans hasseltii*) in Australia and New Zealand. *Medical Journal of Australia,* 2, 620–623.

Thomas, L. K., and Jones, D. N. (1999). Management options for a human-wildlife conflict: Australian magpie attacks on humans. *Human Dimensions of Wildlife,* 4(3), 93–95.

Tilson, D. J., and Stacks, D. W. (1997). To know us is to love us: The public relations campaign to sell a "Business-Tourist-Friendly" Miami. *Public Relations Review,* 23(2), 95–115.

Tourism Queensland. (2003). The Townsville region regional update. http://www.qttc.com.au/research/pdf/regional_updates/tsv_03.pdf. Accessed October 2003.

University of Florida and Florida Museum of Natural History (UFL/FMNH). (2003). *International Shark Attack File.* http://www.flmnh.ufl.edu/fish/sharks/ISAF/ISAF.htm. Accessed January 2004.

Wilde, H., Briggs, D. J., Meslin, F., Hemachudha, T., and Sitprija, V. (2003). Rabies update for travel medicine advisors (travel medicine). *Clinical Infectious Diseases,* 37(1), 96–101.

Wilks, J., Watson, B., and Faulks, I. J. (1999). International tourists and road safety in Australia: Developing a national research and management programme. *Tourism Management,* 20, 645–654.

Williams, A. C., Johnsingh, A. J. T., and Krausman, P. R. (2001). Elephant-human conflicts in Rajaju National Park, northwestern India. *Wildlife Society Bulletin,* 29(4), 107–110.

Woods, B. (2000). Beauty and the beast: Preferences for animals in Australia. *Journal of Tourism Studies,* 11(2), 25–35.

Tourism and Crisis-Management Issues

Yoel Mansfeld and Abraham Pizam

One of the bottom-line conclusions in the theoretical chapter that opened this book was that the negative impact of security incidents on tourist destination, tourists, and host communities is almost inevitable. Two basic questions arise in the wake of these unavoidable outcomes. The first is to what extent is it possible to manage a tourism crisis that emerges in the wake of security incidents and to mitigate the damage by proper policies and appropriate steps during the preventive, reactive (while the crisis is occurring), and recovery levels? The second is obviously what strategies and courses of action are more effective than others and, therefore, should be recommended for affected destinations?

This chapter discusses these dilemmas from various perspectives. Thus, in the first chapter of this section Eli Avraham is concerned with an important aspect of crisis management, that is, the role of public relations and advertising strategies in the efforts to lessen the image damage in affected tourism destinations. His main conclusions are that:

- There is a direct relation between the image of the affected tourist destination and the level of demand generated by tourists to visit it.
- A proper strategic planning is needed in order to develop an effective image-management strategy. Considerable attention in planning long-term image-management campaigns should be devoted to the deflecting of potential crises since such an approach has proved to be highly efficient.
- Media campaigns cannot be utilized on a stand-alone basis when dealing with security induced tourism crises. They have to be part of a comprehensive and concerted effort operating in different yet complementary channels.
- Media campaigns must be accompanied by overall improvement of the tourist product, including the level of security provided to safeguard the life of tourists and the quality of their tourist experience.

David Beirman's chapter looks from a marketing perspective at the issue of recovery campaigns. By adopting a comparative approach to his analysis of three Southeast Asian recovery processes, he manages to detect the most crucial steps to be taken in order to ensure successful recovery through the use of marketing. His main conclusions are that:

- The negative image that emerges in the wake of security incidents is a result of two major forces operating simultaneously: the media and the governmental travel advisories.
- When concerted and well-coordinated efforts took place, involving international organizations, as well as regional-national bodies and national tourism organizations in the 2002–2003 tourism crisis, they proved to yield a fast and effective recovery process.
- Contingency marketing plans can shorten the time lag between tourism crisis situations and actual recovery trends.
- Private-public cooperation with strong governmental commitment to tourism is vital for successful and rapid recovery from security oriented tourism crises.

One of the factors that shape tourists' risk perception, and thus their subsequent travel behavior, is the availability of security oriented information and the type of information source. In a chapter on the role of security information in tourism crisis management Yoel Mansfeld analyzes the importance of crisis communications and of generating balanced, comprehensive, and up-to-date security information, to control the security image of potential and actual tourists. His main conclusions are that:

- The proactive provision of security and risk related information by the host destinations is imperative in order to balance the biased information provided by the media and governments in the generating markets.
- The provision of comprehensive and accurate security related information should not target would-be travelers but be directed toward those who have already made a com-

mitment to travel to affected destinations; tourists who are visiting affected destinations; and tourists who have already returned from a visit to affected destinations.

- Information on current security situations in affected destinations should not be based only on facts, but also on their interpretation in order to assist tourists to make a more balanced evaluation of the risk involved.
- The provision of security information by affected destinations should be both proactive and dynamic to increase the confidence of tourists in this type of information source.

On a more narrow perspective of tourism crisis management, Greg Stafford, Larry Yu, and Alex Kobina Armoo look at how Washington, DC, hotels responded to the September 11, 2001, terrorist attack on the Pentagon. Following an in-depth analysis of the human and financial implications of this crisis on the local hospitality sector, their main conclusions are that:

- A well-coordinated effort from the early stages of the tourism crisis, accompanied by proper objectives, helped hospitality businesses in the Washington, DC, area regain their pre-September 11 business levels.
- This success was attributed to the well-coordinated and centralized industry response which provided a single and reliable information source about the industry status.
- The lesson as to how the DC hospitality sector dealt with the crisis is that contingency plans must be built on solid knowledge of crisis-management skills and refined through workplace practices.
- When emergency conditions have subsided, management must make the transition quickly to the process of recovery.
- Familiarity with crisis characteristics will better enable hotel managers to handle future security oriented crises.

On a similar topic but more related to crisis-management strategies among hotel managers, Aviad Israeli and Arie Reichel discuss how the Israeli hospitality sector tried to mitigate the ramifications of the second period of Palestinian uprising (the *Intifada*). Their main conclusions are that:

- When assessing the performance of hotel managers in times of security induced tourism crises, there is a need to examine both their recognition of the importance of measures that assist the organization in times of crisis and also their level of usage.
- Four crisis management practices were identified among Israeli hotel managers: marketing, infrastructure (or hotel) maintenance, human resources, and governmental assistance.
- With regard to hotel managers in the Israeli hospitality sector, there is considerable correspondence between their perceived practices' importance and their usage.

And lastly Nevenka Cavlek looks at the role of tour operators in crisis management. Being responsible for a large proportion of overall global tourist movements, it is their prime interest to ensure risk-free travel. Also, their level of involvement in both domestic and international demand for tourism makes them one of the most vulnerable sectors within the tourism system once a tourism crisis occurs. Leaning mostly on the Croatian case, her main conclusions are that:

- The crisis behavior of tour operators toward the destination primarily depends on the type of crisis, its dimensions, its predicted length, its consequences in the receiving country, the tour operator's own business interests in a country, and decisions by the governments of the generating countries.

- Since tour operators influence the way a particular destination is viewed by tourists, they have to be highly selective when it comes to promoting destinations that represent a differential level of risk.
- The more a tour operator is vertically integrated with companies in the receiving countries, the more attention it pays to the quick recovery from security oriented crises.
- The return of the large tour operators to an affected destination is a strong indication that the situation is safe enough to visit. Thus, tour operators have a major task in restoring confidence among tourists and other stakeholders in the affected destination.

13

Public Relations and Advertising Strategies for Managing Tourist Destination Image Crises

Eli Avraham

Learning Objectives

- To understand the meaning of destination marketing.
- To become familiar with the concept of destination image management.
- To learn the techniques for creating and promoting destination images.
- To understand media strategies for improving destination images.
- To understand the factors that shape destination images.

Background

Image crises occur frequently to tourist destinations all around the world. Crime waves, terrorist activities, interracial conflicts, epidemics, and natural disasters (earthquakes, hurricanes, mud slides, and floods) are widely covered in the national and international media. As we know, the media plays an important role in travel patterns since it has the power to shape the public's perception of a destination, conflict, and issue (Santana, 2003). Overnight, the bad press of crisis events can empty out hotels, cancel flights, and leave tourist attractions desolate. They have an instant and far-reaching negative effect on the tourism industry of the affected cities, regions, and countries. On the other hand, in contrast to the sudden negative image, tourist destinations can also gradually develop a negative image, as the cumulative result of ongoing, long-term problems, such as a decline in the local tourism industry, low-quality services, hotels that are not well maintained, neglected tourist attractions, lack of general upkeep, rising crime rates, unemployment, and a difficult socioeconomic situation. In such cases, just as in the case of a sudden downturn in image resulting from a particular crisis, an unfavorable image is projected in the media and has a corresponding negative effect on the local tourism industry and on potential tourists' buying behavior.

Whatever the cause of the negative image, be it a sudden crisis or a prolonged period of decline, many destinations suffer from being identified as scary, dangerous, boring, or gloomy; places a tourist would have no interest in visiting. Many decision makers in the local tourism industry stand by helplessly, frustrated by their knowledge that, in most cases, the negative image is not based on facts found on the ground. Given that stereotypes are not easily changed or dismissed, the challenge facing these decision makers is great. An analysis of many tourist crises, however, shows that numerous tourism destinations throughout the world have managed to change a negative image into a positive one and bring back the tourists. To achieve this lofty goal, a number of communications strategies are used. This chapter proposes to analyze these strategies and present those most successful for altering an unflattering public image.

Destination Marketing, Images, and Stereotypes

"Destination promotion" has many different definitions in literature; a summary of most of these can be found in Short et al. (2000), who declare that "Destination promotion involves the re-evaluation and re-presentation of destination to create and market a new image for localities to enhance their competitive position in attracting or retaining resources" (p. 318). This definition emphasizes the re-presentation of destination in order to create and market a new image geared to compete over the retaining and attracting of various resources. Nielsen's (2001) definition of destination promotion is similar, but he stresses the difficulty of achieving the task, especially when dealing with an image-related crisis: "promoting a destination in normal circumstances is a difficult task, but promoting a destination that faces tourism challenges—whether from negative press, or from infrastructure damage caused by natural disasters or man-made disasters—is an altogether more arduous task" (pp. 207–208).

Destination marketing attempts to improve the images and public perceptions of destinations. Kotler et al. (1993) define the image of a destination as "the sum of beliefs, ideals, and impressions people have toward a certain destination." They argue that an image represents a simplification of many associations and assorted information related to a place, and is a cognitive product of the attempt to process large amounts of information.

Many different factors influence a destination's image or perception among tourists. Among these are the characteristics of the destination's population, its status or political power, the population size, its crime rate, the numbers and types of tourist attractions, the numbers and character of national institutions located there, its location and historical background, movies and television series that have been filmed on location at the destination, its media coverage, atmosphere, entertainment options, tourist or cultural value, and physical appearance (Avraham, 2002; 2003b; 2003a).

Various methods may be used in order to evaluate a destination's image among specific target populations, but the most popular of these are attitude surveys, questionnaires, focus groups, and in-depth interviews (Fenster et al., 1994).

Kotler et al. (1993) argue that a destination's image can be positive and attractive, negative, weak (as in the case of peripheral locations that are not well known), mixed (when the image includes both positive and negative elements), or contradictory (when the city has a favorable image among a certain population and a negative image among another population). In addition, destination images can be classified as "rich images" or "poor images"; some destinations have a rich image—that is, a great deal is known about them, usually from a variety of sources as well as from personal visits and knowledge; other destinations have a poor image—very little is known about them, and what is known usually comes from a single source of information (Elizur, 1986).

Image Management and Destination Marketing Strategies

In the scope of the present chapter, it is not possible to describe in detail all the stages of decision making and implementation of marketing campaigns (for such elaboration, see Kotler et al., 1993; Short et al., 1993; Morgan and Pritchard, 2001). One should, however, distinguish between attempting to change a destination's image while changing its actual reality (Beriatos and Gospodini, 2004), and attempting to alter the image without making any concrete changes. In the first scenario, the change must be implemented in stages. The initial stage is diagnosis, followed by building a strategic vision, and finally planning the next steps necessary to implement the determined vision. It should be stressed that prior to marketing a destination, local authorities should make sure that it does in fact supply basic services, that regular maintenance of the existing infrastructure takes place, and that this infrastructure fulfills the needs of patrons in hotels, restaurants, transportation, and at cultural events. If no improvement in basic services is evident, it can be very difficult to succeed in a marketing campaign, and, in fact, in most cases the effort will be rendered useless. Local authorities understand today that more money should be invested in developing new attractions, services, and accommodations to improve a destination's attractiveness with regard to tourists. The nature of these attractions and services is dependent on the type of crowd the destination is interested in drawing (Kotler et al., 1993, 1999; Short et al., 1993).

The necessary first step to successfully change a destination's image is a careful examination of the existing image among the target population, since it is this image that provides the base on which the marketing campaign should be built. Destinations with a positive image need campaigns that reinforce this image, whereas destinations with negative images need campaigns that improve and challenge the existing image. Destinations with weak images need campaigns that focus first on raising awareness about the destination, and only then on constructing the desired image.

The process of constructing an image is not easy, and several different studies have suggested a range of steps: the destination's uniqueness needs to be identified, positioned, and marketed in a way that reflects what is unique about it and how it differs from competing destinations. Special attention must be given to ensuring that all campaign messages are geared toward underlining the unique image chosen by all the participating bodies. In addition, the campaign should not try to cover too many target populations; it should have a succinct, consistent message. Every step agreed upon should be followed, and the various assorted components such as advertising, public relations, and promotion should be well coordinated (Kotler et al., 1993, 1999; Short et al., 1993; Young and Lever, 1997; Morgan and Pritchard, 2001; Avraham, 2003a).

Techniques for Delivering Campaign Messages

A number of different methods for improving a destination's image can be employed by campaign managers to deliver their key points. Choosing between these options depends on the campaign's goals and timing, on the economic situation of the destination and that of competing destinations, on the available budgets, and on many other factors. The most common techniques are: (1) Advertising—this is the most popular tool for marketing destinations, and is based on purchasing advertisements in the media and delivering messages to target populations. According to a CNN poll, $538 million was spent on marketing locations in the United States alone during 1999 (Piggott, 2001, as cited in Morgan and Pritchard, 2002). Transferring advertising messages to the target audience can be accomplished by using several techniques. The ads can highlight existing attractions of a destination, and describe the destination from the viewpoint of the targeted population (such as showing a European tourist in Thailand). They can present the local viewpoint, emphasizing how a visit will affect the tourist, or depicting the destination as suiting a certain lifestyle. In a time of crisis, there is a tendency among advertisers to use celebrities to pitch a destination. For example, movie and television actors were used to promote tourism to New York and Washington after the September 11 events. (2) Direct mailing/marketing—this method involves directly addressing target populations through mail, personal meetings, telephone, or electronic mail. Target audiences are sent brochures, maps, photographs, and information pamphlets in the hope that they will become interested in visiting the destination. (3) Sales promotions—this refers to short-term offers for various destination services, such as reduced prices in local hotels. (4) Public relations—this technique attempts to influence the way the destination is represented in the various media. Destination spokespeople or public relations advisors try to create a favorable image through promoting special events and positive stories, and at the same time attempt to prevent the publication of any unfavorable stories, especially during crises.

In order to achieve success, it is imperative that a high level of coordination exists between the various techniques employed by the campaign and that the messages delivered to the public are of similar content. Marketing and public relations campaigns are complementary and must be mutually supportive. For example, while a marketing campaign attempts to attract tourists to a destination using advertisements and brochures, in the public relations campaign spokespersons and advisors should convince news people to publish positive stories on the destination, such as the growing number of tourists rushing to visit (if this is indeed the case), the opening of new tourist attractions, and reports about celebrities who have visited there.

Dealing with Tourism Crises

Much has been written about the use of media in crisis situations in general (Fearn-Banks, 1995). The above-mentioned points are particularly relevant when dealing with a destination image crisis that is the result of ongoing, long-term problems. In the event of a sudden destination image crisis, a new set of rules comes into effect. In such situations there is a significantly expanded demand for information and, as a result, the role of the media becomes much more important. While it is true that by definition, a crisis is an unexpected event, the proper management of a tourist destination includes a certain degree of potential crisis preparation. Such preparation includes preplanned strategies for crisis handling and the distribution of tasks among the relevant professionals, as well as assigning staff to deal specifically with information requests, including interacting with the media. Regarding the media handling, it must be decided in advance who will be responsible for key jobs, such as issuing press releases, speaking to journalists, and establishing an information center, including an updated web site.

Dealing with Negative Images of Tourist Destinations

Daunting though the task of changing a destination's negative image may be, there are a number of worldwide cases that illustrate that it can be done successfully.

The many strategies that can be used by destination decision makers to deal with a negative image can be divided into two main categories: those that focus on the

Table 1 Public Relations and Advertising Strategies and the Tourist Destinations That Employed Them to Manage a Destination Image Crisis

Category	*Public Relations and Advertising Strategy*	*Tourist Destinations That Employed the Strategy*
Changing the objective reality	1. Tackling the issue	Miami, New York, Syracuse, Egypt
	2. The "come see for yourself" strategy	India, Israel, Nepal, London, Miami, Belfast, Jerusalem, Holon,
	3. Hosting special events	Beijing
	4. Fostering residents' local pride	Syracuse, Miami, Egypt
Managing the symbolic reality	1. The "Crisis? What crisis?" approach	Spain, Manchester, Turkey, New Milford
	2. Employing a "counter-messages offensive"	Miami, Tunisia, Nuremberg, Negev Desert
	3. Acknowledging the negative image	Tulsa, London, Nepal, Israel, Philippines
	4. Geographic isolation strategy	Eilat, Jordan
	5. Spinning liabilities into assets	Minnesota, Be'er Sheva, Bradford
	6. Changing a destination's name	Vichy, North Dakota, Eilat
	7. Creation of a new logo or slogan	Syracuse, Poland, Bradford
	8. Changing the campaign's target audience	Israel, Syria

reality underlying the negative image, and those that fight the negative image by means of advertising and public relations. In other words, strategies in the first category deal with the seen/objective reality while those in the second focus on the symbolic reality.

Category One: Changing the Objective Reality

While strategies in both categories make use of the media, it is only those in the first category (i.e., those that deal with the objective reality) that take steps beyond media representation to change the negative image. Following are several first category strategies used by decision makers.

Tackling the Issue

The first strategy is very clear-cut: if there is a specific problem responsible for the sudden crisis or the destination's prolonged negative image, it needs to be solved. In other words, a destination's negative image is not fictitious but does in fact reflect some real-life problem. Therefore, destinations must solve the problems that led to the negative image associated with them. If, for example, a destination is perceived as violent and unsafe, it must combat crime and violence, as was the case of Miami: "At the same time that the tourism officials were working to rebuild tourism numbers, government and law enforcement officials were getting tough on crime . . . as crime continued to decrease, tourism began to flourish" (Tilson and Stacks, 2001, pp. 159, 162).

A similar case of tackling the crime problem in order to improve a destination's image is that of New York City, where Mayor Rudolph Giuliani halved the city's crime rate. Following citywide campaigns against perpetrators of crime and violence, the feeling of safety in the city increased in unprecedented fashion. Likewise, following a terrorist attack against German tourists, the Egyptian government took a hard line against radical Islamic groups and solved the image crisis (Efrati, 2002; Wahab, 1996). A third example is that of Syracuse, which decided to fix its negative image and embarked on a project of refurbishing the city center through private and public partnerships. Like in Syracuse, tourist destinations the world over are revitalized through the construction of conference and cultural centers, stadiums, shopping malls, theme-oriented areas, museums, agricultural markets, and improved mass transit systems (Short et al., 1993).

The "Come See for Yourself" Strategy

Of those destinations with a cumulative negative image accrued over the years, some suffer as a result of a problematic past and a well-ingrained bad reputation, so that even if positive changes occur, unfavorable opinion persists (Strauss, 1961). Unfavorable stereotypes are firmly established in the public's mind and it is very difficult to overcome them, because no matter how much effort is spent changing the reality, if the negative stereotypes keep the crowds away, no one will see the changes. In such cases, the best course of action is staging events designed to bring in visitors who would not otherwise come. This can be accomplished through holding conferences, exhibitions, tours, and press conferences, among other special events. The main advantage of visiting a destination is that the image holder has a chance to personally experience the objective reality within the destination, without being dependent on mediators or secondary agents. When this happens in destinations associated with negative stereotypes, it may become clear to visitors that these stereotypes are false.

This strategy was successfully employed in the effort to improve the image of the Israeli city of Holon. A poll conducted by local authorities during the early 1990s revealed that although the city in general suffered from a negative image, people who had visited were fond of it. Therefore, as part of the campaign to change the city's image, a decision was made to attract people through special cultural events and performances that would enrich the image. Indeed, this strategy was successful and the city's image greatly improved during the late 1990s (Avraham, 2003a). In addition to bringing visitors, destinations need to convince decision makers and public opinion leaders to come to the destination and "see it with their own eyes."

Journalists are clearly the most important of the visitors that public opinion leaders seek to draw to their destinations. Attracting the journalists to a destination through the efficient use of public relations can be very beneficial for improving the destination's image. This is especially true during times of crisis, when it is advised to bring media people and organize familiarization trips in the hope that they will report that the situation in the destination is "business as usual"; the tourist attractions are open, tourist information booths are operating, cultural events continue, and tourist services are available. Many destinations chose this strategy, among them Israel (during the conflicts with the Palestinians and terror attacks), India (during an outbreak of infectious disease), London (the foot and mouth disease outbreak), Nepal (rebel attacks), and Miami (crime wave against tourists) (Hopper, 2003; Frisby, 2002; Baral et al., 2004; Beirman, 2002).

Local authorities employed the strategy of organizing local visits for public opinion leaders, in order to deal with any negative perceptions held by them, and by potential tourists. It was hoped that these visits would help change negative perceptions and promote a favorable image, which in turn would be passed on to other people with whom the opinion leaders would come in contact (Burgess, 1982). Similarly, the cities of Jerusalem and Belfast hold frequent cultural events and festivals in order to attract visitors and thus address the fear of terrorist activity, which might cause many to avoid these cities (Efrati, 2002).

Hosting Special Events

One of the most famous examples of using special events (often referred to as "spotlight events," "hallmark events," or "mega-events") in order to improve a destination image was the Nazis' use of the 1936 Olympic Games to project a positive image of their regime (Nielsen, 2001). Since then, many cities have used the Summer and Winter Olympics, the World Expo, the Cultural Capital of Europe title, the Eurovision song competition and, in the United States, the Republican and Democratic national conventions as major platforms for massive public relations-led image campaigns (Beriatos and Gospodini, 2004). These events focus attention on a particular location for a short, concentrated period, allowing the destination to promote certain chosen images to the international media, and may be used to improve a negative image. This was the case when Beijing hosted the 1990 Asian Games and used the media attention to improve its image after the Tiananmen Square massacre (Hall and O'Sullivan, 1996). Cities that host international spotlight events such as the Olympics or the World Expo undergo substantial changes in the urban landscape, the result of large-scale investments in the private and public sectors. The resulting development and renewal of tourist attractions, hotels, transportation, and upgrading of destination infrastructures strengthen the destinations' competitive edge and raise their rating in the global hierarchal system.

Inclusion of Residents in Campaign and Fostering Local Pride

A key difference between product marketing and destination marketing is that in the latter it is necessary to consider the people who live in and near the "product" (that is, the destination). It is only natural to involve the residents in the processes that lead to the improvement of the destination's image. Residents of unfavorably perceived destinations often suffer from a lack of local pride and a low self-image. This can lead to indifference toward the destination and an unwillingness to take part in various enterprises, or to volunteer to make it more attractive to tourists. For this reason, dealing with a negative image requires working with the local residents and mobilizing their support for the process of change. This media strategy is internally focused, aimed at residents and the way they perceive their home. The underlying assumption is that a favorable self-image will turn the destination's residents into ambassadors who will speak of its wonders when conversing with residents of other destinations (Tilson and Stacks, 1997). When the city of Glasgow, for example, wanted to change its image, it built public establishments, museums, and cultural and tourist attractions, and all the while received the backing of local residents who became more aware of the city's cleanliness and aesthetics (Paddison, 1993).

It is important to cultivate the pride and enthusiasm of the local residents towards the process of change. In Syracuse, stickers declaring "I have a part in Syracuse" and "We grow together" were handed out, and local residents participated in campaign decisions, such as choosing the city's new logo (Short et al., 1993). Fostering the participation of residents is of major importance, and their opinions and suggestions should be heard when planning campaign strategies and formulating slogans. The role of residents is a key element when attempting to recover from crises in tourism. For example, in order to bring the tourists back to Miami in the beginning of the 1990s, a 10-year program calling for increased promotion and tourism education for the local residents was launched. During the campaign, public service announcements about the importance of tourism ran countywide. The campaign mangers believed that in order to solve the crisis, it was crucial to convince local residents of the importance of tourism to the state economy (Tilson and Stacks, 2001). Similarly, in Egypt, there was an attempt to convince residents to assist in the war on terror, by informing them about the damage done to the country's national image as a result (Wahab, 1996; Nielsen, 2001).

Category Two: Managing the Symbolic Reality

In contrast to the above-mentioned strategies in which changes to the reality play a central role, the strategies in the second category focus almost exclusively on combating image crises by dealing with symbolic changes via public relations and advertising. There are eight strategies in this category.

The "Crisis? What Crisis?" Approach

Some destinations choose to ignore the damage to their images and act as though there never was a crisis, in the hope that new events and the passing of time will cause tourists to forget the crisis. Thus Spain, for example, and in particular Barcelona and Madrid, chose to employ a "business as usual" approach following periodic terrorist attacks (Efrati, 2002). In Turkey, a similar policy was introduced after the terror attacks of 2003–2004. A destination can also ignore its negative image in non-crisis times, and market itself as it wants to appear rather than as it is typically seen. For example, the city of Manchester ignored common stereotypes by marketing itself as an international business center (Young and Lever, 1997). In this context, some destinations' tendency to mention the bright future in their campaigns—while ignoring the more troublesome present—should also be noted. It is in such cases that slogans such as "The future is safe in X," "Think of the future in X," or "X is stepping toward tomorrow" can be found. Holcomb (1994) cites the example of the American city of New Milford, Connecticut, which chose the slogan "A great past, and a greater future," but remained surprisingly silent regarding the present.

Employing a "Counter-Messages Offensive"

A destination can produce messages contrary to those that led to the crisis and the negative image associated with it, or in other words, launch a counter-messages offensive (Kotler et al., 1993). There are two main methods to accomplish this strategy. The first includes specific references in public relations and advertising campaigns to the source of the negative components of the image, while the second attempts to reposition the destination. When the first possibility is employed, counter-messages are sent that are geared toward changing the negative components of the image so that the destination is no longer perceived as, for example, unsafe, dirty, or boring. If several high-profile crimes occurred in a destination and received a great deal of media attention, the destination can publish data concerning its actual crime rate, which may be much lower than that of other similar-sized cities. If a destination is perceived as unsafe, advertisements can be used in which visitors say how much they enjoyed their visit and how safe they felt.

An example of the above strategy was used in Miami where, in the wake of the tourism crisis, the convention bureau unveiled an advertising campaign addressing tourist safety and portraying the city as a safe destination (Tilson and Stacks, 2001). Similarly, Tunisia's tourism office adopted, after the terrorist attack in 2002, an ad campaign aiming at inspiring "peace and tranquility" in visitors (Media Line, 2004).

When using the second possibility, the destination is positioned in a manner directly opposed to the existing stereotype. Thus, for example, the German city of Nuremberg, one of the prime symbols of the Nazi regime whose name is linked to the post war trials of Nazi leaders, has been steadily positioning itself over the past decade as a center of "peace and human rights." This repositioning has been successful, as Nuremberg won international recognition by UNESCO. The chosen position was the direct opposite of the negative image conferred upon the city by its Nazi past: a place of war, racism, destruction, and nationalism, Nuremberg now stands for justice, freedom, peace, and equality. This change in repositioning was accomplished by the creation and promotion of museums, monuments, art exhibits, cultural events, and conferences dedicated to human rights and social justice (Ha'aretz, May 12, 2003).

Repositioning was also used in marketing the Negev Desert in southern Israel. The "Negev Action" campaign, for example, challenged the perception of the region as boring, remote, uneventful, and monotonous. As part of the repositioning, the campaign promoted trips, sporting events, and family-oriented activities in the Negev, and the advertisements focused on active recreation photographs. Another campaign in the same region used the slogan "Green in the Negev," which challenged the perception of the area as desert-like, arid, barren, and distant from the country's center (Avraham, 2003a).

Acknowledging the Negative Image

Understandable as the urge to deny might be, sometimes acknowledging the negative image directly is the most effective course of action. This might be done during the crisis, or immediately after it has passed. For example, the American city of Tulsa, Oklahoma, used advertisements portraying local sites and residents under the slogan "This should fill in the blank about Tulsa," and with the acknowledgement that "When someone says 'Tulsa' some people draw a blank." While admitting that the city is not well known and has a weak image, the campaign provides the solution—pictures of the city and information about it, attempting to enrich its image (Holcomb, 1994). After the crisis is over, an advertisement may be placed presenting the message that "the image of the city as gray and covered with smoke has disappeared, and now the city is attractive" (Burgess, 1982). Such a slogan acknowledges the difficult past and thus creates a feeling of trust between the advertisers and the external audience. This strategy also includes the initiative employed by some places to acknowledge that a problem exists in a specific regional area and to frankly advise tourists not to go there. A prime example is the approach taken by the London Tourist Board during England's much-publicized outbreak of foot and mouth disease, which emphasized that the problem was only in rural areas (Hopper, 2003; Frisby, 2002). Similar approaches have been used by Israel, Nepal, and the Philippines (Beirman, 2002; Baral et al., 2004).

Geographic Isolation Strategy

In his analysis of safety and security issues, Santana (2003) notes that "as far as tourism is concerned, security issues (real or perceived) always have a spillover effect. That is, tourists tend to associate a security incident with an entire region" (p. 305). This strategy comprises a destination's attempt to distance itself from a problematic area or region with which it is identified. Examples of separation can be seen in the marketing of the Israeli city of Eilat, which during the late 1990s was presented in Europe as "Eilat on the Red Sea." The advertising campaign did not mention the fact that the city was located in Israel, which was perceived at that time as unsafe for tourists due to security issues. The campaign managers apparently believed that not mentioning the exact location of the city would help in its marketing and prevent European tourists from avoiding it due to perceptions of Israel as unsafe. This strategy, which Pizam and Mansfeld (1996) call "destination-specific" or "isolation strategy" (Beirman, 2002), is recommended when promoting destinations located in countries suffering from ongoing image crises. Beirman (2002) brings an additional example when he mentions that Jordan aimed "to differentiate destination Jordan from destination Israel" (p. 174), after the start of the latest outbreak of conflict between Israel and the Palestinians at the end of 2000.

Spinning Liabilities into Assets

This strategy also takes the key point of a previous strategy—acknowledging the negative image—a step further by acknowledging a negative factor responsible for the image and spinning it into a positive trait. For example, Minnesota winters are known as extremely cold, a fact that harmed the state's image and drove people away. Today, however, these winters are marketed as unique and have become a tourist attraction, with various winter cultural events and festivals catering to many (Kotler et al., 1993). Similarly, one of the causes of the negative perceptions of the city of Be'er Sheva is its location in the Negev Desert in southern Israel, which is associated with extreme heat and the "backward" Orient. Several researchers (Fenster et al., 1994) suggested to local authorities that they should market the city as a modern and unique (in Israel) embodiment of an Eastern, exotic, desert city. This suggestion made use of the positive perceptions linked to the desert, and these were associated with the city. The researchers suggested the use of Eastern architecture (a unique oriental shopping center, a Bedouin market) and other activities that would promote its image as an Eastern, modern desert city. The Eastern character suggested refers to perceptions of the East as innocent, unspoiled, hospitable, spirited, and lively. Another similar case is that of the British city of Bradford, which for many years suffered from a negative image due to the many foreign immigrants who had settled in it and the ethnic and racial clashes that ensued. Over the past few years, the city has been trying to turn this characteristic into an advantage by marketing itself as a multicultural oasis (their slogan is "Flavors of Asia"), where different social groups and races coexist in harmony and a spirit of cooperation (Bramwell and Rawding, 1996). In all of these examples, those characteristics seen as responsible for the negative image are presented with a new twist, so as to turn them from liabilities to assets.

Changing a Destination's Name

A destination's name, like that of a person or an organization, is part of its identity and, as such, has considerable influence on the way it is perceived by the public. Therefore, in order to effect a change in image, it may prove prudent to change the name as well. Some destinations have names that are beneficial to marketing, whereas others have names associated with negative stereotypes and perceptions that harm their attractiveness. Thus, for example, a French member of Parliament attempted to pass a law that would prohibit the use of the term "Vichy Government," claiming that it was harmful to the development of the city of Vichy.

This city, which was once the world capital of healing-related tourism, was also capital to the Nazi-collaborating World War II government that has since been known as the "Vichy Government." Although many years have passed since that difficult time, the city's name is still problematic, and tourists, conference organizers, and visitors refrain from including it on their itineraries.

There are cases in which a destination's image is so unfavorable that local authorities have given up hopes of rectifying the situation. Some of these places simply change their names in the hope that the negative image associated with the old name will disappear along with it (Avraham, 2003a). An interesting example is that of the US state of North Dakota, which has begun taking measures to rid itself of the prefix "North" and be known simply as "Dakota," since the term "North" leads to perceptions of the state as cold, isolated, and unattractive (Singer, 2002). Similarly, in order to combat the tourism crisis in the city of Eilat in Israel, a decision was made to change its name to "Eilat City." The city's decision makers believe that with its new name, they will be able to market Eilat as a lively, international, and borderless city.

Creation of a New Logo or Slogan

Along with names, destinations' symbols and logos are also important components of marketing campaigns. Thus, a destination in transition may also need to change its symbols. Industrial cities moving on to the post-industrial era often have symbols that are no longer relevant to the city's new spirit. The city of Syracuse in New York State, for example, strove to move toward the post-industrial era but had a symbol containing chimneys, industrial plants, and smoke that was incongruent with the new image. Indeed, after lengthy debate a new logo was adopted, one displaying skyscrapers, a modern skyline, open skies, and a lake (Short et al., 1993). In another example, a Polish marketing agency recently suggested making a kite the new national logo, in order to promote tourism and to improve the nation's general image. According to the agency, kites symbolize a plethora of positive attributes, including freedom, youth, love of life, and hope. The idea is that associating Poland with these concepts via the logo will help to undo the stereotype of Poland as gray, boring, cold, conservative, and poor (Boxer, 2002).

Like logos, slogans are effective methods of delivering messages. One of the classic slogans used to foster fondness for a destination is the well-known "I ♥ New York." A good slogan may be used for many years and through several different campaigns. Good slogans lay out a destination's vision, reflect its spirit, and create enthusiasm and momentum. Formulating a slogan depends on the target population and on the goals of the campaign, but the slogan must also be at least somewhat congruent with reality (Kotler et al., 1993). A destination undergoing a change should change its slogan, emphasizing the change and its new look, as in the case of the British city of Bradford, whose new slogans include "Bradford's Bouncing Back," and "A Surprising Place" (Bramwell and Rawding, 1996).

Changing the Campaign's Target Audience

An additional strategy that can be used by destination decision makers to deal with an image crisis is to change the target audience of their advertising and public relations campaigns. Sometimes, the destination's image is so problematic among the existing target audience (past visitors, for example) that there is virtually no chance of overcoming it. This scenario is especially severe when a destination suffers frequently from negative coverage in the international media. When destinations in this situation change their target audience or their type of tourism, they begin to concentrate on a different market segment that is less affected by the issues raised in the negative coverage. In the beginning of the new century, several countries in the Middle East employed this strategy. In Israel, for example, as a result of the damage caused to general tourism, advertising campaigns began to concentrate on religious tourists in the United States and Europe, with the assumption that tourists of this type would be less sensitive to security issues. Jews and Evangelical Christians became a prime target audience, while domestic tourism was also encouraged at new levels. The new campaign attempted to use the potential tourists' religious identity to convince them to visit Israel, using the slogan "Don't let your soul wait any longer. Come visit Israel" (Ha'aretz, May 6, 2003; Media Line, 2004).

Another example of this strategy can be found in Syria, another country that suffered a tourism crisis after the terrorist events of September 11. As a result of the crisis, Syria began a campaign based on the slogan "Syria, Land of Civilizations" in order to attract tourists from throughout the Arab world who no longer felt comfortable traveling in the West (Ha'aretz, July 30, 2002). Jordan initiated a similar approach, and began concentrating on regional tourism, attracting visitors from Gulf nations and neighboring countries (Media Line, 2004).

Choosing the Right Strategy to Solve the Image Related Crisis

Having surveyed the various strategies used by destinations to change a negative image, the issue that remains is how to select the correct strategy for a given situation. Some of the many factors behind this decision include the specific circumstances and nature of the crisis, the amount and nature of negative press received by the destination, and the feasibility of hosting mega-events. If, for example, the actual problems that led to a crisis and the formation of a negative image have not been solved—such as crime against tourists and a perceived lack of safety—it will not be productive to select a strategy that involves attracting visitors or public opinion leaders to the destination in order to nullify stereotypical perceptions, or to deliver counter-stereotypical messages.

It is also crucial to address the issue of how widespread the negative image is among the target population, and if it is at all possible to change this image. If the image is not extremely unfavorable, is not very prevalent, and is relatively easy to change, the best strategy is probably to ignore it. Spanish cities can ignore terrorist attacks because the attacks are perceived as being targeted at specific politicians or occurring outside the tourist demand area (Nielsen, 2001); cities in Israel, in which terrorists target the civilian population, cannot disregard such attacks, especially since extensive media coverage is given to them. In any case, the choice of this strategy depends on the strength of the place as central tourist destination: "A 'flagship destination'—because of its inherent popularity—recovers quickly from adverse publicity" (Nielsen, 2001, p. 219). In other words, the decision to choose the "business as usual" strategy depends on the size and amount of the crisis's media coverage, as well as the specific nature of the crisis. Employing a strategy of counter stereotypical messages is likely to be the best option if a crisis is not very serious and the negative image is prevalent but relatively easy to change. On the other hand, in cases where the problem is very well known and difficult to solve through counter-stereotypical messages, it is more logical to first address the problems with the destination's reality, for example, fighting crime or improving tourist infrastructure, and then to market the change to the destination.

Conclusion

A tourist destination's perceived image can affect—positively or negatively—its demand. A positive image can bring a multitude of tourists, while a negative image can keep them away. This chapter has presented a summary of the 12 relevant media strategies chosen by decision makers around the world who were interested in improving their destinations' negative image, both in the case of a sudden negative event or a prolonged decline. The analysis revealed two categories of media strategies: those that involve changing the reality and those that occur only in the symbolic reality. Favorable images are becoming more and more vital due to the growing competition between destinations. Due to this competition, destinations can no longer remain indifferent to the way they are perceived (Kotler et al., 1999).

It should be emphasized that improving a destination's image is a complicated task and may take many years (Fenster et al., 1994, suggest a time frame of 6–8 years). The most important step toward an effective management of the crisis and changing the negative image is proper long-term strategic planning. When done correctly, this planning devotes considerable attention to deflecting potential crises, and has been shown to be very successful when implemented.

One must remember, however, that although proper handling of the relevant media aspects is important, an advertising campaign alone is not enough to bring about significant change. Real-life problems faced by those who need to use a destination's tourist services—including accommodations, restaurants, and transportation—must also be dealt with. Actions must also be pursued in different and complementary channels. Members of the former Yugoslav republics, for example, use media advertisements, but at the same time have also launched an extensive sales campaign in tourist fairs around the world, and regularly meet travel agents and tour operators—all this in order to bring back those tourists who used to flock to the area prior to the Balkan Wars (Efrati, 2002). In other words, several techniques should be used in unison in order to achieve the desired end result: advertising, public relations, sales promotions, direct mailing, and marketing. In addition to the need to use all of these techniques, it is important to unite all the players who are connected to the crisis: local residents, the tourist industry (hotels, attractions, tour-operators), local government, national government, local media, public relations, and ad people (Tilson and Stacks, 2001). In this way, for example, a massive advertising campaign sponsored by the local government and hotel industry needs to be accompanied by promotional discounts in local hotels.

Having presented the various media strategies, it is crucial to point out the obvious: the analysis in this chapter demonstrates clearly that a given location's actual real-life situation is more important than any media strategy employed. Spain did not only redefine and reposition itself, but also employed a general program of improving tourist attractions, facilities, and services (Morgan and Pritchard, 2002). The Spanish city of Bilbao—which used to face severe problems stemming from poverty, unemployment, and terrorist attacks—would not have become an attractive tourist destination through media campaigning alone, had it not at the same time developed real attractions (establishing the Guggenheim Museum and other tourist sites and services). Media strategies are helpful in raising awareness among target populations to what a destination has to offer, and can thus assist in making the most of these attractions, but they are not enough by themselves to bring about a much longed-for change.

Concept Definitions

Public relations Technique that aims to influence the destination's coverage patterns in the mass media. Public relations professionals try to build a positive image of the destination among a target audience by spreading positive stories and preventing the publishing of negative articles.

Advertising By buying space in the media, advertisers send positive messages about the destination to the target audience. Advertisers can use various strategies to convey their message and make the destination desirable to target audiences.

Destination marketing The attempt to promote a positive image for a destination among a specific target audience by using techniques such as advertising, public relations, sale promotion, and direct marketing.

Destination image crisis Destination images can suffer from one of two kinds of crises. The first occurs after a sudden negative event like a terrorist attack or an earthquake; the second is the gradual result of a prolonged period of decline.

Review Questions

1. What factors affect the decision of which strategy to choose when managing a destination crisis image?
2. What is the correct way to manage a destination's image in general?
3. Which techniques are available in order to build a positive destination image?
4. What are the possible advertising and public relations strategies available in order to handle a destination image crisis?
5. What factors affect destination image and what are some of the types of images that exist?

References

Avraham, E. (2002). Cities and their news media images. *Cities,* 17, 363–370.

Avraham, E. (2003a). *Promoting and Marketing Cities in Israel.* Jerusalem: Floersheimer Institute for Policy Studies (in Hebrew).

Avraham, E. (2003b). *Behind Media Marginality: Coverage of Social Groups and Places in the Israeli Press*. Lanham, MD: Lexington Books.

Baral, A., Baral, S., and Morgan, N. (2004). Marketing Nepal in an uncertain climate: Confronting perceptions of risk and insecurity. *Journal of Vacation Marketing,* 10, 186–192.

Beirman, D. (2002). Marketing of tourism destinations during a prolonged crisis: Israel and Middle East. *Journal of Vacation Marketing,* 8, 167–176.

Beriatos, E., and Gospodini, A. (2004). "Globalising" urban landscapes: Athens and the 2004 Olympics. *Cities,* 21, 187–202.

Boxer, S. (2002, December 12). New Poland, no joke. *New York Times*.

Bramwell, B., and Rawding, L. (1996). Tourism marketing images of industrial cities. *Annals of Tourism Research,* 23, 201–221.

Burgess, J. (1982). Selling places: Environmental images for the executive. *Regional Studies,* 16, 1–17.

Efrati, B. (2002, September 6). Not welcoming. *Kol Ha'ir*- (in Hebrew).

Elizur, J. (1986). *National Images*. Jerusalem: Hebrew University.

Fenster, T., Herman, D., and Levinson, A. (1994). *Marketing Beer Sheva: Physical, Social, Economic and Administrative Aspects*. Beer Sheva, Israel: The Negev Center for Regional Development, Ben Gurion University (in Hebrew).

Fearn-Banks, K. (1996). *Crisis Communication: A Casebook Approach*. Mahwah, NJ: Lawrence Erlbaum.

Frisby, E. (2002). Communication in a crisis: The British Tourist Authority's responses to the foot-and-mouth outbreak and 11th September, 2001. *Journal of Vacation Marketing,* 9, 89–100.

Hall, C. M., and O'Sullivan, V. (1996). Tourism, political stability and violence, in A. Pizam and Y. Mansfeld (eds.) *Tourism, Crime and International Security Issues,* Chichester, UK: John Wiley & Sons, pp. 105–126.

Holcomb, B. (1994). City Make-Over: Marketing the Post-Industrial City, in J. R. Gold and S. V. Ward, (eds.) *Place Promotion: The Use of Publicity and Marketing to Sell Towns and Regions*. Chichester, UK: John Wiley & Sons, pp. 115–121.

Hopper, P. (2003). Marketing London in a difficult climate. *Journal of Vacation Marketing,* 9, 81–88.

Kotler, P., Haider, D. H., and Rein, I. (1993). *Marketing Places*. New York: Free Press.

Kotler, P., Asplund, C., Rein, I., and Haider, D. H. (1999). *Marketing Places, Europe*. Edinburgh, UK: Financial Times, Prentice Hall.

Media Line. (2004). *Tourism in the face of terror*. http://themedialine.org/news/news_detail.asp ?NewsID=5768. Accessed June 14, 2004.

Morgan, N., and Pritchard, A. (2002). Contextualizing destination branding, in N. Morgan, S. Pritchard, and R. Pride, *Destination Branding: Creating the Unique Destination Proposition*. Woburn, MA: Butterworth-Heinemann.

Morgan, N., and Pritchard, A. (2001). *Advertising in Tourism and Leisure*. Oxford, UK: Butterworth-Heinemann.

Nielsen, C. (2001). *Tourism and the Media*. Melbourne, Australia: Hospitality Press.

Paddison, R. (1993). City marketing, image reconstruction and urban regeneration. *Urban Studies,* 30, 339–350.

Pizam, A., and Mansfeld, Y. (eds.). (1996). *Tourism, Crime and International Security Issues*. Chichester, UK: John Wiley & Sons.

Santana, G. (2003). Crisis management and tourism: Beyond the rhetoric. *Journal of Travel and Tourism,* 15, 299–321.

Short, J. R., Benton, L. M., Luce, W. B., and Walton, J. (1993). Reconstructing the image of an industrial city. *Annals of American Geographers,* 83, 207–224.

Short, J. R., Breitbach, S., Buckman, S., and Essex, J. (2000). From world cities to gateway cities. *City,* 4, 317–340.

Singer, M. (2002, February 18, 25). True north. *The New Yorker,* 118–123.

Strauss, A. L. (1961). *Image of the American City*. New York: The Free Press.

Tilson, D. J., and Stacks, D. W. (1997). To know us is to love us: The public relations campaign to sell a "business-tourist-friendly" Miami. *Public Relations Review,* 23, 95–115.

Tilson, D. J., and Stacks, D. W. (2001). Paradise lost and restored: Florida and the tourist murders, in D. Ross and B. DeSanto (eds.) *Public Relations Cases: International Perspective*. London: Routledge, pp. 155–166.

Wahab, S. (1996). Tourism and terrorism: Synthesis of the problem with emphasis on Egypt, in A. Pizam and Y. Mansfeld (eds.) *Tourism, Crime and International Security Issues,* Chichester, UK: John Wiley & Sons, pp. 175–186.

Young, C., and Lever, J. (1997). Place promotion, economic location and the consumption of image. *Tijdschrift voor Economicische en Sociale Geografie,* 88, 332–341.

14

A Comparative Assessment of Three Southeast Asian Tourism Recovery Campaigns: Singapore Roars: Post SARS 2003, Bali Post–the October 12, 2002 Bombing, and WOW Philippines 2003

David Beirman

Learning Objectives

- To understand the concept of tourism recovery.
- To understand the role of marketing in recovery from tourism crises.
- To understand the role of different components of the tourism industry in recovery processes and their interrelations in recovery processes.
- To understand the meaning of ad hoc and post-crisis marketing concepts.
- To understand the importance of alliances in facilitating centralized, fast, and effective recovery.
- To understand the importance of a multinational regional approach to tourism recovery.
- To understand the growing significance of government travel advisories and global media coverage as image determinants for destinations.

Introduction

The advent of mass international tourism since the 1970s stimulated rapid growth for inbound and stopover tourism in Southeast Asia (SE Asia). Tourism represents one of the most lucrative economic segments for most SE Asian countries. The tourism industry is a major source of foreign currency earnings, creating millions of service industry jobs throughout the region. Increased per capita earnings within Eastern and SE Asia combined with the reduction of tourism prices in real terms within Asia are key factors in accelerating SE Asia's development as one of the world's fastest growing regional tourism destinations.

SE Asia's tourism industry was spared the most severe initial repercussions of September 11, the 2001 attack against the United States and the launch of the US-led global "War on Terror" against Islamic political movements. During the 6 months following September 11, 2001, international tourism experienced a decline predicated on safety threats to long-haul air travel. SE Asian tourism remained relatively buoyant throughout most of 2002 partially due to the perceived isolation of SE Asia from the Afghan and Middle East trouble spots. Equally relevant is the fact that most travel to SE Asia originates from within or near the region (Koldowski, 2003).

The bombing of the crowded Sari Nightclub in Denpassar, Bali, on the evening of October 12, 2002, resulted in the murder of 200 tourists and wounded 300 from 24 countries (Henderson, 2003). The world's deadliest terrorist attack in recent history specifically targeting tourists placed SE Asia at the epicenter of the War on Terror. The convicted perpetrators were members of the radical Islamic group Jemal Islamyia with affiliated cells throughout Indonesia, Malaysia, Thailand, Singapore, and the Philippines. The Sari Club bombing had a devastating impact on tourism to Bali and throughout Indonesia, causing collateral damage to the security perception of many neighboring SE Asian countries. Although Islamic organizations have been active in SE Asia for many years, the Bali attack exacerbated fear of Islamic violence in SE Asia. The governments of many tourist generating countries including Japan, United States, United Kingdom, and Australia issued cautionary travel advisories which encompassed Indonesia, Malaysia, Singapore, and the Philippines. The combination of negative media coverage and negative government travel advisories was perceived as a major threat to the viability of the tourism industries in the region.

Southeast Asian tourism was forced to respond to a further challenge in March 2003 when reports surfaced about the outbreak of a condition known as Severe Acute Respiratory Syndrome, which became known worldwide as SARS. This illness has similar symptoms to pneumonia and was initially diagnosed in Southern China but by March and April cases appeared in Hong Kong, Singapore, Vietnam, and Canada (Singapore Tourist Board statistics, 2003). According to the World Health Organization, between March and July 2003 there were over 7,000 confirmed cases of SARS. The deaths of 774 people were attributed to SARS.

The outbreak of SARS was regarded as a potentially serious public health risk. However, the actual extent of the problem was relatively minor in comparison with ongoing global epidemics including AIDS, malaria, cholera, and typhoid, which cause the death of millions annually. The blanket international media coverage of SARS, the mysterious nature of the disease, and the absence of a known cure combined to ignite panic in the countries affected and among international travelers.

Media footage of people in the streets and airports of Hong Kong and Singapore wearing surgical masks created a sense of fear among travelers who sought to avoid even transiting in cities known or even thought to have contacted the disease. Although tourism to Singapore and Hong Kong was most heavily affected by the SARS scare, the tourism industries of most SE Asian countries were collateral victims of SARS. Thailand, Malaysia, and the Philippines had minimal exposure to SARS but tourism to all these destinations was negatively affected.

Fear of terrorism arising from the Bali bombing and the outbreak of SARS in SE Asia, much of it magnified by media coverage in the West, was exacerbated by a series of negative travel advisories issued by the governments of many economically significant tourism source markets including the United States, United Kingdom, Australia, and Japan. The advisories suggested that citizens defer travel to several countries in SE Asia, notably Indonesia, the Philippines, Malaysia, and, for brief periods, Hong Kong, Thailand, Vietnam, and Singapore (Henderson, 2003).

Although there were varying impacts on tourism within each SE Asian country, the regional extent of the threat to tourism industry spurred SE Asian nations to respond regionally. On November 4, 2002, the heads of state of ASEAN (Association of South East Asian Nations) met in the Cambodian capital Phnom Phenh to pledge their support for an ASEAN tourism recovery marketing campaign (Association of South East Asian Nations, 2002). An ASEAN tourism marketing organization has theoretically been in existence since 1988; however, regional tourism marketing was rarely practiced. The crisis engendered by the Bali bombing motivated the region's political leadership to swiftly reactivate ASEAN tourism marketing cooperation.

At the private sector level, the Bangkok-headquartered Pacific Asia Travel Association (PATA), the main travel industry body representing the Pacific Rim, placed the restoration of SE Asian tourism as its top priority project from the time of the Bali bombing and was supported by national tourist offices and peak travel industry bodies of all SE Asian nations and the International Air Transport Association (IATA). In 2003 PATA launched its innovative recovery campaign, Project Phoenix, designed to stimulate tourism recovery throughout SE Asia. In essence, Project Phoenix involved cooperation between specific media organizations, airlines, national tourist offices, tour operators, and hotel chains. The campaign involved SE Asian regional and destination specific marketing promotions and a range of incentive travel offers to jump-start consumer recovery post- SARS.

This chapter will examine the regional recovery campaigns and issues while focusing on three specific recovery campaigns representing the major challenges encountered by the tourism industry in SE Asia during the final quarter of 2002, the first half of 2003, and the varying successes of recovery strategies employed.

The first specific recovery campaign discussed is Bali's attempt to restore inbound tourism following the Sari Club bombing of October 12, 2002. The restoration campaign was unusual in that it was dominated by the private sector.

The Balinese provincial tourism authorities were professional in their approach but there was considerable evidence that Indonesian government actions did more to hinder, rather than to help, the recovery processes. Bali enjoyed an unusually high degree of support from foreign tour operators and from PATA in its market recovery program.

The Singapore Roars marketing campaign was a highly professional, government initiated, and generously funded restoration marketing campaign designed to stimulate a rapid recovery of tourism to Singapore following the end of the SARS scare in July 2003. An unusual feature of the Singapore Roars campaign was that it was conducted in selected source markets in conjunction with similar campaigns promoted by Thailand and Malaysia. Although it emphasized Singapore, the Singapore Roars campaign was not conducted in isolation. Singapore, as a highly centralized city-state, marshaled a high level of coordination in its marketing campaign. Singapore's tourism marketing is characterized by close cooperation at the leadership level between the public and private sectors of the nation's tourism industry.

The Philippines tourism industry has had a long history of navigating its way through a variety of crisis events since the mid-1980s. Islamic terrorism has posed a background problem in the Philippines since the early 1990s, although occasional attacks against foreign tourists have raised the media profile of this problem. The Philippines was a collateral victim of the fear of terrorism and SARS. The Bali attack focused global attention on the endemic issue of Islamic terrorism in the Philippines. Despite rare attacks in Manila and several well-publicized incidents of abduction and killing of tourists, Islamic terrorism is primarily confined to southern Mindanao. Although there were no SARS cases reported in the Philippines, tourism suffered to a limited extent from the SARS scare.

From 2002, the Philippines Department of Tourism, under the leadership of its energetic secretary Richard Gordon, decided to conduct a highly visible and positive marketing campaign which directly challenged the negative stereotypes prevalent about the Philippines. The WOW Philippines campaign was carefully targeted and designed to re-image destination Philippines and to spread the benefits of tourism as widely as possible across the 7,000 island archipelago nation. While not designed as a campaign to recover from specific events such as was the case with Bali, local terrorist incidents, or SARS, the WOW Philippines campaign is designed with the longer-term goal of changing consumer and industry perceptions of the Philippines and stimulating real growth for the country's tourism industry. Gordon, elected chairman of PATA in 2003, also sought to incorporate the Philippines recovery marketing campaign within a broader SE Asian tourism recovery campaign. An unusual aspect targeted the 8 million Filipino nationals living overseas to act as tourism marketing ambassadors of the Philippines in their respective countries of residence. They were encouraged to make return visits and incentives were offered as motivation to nationals of their country of residence to visit the Philippines.

SE Asian Tourism: A Regional Approach to Tourism Recovery

Regional marketing of SE Asia, especially to long- and medium-haul source markets including North America, Europe, Australasia, and Japan was an important feature of tourism marketing during the 1980s. However, as the economies of SE Asian countries modernized and diversified, the major tourism destinations and stopover points in SE Asia increasingly sought to differentiate their marketing identities, resulting in a tendency toward competition and limited cooperation. Intense competition ensued among Hong Kong, Bangkok, Kuala Lumpur, and Singapore as airline hub and spoke points that influenced their relative positions for stopover traffic. Malaysia, Thailand, Indonesia (especially Bali), the

Philippines, and Vietnam all regarded tourism as strategic fulcrums for national economic expansion.

Consequently, their government-dominated tourism marketing bodies sought inbound tourism at the expense of neighboring nations. The highly competitive nature of tourism during the 1990s was simply not conducive to regional tourism cooperation as each destination assiduously sought to maximize its own share of a growing international market. The ASEAN tourism marketing body, a product of late 1980s government policies, technically remained in existence, but in every practical sense was moribund by 2000.

During the 1990s the tourism industry's private sector had a very different set of priorities. Airlines, multinational hotel chains, cruise and tour operators servicing multiple destinations, and global financial institutions were far more amenable to regional destination marketing than the private sector during the 1990s. The major regional travel industry association PATA was, until the 1990s, primarily oriented towards America. The relocation of its headquarters from San Francisco to Bangkok in 1998 represented the realization of its organizational evolution from an America-centric association in its early years to an Asia-centric organization by the mid-1990s. PATA took over the role of regional SE Asian tourism marketing abandoned by ASEAN during the 1990s (Chuck and Lurie, 2001). Although IATA's global headquarters remained in Switzerland, it maintained a high-profile presence in Asia (interview: Concil, 2003). By 1997, SE Asian tourism growth was increasingly sourced from within Asia. During the 1990s Japan, South Korea, Hong Kong, and Taiwan (the so-called Tiger Economies) had become significant SE Asia source markets. The Asian economic collapse resulted in a 2-year hiatus to this growth. The emergence of India and the People's Republic of China as rapidly growing tourism generating markets promised long-term tourism growth for SE Asian tourism destinations. The quantum growth of mass Chinese outbound tourism exacerbated competitive pressures largely due to the PRC government policy of limiting approved destinations for Chinese nationals. The criteria for a country to become an approved destination for Chinese citizens included operational and political elements. Murray Bailey's report on the Chinese outbound market commissioned by PATA in 2001 provides a detailed guide to the growth forecasts and the specific qualifications for approved destination status (Bailey, 2001).

The prevailing dissonance between the private and the government sector's approach to regional tourism marketing in SE Asia during the 1990s and the early years of the twenty-first century was radically affected by the Sari Club bombing of October 12 , 2002, and the SARS scare of March–June 2003. The two events put a sudden end to the divergence between the two approaches and resulted in a rapid new convergence.

Regional focus on the actual and potential threat of terrorism directed at tourists highlighted by the Bali bombing presented a common threat to the regional tourism industry requiring a coordinated response. The perceived threat of terrorism was compounded by a sudden upsurge in media-driven speculation concerning the potential threat to tourist security. Global media networks including CNN, BBC, Deutsch Welle, and Australian ABC covered this issue extensively. Simultaneously the governments of many economically significant source markets including the United States, Canada, United Kingdom, Germany, and Japan issued travel advisories recommending their citizens defer travel to several

SE Asian countries. From the perspective of these governments, most of which had citizens who were victims of the Bali attack, enhanced caution was not only deemed justified but their own societies demanded it. SE Asian countries, including Singapore, Malaysia, Indonesia, the Philippines, and Thailand, publicly criticized these negative travel advisories and treated them as a diplomatic affront. Although government travel advisories have been issued since the September 11, 2001, attack their global media profile and public visibility has been substantially enhanced. Outspoken former Malaysian Prime Minister Dr. Mahathir Mohammed, Indonesian President Megawati Sukarnoputri, and Philippine Tourism Secretary Richard Gordon (who would be elected as PATA Chairman in mid-2003) were scathing in their condemnation of negative Western and Japanese government travel advisories. The Australian government was the most frequent target of criticism partially due to Australia's position as a country that assiduously courted inclusion within Asia while being perceived as a hostile mouthpiece of the United States (Wilks, 2003).

A relatively unrecognized development that arose from the Bali bombing involved the international travel insurance industry. With few exceptions, prior to the Bali bombing it was almost axiomatic that commercial travel insurance coverage for passengers was linked to the wording of government travel advisories. Most insurance providers automatically excluded coverage for loss or injury from such acts as "politically motivated violence, civil disobedience, war or terrorism." The insurance industry's term for this exclusion was "the general exemption." After September 11 there was growing consumer and stakeholder pressure placed on insurers to provide coverage for unforeseen events (exemplified by the September 11 attack) and some insurers relaxed their adherence to the general exemption to provide coverage on a case-by-case basis. The high number of fatalities in the Bali bombing placed the spotlight on insurance companies. There was intense media pressure applied on travel insurers to assist victims of the bombing. Since the Bali attack several major travel insurers gradually abandoned the general exemption and either provided coverage on an individual basis or factored in terrorism as optional premium coverage. Concurrently there has been an erosion of the linkage between government travel advisories and insurance coverage. Until the September 11, 2001, attack it was a common (though not universal) practice for insurers to deny regular travel insurance coverage for travelers visiting destinations for which governments had issued an advisory to defer travel. The increased incidence of governments issuing negative travel advisories since the advent of the "war on terrorism" has led insurers to treat them less seriously as a commercially valid assessment of tourist risk and determine their coverage based on independent statistical risk assessments.

The combination of negative media coverage from the West and the need to respond to what was widely perceived in SE Asia as diplomatic attacks (in the form of travel advisories) from the West spurred the governments of SE Asia to present a united front to defend regional tourism. On November 4, 2002, just over three weeks after the Bali incident, a summit of ASEAN heads of states gathered in Cambodia to sign a tourism marketing accord pledging a coordinated ASEAN approach to tourism marketing, effectively marking the rebirth of ASEAN tourism. ASEAN tourism cooperation following the Bali attack would also be reinforced and amplified during the 2003 SARS crisis.

By sheer coincidence, the executive of PATA was in Bali on the night of the bombing (October 12, 2002) for the wedding of PATA Vice President Peter Semone. The reception was due to be held at the Sari Club but the explosion occurred while they were en route to the club. The presence of the senior executives of PATA at the scene of the bombing meant that PATA was able to directly assist with the recovery and crisis communication process. PATA played a significant role in mobilizing private tourism industry support for the victims of the bombing and for the Balinese tourism industry. PATA also pledged its cooperation with the Indonesian tourism industry to assist with marketing recovery and adopted a supportive approach to regional tourism recovery in SE Asia (interview: Semone, 2004).

In early 2003 PATA sponsored a task force of tourism recovery experts to assist and report to the Indonesian government and the Balinese tourism authorities. In April 2003, during a period in which SARS was already exerting a significant impact on tourism in SE Asia, the annual PATA conference was held in Bali. Although venues for PATA's annual conferences are determined at least 3 years in advance, the 2003 conference was a most propitious opportunity for the Asia/Pacific travel industry to express solidarity with their Indonesian and Balinese colleagues. PATA prepared a crisis-management manual for the use of its members and affiliates (Winning Edge, 2003). In July 2003 following the SARS scare, PATA launched their major tourism market recovery campaign, Project Phoenix, which included several major elements. It involved the Tourism Ministries of most SE Asian countries, media organizations (notably CNN), airlines, hoteliers, and major tour operators. The campaign's message was intended to reassure travelers that SE Asia was clear of SARS and welcoming visitors. The positive messages were reinforced by a series of pull marketing incentive programs designed to lure tourists. Project Phoenix included broad regional promotion of SE Asia combined with destination-specific promotion. Effectively, Project Phoenix was a private sector initiative involving a high level of public sector support from the tourism ministries of the major SE Asian tourism destinations.

During the SARS scare, IATA relocated its Asia Pacific Crisis communications headquarters from Tokyo to Singapore where it coordinated and communicated the preventative activities of airlines servicing and transiting the region in response to the SARS crisis.

PATA, in conjunction with the World Tourism Organization and the Asia Pacific Economic Cooperation, jointly commissioned an extensive report entitled *Tourism Risk Management for the Asia Pacific Region* written by Professor Jeff Wilks and supported by several other tourism scholars. Wilks heads the Centre for Tourism and Risk Management at the University of Queensland. Professor Wilks' report adopted a broad and well-structured approach to the major risk management and security issues affecting tourism in SE Asia.

During the second half of 2003 it was clear that the convergence between the priorities of SE Asian governments and the tourism private sector was manifested by an increasingly cooperative approach to the restoration marketing of tourism in the region. From July 2003 the overwhelming regional priority for SE Asian tourism was the rapid implementation of demand recovery for the inbound tourism market. This regional priority would be complemented by the market recovery campaigns of the individual countries within SE Asia.

Case Study 1. Singapore Roars 2003

Singapore has earned an outstanding reputation for success in SE Asian tourism since the 1980s. The opening of Changi International Airport 22 km north of the center of Singapore in 1982 greatly enhanced Singapore's capacity to handle a massive growth in airline capacity. Singapore's inbound tourism numbers in 2000 with 7.69 million arrivals represent an all-time record. By comparison, Singapore's inbound tourism level was only 6 million arrivals in 1992. However, it must be noted that tourism in Singapore is dominated by short stays and a high proportion of stopover tourism. The average length of stay in 2001 according to the Singapore Tourism Board was only 3.19 days. Although not counted in tourism statistics, transit passengers remaining at the airport or passing through Singapore en route to other destinations may contribute financially to Singapore, but not to its hotel revenues (Singapore Tourist Board Statistics, 2004).

Singapore's reputation as a tourist attraction is based on its shopping facilities, a concentrated range of tourist attractions, the high quality of its cuisine, hotels, and its image as being safe, clean, law abiding, free of health concerns, and tourist-friendly, are among the key attractions for visitors. According to Joan Henderson, tourism to Singapore in 2001 contributed US$5.3 billion or approximately 10% of Singapore's GDP. Consequently, tourism's contribution to Singapore's economy and employment creation is significant by world standards. Singapore's tourism experienced a minor downturn of 2.2% in 2001 compared with 2000, primarily due to the global downturn in long haul flights following September 11, 2001, which affected inbound tourism during the final quarter of 2001. Most of Singapore's major markets remained stable during that year with the only notable downturn from Japan, the United States, and Taiwan. Singapore was largely perceived as relatively immune from the threat of Islamic terrorism and geographically removed from the epicenter of America's "war on terrorism" being fought in Afghanistan during 2001 and early 2002 (Henderson, 2003).

During 2002 inbound tourism to Singapore continued to decline, although this was concentrated within the period immediately following the Bali bombing in October 12, 2002. Singapore became a classical collateral victim of the Bali terrorist incident. As discussed earlier in this chapter, the Bali attack focused world attention on the potential threat of Islamic terrorism in SE Asia. This was not a new phenomenon; but the scale of the Bali attack, the fact that the majority of victims were tourists from Western countries, and the intense media coverage of the incident had regional implications. In Singapore, days after Bali, the US, British, and Australian intelligence services claimed to identify a Jemayal Islamiya/Al Qaeda plot to attack several Western diplomatic legations in Singapore. Several suspects were arrested by Singapore police and documents and videotapes were produced to support these allegations. The foreign ministries of several of Singapore's main source markets including Australia, United Kingdom, United States, and Japan issued travel advisories to defer nonessential travel to Singapore. Although this level of warning remained in force relatively briefly, the advisories aroused outrage throughout Singapore, especially in its tourism industry. Singapore was not the only SE Asian country subjected to negative travel advisories following the Bali bombing. However, the very concept of a highly regulated and ordered society such as Singapore being defined as a tourism security risk was considered by the political and business elite of Singapore as a diplomatic insult and a loss of face.

The first confirmed case of SARS in Singapore was identified in mid-March 2003. According to the Singapore Ministry of Tourism, a Chinese doctor visiting Hong Kong in late February unwittingly infected several hotel guests, including a number of Singaporeans who carried the disease home to Singapore. At its zenith, by late May, a total of 238 people in Singapore were diagnosed with the virus, of whom 33 died. A further 930 people were placed under preventive quarantine. As Joan Henderson correctly observes, the crisis surrounding SARS was one of fear rather than of substance. This is not to suggest the very real concern generated by SARS, especially among those unfortunate enough to have contracted the condition. In global epidemiological terms, the SARS outbreak in Singapore was relatively minor compared to the 1980s Hong Kong Flu. The overwhelming media coverage of SARS accompanied by the widespread dissemination of World Health Organization warnings in the media and government travel advisories created a global impression that Singapore was a latter day leper colony. The psychological fear of a looming pandemic quickly descended on neighboring countries. Travelers were instilled with dread about visiting or transiting Singapore or even flying in aircraft that might travel in the general direction of Singapore (Henderson, 2003a).

IATA temporarily relocated its Asia Pacific Crisis Communications headquarters from Tokyo to Singapore in order to focus its attention on reassuring airline passengers of the negligible risk of contracting SARS on airlines and in transit lounges in Singapore or other SE Asian transit points. Tony Concil, IATA's Asia Pacific Manager of Crisis Communications, sent daily media releases about the SARS situation as it impacted on airlines and airports. Concil recalled the many occasions he fielded journalists' questions and issued media releases detailing the precautions airlines had taken to minimize the risk of the spread of SARS among airline passengers (interview: Concil, 2003).

During the height of the SARS crisis in the period of March–June 2003, Singapore's government committed SID$230 million to support the travel and tourism industry during a period in which tourist arrivals to the island state fell by 70% over the corresponding months of 2002. A portion of these funds was devoted to maintaining employment levels in a wide range of tourism enterprises including hotels, attractions, tour operators, and the government-owned airlines. A push marketing campaign was initiated to inform and reassure the travel industry and consumers in key markets that Singapore had adopted a range of measures to minimize the threat of SARS. Many hotels and attractions were identified as complying with a series of measures which declared them "SARS safe." This was conducted under the slogan COOL Singapore Campaign (Henderson, 2003a).

By June 2003 Singapore was declared SARS free by the WHO, enabling the Singapore Tourism Board to instigate its major post-SARS marketing recovery campaign. Singapore Roars referred to the national Merlion symbol and the explicit connection with tourists "roaring" back to Singapore. The Singapore Tourism Board was granted a SID$50 million budget increment devoted to marketing Singapore toward a post-SARS recovery. The primary message communicated, of Singapore being SARS-free, incorporated a range of inducements and discounts designed to encourage travel agents to sell and promote Singapore to their clients. Travelers visiting Singapore were made aware of an extensive range of discounted and value-added deals on tours, hotel accommodation, access to attractions, etc. Promotions targeted at the travel industry in specific source markets incorporated the Singapore Roars campaign as part of a cooperative marketing exercise with campaigns to visit Malaysia, Thailand, Hong Kong, and the Philippines. This strategy was part of a broader ASEAN and PATA approach to tourism recovery. The Singapore Roars campaign marked the first occasion for Singapore's tourism industry in which a destination-specific recovery campaign was conducted in conjunction with a broader regional tourism recovery campaign. The significant downturn of tourism to most of SE Asia irrespective of the prevalence or even presence of SARS was a common factor prompting a cooperative approach. Operational and budgetary benefits derived from sharing the costs between government and private sector stakeholders communicating a common regional message to the trade and consumers in key source markets were persuasive reasons to cooperate.

The Singapore Roars campaign involved concerted media coverage in major source markets, some of it subsidized by CNN as part of a broader SE Asian regional Project Phoenix campaign coordinated along with PATA. A consortium of 270 private tour operators, hoteliers, attractions, shopping outlets, and airlines contributed marketing funds and offered a range of special deals to expand incentive arrangements designed to reactivate tourism demand for Singapore. By November 2003, it was clear statistically that the recovery process was achieving success. November 2003 arrival figures for Singapore were 7.3% up on the preceding year. However, it must be recalled that November 2002 figures were historically low due primarily to the impact of the Bali incident on SE Asian regional tourism. The impact of SARS was pronounced on the annualized figures for Singapore's inbound tourism, which suffered a 19.04% drop in 2003 compared to 2002 or 6.127 million tourists in 2003 compared to 7.293 million in 2002 (Singapore Tourist Board Statistics, 2003).

The relative speed and success of Singapore's recovery exemplify the best elements of a professionally managed restoration marketing campaign. Singapore Roars was coordinated and centralized by the Singapore Tourism Board involving a high level of cooperation with national carrier Singapore Airlines, the private sector of Singapore's tourism industry, especially hoteliers, major attractions, and inbound tour operators. Singapore's advantage in this achievement is largely due to the fact that it is a highly regulated and compact city-state and there is universal recognition in Singapore that tourism is a strategically and economically significant industry.

Case Study 2. Bali's Recovery after the October 2002 Sari Club Bombing

The eastern Indonesian island of Bali is traditionally promoted as a destination with little reference to the fact that it remains politically an integral part of Indonesia. In the period of the late 1990s and the beginning of the twenty-first century, the Indonesian nation, which gained independence in 1949 under the slogan "Unity in Diversity," has experienced an upsurge of political instability accompanied by outbreaks of ethnic, religious, and nationalist separatist movements. The most significant campaign resulted in the creation of an independent East Timor in 2002 following a bloody 25-year guerrilla war. However, from Aceh in the northwest extremity of Sumatra to Irian Jaya in Western New Guinea (Indonesia's eastern extremity), separatist movements are now an integral part of Indonesia's complex political landscape in a country with a population of 210 million people and

12,000 islands, the sprawling archipelago dominating the region between the Malay Peninsula and Australia.

Until October 2002, Bali, especially from a tourism perspective, was a superficially idyllic anomaly in Indonesia. Bali's population and culture are predominantly Hindu, whereas 90% of Indonesia's people are Muslim. Bali's distinctive culture, as the last surviving remnant of India's former cultural hegemony over the Indonesian archipelago, is unique in Indonesia. Over the past 20 years Bali has become a popular "paradise" resort island for tourists from all over the world. Beautiful scenery, a pervasive spiritualism among the traditional Balinese, excellent surfing beaches, friendly people, and relatively relaxed attitudes to cultural pluralism are the elements that have resulted in Bali's popularity as one of the most visited single tourist and resort destinations in SE Asia. Bali is the most popular destination for foreign tourists in Indonesia.

Bali's tourism infrastructure has attracted a vast amount of foreign investment and contributed to its comprehensive and diverse tourism infrastructure, with resorts ranging from the ultimate in luxury to the basic. The influx of tourists has brought a mixture of economic benefits and social and economic problems. Bali's economy is heavily dependent on tourism but many Balinese are exploited by the tourism-based economy. Many of the resorts and hotels are owned by multinational companies and Australian or Western entrepreneurs. The dark side of Bali's tourism success has resulted in the growth of a subculture of corruption, drug dealing, and a high rate of petty crime. The combination of unregulated capitalism and the daily influx of ten jumbo loads of foreign tourists, some of whom are totally ignorant and insensitive to traditional Bali's social mores, has undermined Bali's traditional society (interview: King, 2003).

A minority of Western tourists, especially from nearby Australia, treated Bali as an exotic tropical "pub" and a relatively inexpensive resort island paradise to let off steam after a football season or to celebrate personal or communal events. The alcohol- and drug-induced hedonism of this section of the Western market has caused offense to the traditional local Hindu Balinese. For the growing numbers of Indonesians who have embraced Islamist fundamentalism, these tourists symbolize Western decadence inimical to their beliefs. Since the mid-1990s, some Islamic groups in Java have attacked nightclubs and liquor outlets as part of a wider demand to impose (Islamist) Sharia law in Indonesia. Until 2002, Bali had been spared this form of violence primarily because most Balinese are non-Islamist and such acts would harm the lucrative tourism industry.

Inbound tourism to Indonesia statistically peaked in 2001 with a total of 5,163,620 arrivals. The main source markets in 2001 were Singapore, 1.477 million tourists; Japan, 611,000; Malaysia, 485,000; Australia, 397,000; Taiwan, 391,000; South Korea, 212,000; United Kingdom, 189,000; United States, 177,000; and Germany, 159,000. Malaysian and Singaporean numbers were heavily inflated by their proximity to Indonesia and the fact that many of these visits were essentially cross-border short trips. NE Asia and Japan, Europe, Australasia, and North America were the primary sources of high-yield tourism for Indonesia and Bali specifically. Bali's capital, Denpassar, was the port of entry for 27.71% of foreign visitors to Indonesia and the principal port of entry for inbound tourists (Badan Pusat Statistik, BPS-Statistics Indonesia, 2004).

The October 12, 2002, Bali bombing had a profound impact on Balinese tourism but it would impact tourism broadly throughout Indonesia and SE Asia.

On that evening the crowded Sari Club and nearby Paddy's Bar were destroyed by powerful truck bombs. The convicted perpetrators, members of Jemaal Islamiya, a radical Islamic group allegedly linked to Al Qaeda with a substantial presence in Indonesia, readily admitted responsibility for the attack. The bombing was the deadliest attack directed against tourists in the world for many years. Of the 200 deaths, 88 were Australian, 24 British, and 10 Indonesian, and the victims included citizens from 21 other countries (Wilks, 2003).

The impact of the Bali bombing on tourism to the island was instantaneous. Tourism demand fell and forward booking dried up. The bombing exposed a number of serious infrastructure problems, the most obvious of which was the lack of medical and hospital facilities to cope with a disaster on this unprecedented scale. The lack of mortuary facilities was a factor that contributed to the major difficulties in identifying many of the deceased. The bombing also exposed the serious security shortcomings of the entire tourism infrastructure in Bali. The Indonesian government does not make provisions for adequate medical facilities to cope with the general needs of locals or tourists. The Balinese were unprepared and ill-equipped for an act of terrorism so alien to their way of life.

Many of Bali's major tourism source countries immediately imposed upgraded travel advisories. Australia, the United States, United Kingdom, and Japan advised their citizens to defer travel. Garuda Airlines and Qantas facilitated the evacuation from Bali of many travelers, especially Australians. The Australian government offered medical assistance to the Indonesian government and sent medical and rescue teams, providing Australian medical facilities and personnel to all victims of the attack.

As discussed in the introduction, the executive members of the PATA were in Bali on the night of the Sari Club bombing to celebrate the wedding of PATA Vice President Peter Semone. The PATA executives were to play a significant role in galvanizing regional travel industry support for Bali's beleaguered tourism industry. They immediately assisted local tourist authorities in establishing a crisis communications unit. PATA recruited a team of crisis management experts to assist the Balinese and Indonesians in planning a recovery strategy for the tourism industry.

The Indonesian government contracted Australian tourism crisis management specialist John King, CEO of Tourism and Leisure and the Chairman of Tourism Tasmania, to assist them in developing a marketing recovery strategy. King's significant involvement in Tourism Tasmania's highly successful campaign to restore tourism to Tasmania after the 1996 Port Arthur massacre of tourists eminently qualified him to assist in Bali catastrophe. Mr. King observed that the Indonesian government demonstrated minimal commitment to implementing the advice provided by him, his colleagues, or any other foreign experts.

The private sector of the tourism industry within Bali and regionally was characterized by a strong sense of unity of purpose. Conscious of the need to reassure tourists, hoteliers, tour operators, and local attractions upgraded security. Tour operators and airlines combined to introduce a range of incentive-based marketing campaigns and familiarization tours for media and travel industry leaders from source markets. PATA maintained its long-term plan to hold its annual conference in Bali in April 2003. Australian tourism industry organizations, including the Australian Institute of Travel and Tourism and the Council of Australian Tour Operators, chose to hold major conferences in Bali as gestures of support for the Balinese tourism industry. The Balinese regional tourism authority produced a highly professional promotional film that illustrated the distinction between negative perception and the actual experience of travelers in 2003. The film featured tourists from the main source markets telling viewers how much they were enjoying their visit and showing there was no cause for concern. The underlying theme was to communicate a message that tourism to Bali was safe.

Major obstacles to a fully integrated tourism recovery in Bali included the maintenance of negative travel advisories, negative media publicity mostly emanating from international media, and the Indonesian government's decision to conduct the trial of the Bali bombers in Bali. According to Australian, British, American, and Japanese travel industry professionals invited by the Indonesian government to assist in the recovery, the greatest barrier to implementing a recovery was the ineptitude and incompetence of the Indonesian government and the Indonesian Ministry of Tourism. The Indonesian government ignored professional advice and implemented policies counterproductive to an effective tourism recovery campaign. The Indonesian government compounded its problems from the very beginning of the Bali crisis by its failure to show due concern for the welfare of the victims and launching into attacks verging on the paranoia against governments that had issued negative travel advisories after the bombing. Indonesian government attacks against the Australian government were especially vituperative. Considering that Australians accounted for almost half the victims of the Sari Club attack, the Indonesian response to the advisories demonstrated insensitivity at best, and at worst a petulant abandonment of diplomacy. The attitude toward Australia, in particular, was colored by wider strains in bilateral relations including Timor's struggle for independence from Indonesia in 1999 and Australia's prominent military and diplomatic involvement as a guarantor of this process.

The Indonesian government was a vocal supporter of reconstituting the ASEAN tourism marketing organization. During the November 2003 summit of ASEAN leaders in Phnom Phenh, Indonesian President Megawati Sukarnoputri joined many of her fellow ASEAN leaders in criticizing Western government negative travel advisories. The Bali attack had led to the imposition of security-based negative travel advisories applying to several SE Asian countries. Indonesia was strongly supported in this forum by all SE Asian leaders, especially Malaysian Prime Minister Dr. Mohathir Mohammed.

The Indonesian government officially committed US$9 million toward Bali's tourism recovery but there was little evidence of spending on any identifiable program. Tourism, Security and Finance Ministry officials undertook "fact finding marketing missions" to source markets; two were conducted in Australia. The Indonesian delegates involved in the second visit in November 2003 were unaware of the first one, which had taken place in May 2003. One policy initiative widely condemned by local and overseas tourism industry stakeholders was the Indonesian government's decision to impose a series of visa fees. Whatever the merits of the policy, the timing of its release several weeks after the Bali bombing was condemned by private sector stakeholders who viewed it as a major disincentive to tourism recovery.

The decision to try the alleged bombers in Bali was controversial. There were legitimate legal reasons to conduct the trial at or near the scene of the crime. However, from a tourism recovery perspective, the trial served as a painful and protracted reinforcement of the Bali bombing. Understandably, the trial focused far more global media coverage on Bali than the issue of restoring tourism.

Private industry marketing campaigns led to a gradual recovery of tourism to Bali from April 2003. PATA's conference during that month focused on positive tourism attention on Bali. Bali also figured in the broad scheme of PATA's Project Phoenix. Additionally, a range of cut-price holiday deals was successful in enticing many tourists to return to Bali. The Balinese tourism authorities successfully cooperated with the private sector of the tourism industry. The anniversary of the Bali bombing was conducted in a manner intended to end to the formal mourning period (if not for the survivors and the families of victims) to this terrible event at least to non-Balinese victims. The Balinese tourism industry took almost a year to recover. Australian and Japanese markets have been slower to recover than SE Asia, the United Kingdom, and the United States.

Case Study 3. WOW Philippines Campaign

The Philippines tourism industry has a long history of overcoming a range of events that have restricted its potential to become a major tourism destination in SE Asia. Political instability, AIDS, localized terrorism, especially in the large southern island of Mindanao, natural disasters ranging from seasonal typhoons to the 1991 eruption of Mount Pinatubo, and the 1997 Asian economic crisis all contributed to acting as a brake on tourism growth. Considering all these problems, the overall trend for tourism in the Philippines during the 1990s was positive. By 2000, inbound tourism numbers to the Philippines passed 2 million, double the 1990 level (Beirman, 2003).

In 2001 the forced resignation of President Joseph Estrada led to the appointment and later election of Gloria Arroyo as President of the Philippines. Arroyo appointed Richard Gordon as Secretary of Tourism. Gordon took on the position with an almost missionary zeal to advance tourism to the Philippines as a major industry. An energetic and ambitious man, Richard Gordon sought to radically alter the tourism infrastructure and the marketing of the Philippines. He also took on a high profile in supranational tourism organizations. In 2003 he was elected as Chairman of PATA and rapidly became a powerful regional voice within the World Tourism Organization.

Gordon rallied the Philippines tourism industry by calling the first TRICON (Tourism Related Industry Conference) a few months after his appointment as Tourism Secretary in 2001. The underlying strategy to revitalize Philippines tourism was to gather an alliance of all sectors of the tourism industry and the local and provincial political leadership. The second and third TRICONs in 2002 and 2003 developed the concept of a marketing campaign called WOW Philippines. The core strategy of the campaign was to diversify both the image and benefits of inbound tourism to encompass a wide geographical area of the country. The campaign sought to appeal to a variety of niche markets, including golfing, resorts, ecotourism, meetings and conventions, business and incentive tourism, adventure tourism, diving and water sports, cultural tourism, and to identify specific regions of the Philippines as representative destinations for these activities. In the capital, Manila, the Department of Tourism actively participated in restoring specific districts of the city, most notably Intramuros (the old walled Spanish quarter of Manila) and Manila Bay's waterfront.

The Philippines Department of Tourism, in close association with major tour operators, Philippine Airlines, and other carriers servicing the Philippines, offered a range of cut-price and value-added incentives for tourists. It also sought to encourage the estimated 8 million Filipinos living abroad to act as tourism ambassadors. The Department of Tourism, supported by the national government and the local media, instituted educational and promotional programs designed to encourage locals to adopt a welcoming attitude toward tourists and to encourage hospitality and cleanliness in the county's towns and cities. The Philippines Department of Tourism claimed a high level of success for this program.

Following the Bali bombing in October 2002, there was renewed focus on Islamic activists in the southern Philippines by the global media and by the governments of some of the Philippines' most lucrative tourism source markets, notably Japan, the United States, Western Europe, and Australia. Tourism Secretary Richard Gordon was a vocal critic of negative media coverage and government advisories. The primary basis of his criticism was that the reputation of the entire country was under attack because of activities that took place in one particular locality. Gordon demanded that the media and government advisories specify the problem areas as opposed to calling the entire destination into disrepute. Gordon's demand for specificity in both media coverage and government travel advisories resonated with the governments and tourism officials in many other countries in SE Asia that shared similar problems after the Bali bombing. The Philippines inbound tourism industry was a collateral victim of the 2003 SARS outbreak despite the fact that the country was virtually free of the disease. Media-magnified fears of SE Asia as a potential risk area for SARS impacted SE Asian countries irrespective of the incidence or absence of SARS.

Richard Gordon's appointment to the chairmanship of PATA in 2003 gave him a powerful voice well beyond the Philippines. He used his regional power base in both PATA and ASEAN to exert pressure on the World Tourism Organization to take a global tourism industry approach to the issue of government travel advisories and received support from many countries that shared the problems the Philippines itself had encountered. In early 2004 the WTO's Secretary General, Francesco Frangialli, called on the WTO to initiate discussions through the UN for a global approach to realistic government travel advisories. The WTO obtained UN status in 2003, which enabled them to officially propose a global set of standards on travel advisories to the UN.

The WOW Philippines campaign involved a mixture of marketing campaigns, travel industry training and education, media and travel agency familiarization visits, and the incorporation of WOW Philippines within PATA's Project Phoenix. A unique feature of the campaign was the incentive scheme offered to expatriate Filipinos to make a home visit accompanied by citizens of their host nation. In effect, the more citizens Filipinos could identify as visiting because of their individual promotion, the cheaper the home visit would be. The expatriate program was designed as an attractive bonus/incentive scheme for the many Filipino citizens working abroad on low incomes with families in the Philippines. Provided they could identify a given number of tourists, the home visit for the expatriate would be free of charge. As Filipino nationals traditionally represented 5% of the inbound tourist market during the 1990s, the bonus scheme was a potentially effective program to bolster inbound tourism numbers. In the period of 2002–2003 overseas Filipino arrivals represented the second fastest growing sector (up 19.8%) of an inbound tourism market that as a whole actually dropped by 2.3% between 2002 and 2003 (Philippine Department of Tourism, 2003).

The overall concept of WOW Philippines was expected to be statistically significant during 2004 and early indications for 2004 already indicated a strong return to inbound tourism growth. The overall market for the Philippines remained steady during 2003 despite the collateral problems of SARS and the more pressing problem of terrorism-focused negative travel advisories. The total inbound tourism figures for 2003 were 1.907 million tourists compared to 1.932 million tourists for 2002, representing a drop of 1.3%, considerably less than most other SE Asian countries.

Of the Philippines major source markets, slight declines in US and Japanese tourists were offset by growth from South Korea. However, in 2003 the Philippines suffered drops in arrivals from significant SARS-affected source markets: Hong Kong, Taiwan, and Singapore. Outbound tourism from these three countries was heavily reduced during the period of the SARS outbreak period (March–July 2003) due to entry restrictions imposed by tourism receiving countries including the Philippines.

Conclusion

The tourism industry in SE Asia was subjected to an unprecedented degree of volatility during late 2002 and the first half of 2003. The Bali terrorism attack in October 2003, while disastrous for Bali, exerted a powerful ripple effect which impacted most SE Asian destinations, especially those with substantial Muslim populations. The Bali bombing placed an unwelcome global spotlight on the perceived and actual threat from Islamic organizations on the overall political stability and tourism safety in some SE Asian nations.

The combination of negative media attention and negative government travel advisories from major source markets in the West and Japan spurred regional and national responses in SE Asia. The outbreak of SARS in the period of May–June 2003, while largely confined to SE China, Hong Kong, and Singapore, was magnified by the media and to some extent by the World Health Organization as a public health threat, grossly disproportionate to the number of people affected by the disease. On the scale of global epidemic outbreaks, SARS was a minnow compared to AIDS, malaria, cholera, typhoid, and historical manifestations of influenza. The impact of SARS in SE Asia was a severe curtailment of tourism arrivals during the identifiable danger period.

The tourism industry response in SE Asia to SARS and terrorism involved an unprecedented degree of regional government and private industry cooperation in recovery marketing programs. The previously moribund ASEAN tourism agreement was rapidly resurrected by the initiative of ASEAN heads of state. PATA, IATA, and the WTO worked closely with the global media and government tourism bodies in SE Asia to implement a regional marketing recovery program revitalizing tourism it the region. Each country in SE Asia developed and implemented destination-specific marketing and confidence-building programs to restore tourism. The Singapore Roars campaign was highly organized, centralized, and professional and achieved rapid results. The WOW Philippines campaign successfully minimized deterioration in tourism during a challenging period for the country. The Indonesian government's campaign to restore tourism was hampered by poor leadership, incompetence, a failure to heed professional advice, and a lack of national direction. Conversely, the Balinese regional authorities implemented a successful marketing recovery campaign heavily supported by local and foreign private sector stakeholders. Consequently, Bali was less dependent on Indonesian government support than other regions in the country.

The events of late 2002 and the first half of 2003 demonstrated the vulnerability of the SE Asian tourism industry to events beyond direct managerial control. SARS and terrorism demonstrated the importance of contingency marketing and management practices to optimize recovery from crisis events. The tourism industry's private sector demonstrably and quickly learned from the many crisis events of the early twenty-first century. Increasingly, SE Asian governments with a powerful vested economic interest in maintaining a growing tourism sector have taken the issue of recovery marketing seriously. A minority of SE Asian countries, especially Indonesia, have a long way to go before they are in a position to recover rapidly from tourism crises without a high level of dependence on outside aid. The majority of ASEAN countries at both government and the private sector level are demonstrating a high level of professionalism in their strategic marketing approaches to the protection and recovery of their tourism industries and infrastructure from crisis events.

Concept Definitions

Crisis (as applied to this chapter) A situation requiring radical management action in response to events beyond the internal control of the organization necessitating urgent adaptation of marketing and operational practices to restore the confidence of employees, associated enterprises, and consumers in the viability of the destination.

Restoration alliance The working alliance between all private and public sector segments of the tourism and hospitality industry and destination marketing authorities designed to restore the marketability and security of the tourism industry. The alliance works on both restoring the image and marketability of tourism destinations and regions concurrently to sellers and distributors of tourism product and to consumers.

Source markets Markets that are the main sources from where tourists are attracted to the destination or region. This is usually determined by country of origin and measured by counting the citizenship of tourist arrivals.

Government travel advisories Assessments determined by ministries of foreign governments that focus on the security and safety of destinations. These are designed to inform citizens of the issuing country of the likely threats that citizens of that country may encounter if visiting a given destination. In most instances advisories are graded in accordance with a perceived level of threat.

General exemption Insurance companies frequently deny cover to insured travelers who incur loss, injury, or death resulting from politically motivated violence. In insurance parlance, denial of such coverage is labeled as "the general exemption." The general exemption is currently undergoing review as a result of consumer pressure.

Regional marketing In the context of this chapter, regional marketing involves the coordinated promotion and marketing of a group of countries to a commonly targeted source market or set of source markets. PATA and ASEAN employed regional marketing of several SE Asian destinations to specific source markets including China, Japan, Australia, Europe, and the United States with the aim of encouraging tourists to visit more than one of the destinations in a single trip. It also recognized that the crisis indicators that led to the coordinated recovery campaign were common to the region, as opposed to being confined to a single destination.

Push marketing Targeting the sellers, promoters, and suppliers of the tourism product in source markets including wholesalers, travel agents, airlines, hotel chains, and the media.

Pull marketing Marketing to consumers either directly or in cooperation with stakeholders such as wholesalers, the media, airlines, hoteliers, and other principals that sell or have product in the targeted destination.

Review Questions

1. Explain the recent upsurge in the significance of government travel advisories as a factor impacting on tourism to SE Asia and its broader implications for global tourism.
2. PATA's Project Phoenix represented a new approach to regional restoration marketing in SE Asia. Assess its success as a strategy and identify the key elements incorporated in the campaign.
3. In what ways has media coverage impacted the perception of tourism destinations since the development of global media broadcasting through networks such as CNN.
4. Analyze and compare the effectiveness of the three campaigns examined in this chapter: Singapore Roars, the Bali recovery program, and WOW Philippines.
5. To what extent to you attribute the reconstitution of ASEAN Tourism Marketing as a reactive measure to tourism crisis or a proactive approach to recovery?
6. How does the marketing recovery campaign in SE Asia compare to other recovery campaigns in the world, such as the US recovery campaign post–September 11, Turkish recovery after the Izmit earthquake, Britain post–foot and mouth outbreak, Sri Lanka post–civil war or others.
7. If you were a destination marketer in SE Asia how would you have approached the management of SARS and the Bali bombing? Are there approaches you would use that would be different from those undertaken in the case studies?

References

Association of South East Asian Nations. (2002, November 4). *Agreements on Tourism,* Phnom Penh. www.aseansec.org.

Badan Pusat Statistik (BPS-Statistics) Indonesia. (2004). http://www.bps.go.id/sector/tourism/tables.shtml.

Bailey, M. (2001). *China Outbound Market Report*. Bangkok, Thailand: PATA.

Chuck, Y. G., and Lurie, M. (2001). *The Story of the Pacific Asia Travel Association*. San Francisco: PATA, pp. 250–253.

Beirman, D. (2003). *Restoring Tourism Destinations in Crisis*. Sydney: Allen & Unwin, pp. 245–264.

Henderson, J. C. (2003). Terrorism and tourism. Managing the consequences of the Bali bombings. *Journal of Travel and Tourism Marketing,* 15(1), 42–43.
Henderson, J. C. (2003a). Managing a health related crisis: SARS in Singapore. *Journal of Vacation Marketing,* 10(1), 67–77.
Koldowski, J. (2003). Inbound story. Strategic intelligence group. *PATA Compass,* July/August.
Online Asia Times. (2003, July 16). Singapore girds for SARS. *Online Asia Times.* www.atimes.com.
Philippines Department of Tourism. (2003). www.wowphilippines.com.ph/dot/statistics.asp.
Sawyer, M. M. (2003). Flight of the Phoenix. *PATA Compass.* September/October, 10–11.
Singapore Tourist Board Statistics. (2003). www.stb.gov.sg.
Singapore Tourist Board Statistics. (2004). www.stb.gov.sg.
Wilks, J. (2003). *Tourism Risk Management for the Asia Pacific Region.* Bangkok: World Tourism Organization (WTO) and PATA, pp. 151–157.
Winning Edge. (2003). *Crisis—It Won't Happen to Us.* Bangkok: PATA.

Interviews

Concil, Anthony. Director, Corporate Communications, IATA Asia Pacific. Interviewed in Sydney, August 13, 2003.
King, John. CEO Tourism and Leisure, Coffs Harbour Australia. Interviewed October 2003.
Semone, Peter. Vice President of PATA. Interviewed in Sydney, May 17, 2004.

15

The Role of Security Information in Tourism Crisis Management: The Missing Link

Yoel Mansfeld

Learning Objectives

- To understand the "risk" construct in travel behavior.
- To understand the role of travel information in general, and security information in particular, in shaping destination choice and travel behavior.
- To understand the role of security and risk oriented information in crisis communication strategies.
- To understand the role of various information sources in shaping tourists' perception of risk.
- To understand the concept of travel behavior sequence and its relation to tourists' acceptable risk levels.
- To understand the risk reduction strategy based on a security information platform that is operated by the private and public sectors of affected countries and destinations.

Introduction

Risk-free travel is not just the aspiration of every tourist. In order to avoid, or at least minimize, tourism crises, it is also in the tourism industry's interest to ensure safe trips for each and every traveler worldwide. Since September 11, 2001, this fact is no more just a slogan or a minor travel consideration, but rather an uncompromising prerequisite to which each tourism destination must strictly adhere.

Since security situations and potential security hazards nowadays are not endemic to specific countries or specific regions around the world, this prerequisite has become more and more global in nature.

In pursuit of risk-free travel, potential tourists use information sources to facilitate their constructs of perceived travel risk, which they subsequently employ in their destination choice behavior. Similarly, efforts to promote tourist products and tourist destinations are normally accompanied by travel information that is made available to tourists prior to their destination choice. This cross-functional role of tourism information is quite extensively dealt with by the tourism, marketing, and consumer behavior literature. However, the literature on the specific cross-functional role of security information with regard to both consumers' perception of risk and risk management at affected tourist destinations is still very limited and undeveloped.

Hence, the aim of this chapter is threefold: first, to discuss theoretically the construct of risk in travel behavior and to assess the role of travel information in mitigating risk perception; second, to critically evaluate how and to what extent security information as a risk management tool is used or misused by the major travel information providers; and third, to develop an alternative conceptual framework for the effective utilization of security oriented information in the management of security induced tourism crises.

The Risk Construct in Travel Behavior

Consumer decisions are almost always made under a certain level of risk (Weber and Bottom, 1989; Roehl and Fesenmeier, 1992). This risk becomes much more evident in cases where consumers seek to purchase invisible or non-tangible products such as tourism (Sönmez and Graefe, 1998b; Fuchs and Reichel, 2004). Moreover, many previous studies of tourist behavior proved that destination choice is to a large extent a result of subjective evaluation of perceived destination benefits and destination detriments (Mayo, 1973; Mansfeld, 1992; Oppermann and Chon, 1997). An erroneous tourist decision might be regarded as such if at least one expected destination benefit turned out to be a destination detriment. From a consumer behavior perspective, the chance of making a wrong decision becomes a perceived risk (Fuchs and Reichel, 2004). Therefore tourists, like any other consumers, would be interested in making an effort to minimize the risk, thus helping to maximize the value of their travel experience (Mansfeld, 1992; Fodness and Murray, 1997; Lepp and Gibson, 2003). However, apparently, tourists differ in the way they conceive risk perception (Fuchs and Reichel, 2004).

Roehl and Fesenmaier (1992) defined three types of risk-taking tourists: risk neutral, functional risk, and place risk. While the risk neutral category refers to tourists who do not associate tourism and travel destinations as involving any risk, the functional risk category refers to tourists who place more emphasis on the operational side of their tour. The last category, place risk, refers to tourists who develop a risk perception based on the risk factors related to the destination and its political, social, and/or security situation. This classification is interesting since it effectively differentiates between two groups of tourists: those who develop some kind of risk perception and will probably try to mitigate it or to avoid it altogether regardless of whether the perception is "place" or "functionally" induced. The other group looks for some level of risk in their travel experience. For this group, such risk is regarded as an added value that contributes to the overall fulfillment of their travel motivations. It may be assumed that even if risk is perceived as a positive ingredient in tourists' destination choice, it is still a calculated risk. When assessing the proportion of these two groups it would seem that the majority of tourists belongs to the functional and place related risk categories. Those residual risk seekers tend to share common sociodemographic and cultural characteristics that have a relatively short life span and are mainly novelty seeking, for example, young backpackers (Elsrud, 2001; Uriely et al., 2002; Lepp and Gibson, 2003; Pizam et al., 2004). Lepp and Gibson (2003) examined a less obscure type of tourist. In their study of risk perception among young American tourists, they found that those preferring either the organized or independent mass tourism type of travel are more apt to react to travel risk hazards. Their conclusion was that when confronted with a given travel risk, they would be the first to cancel their travel plans (Lepp and Gibson, 2003, p. 620). Since mass tourism is the major driving force that sets in motion the wheels of this industry, it is vital to further explore not just how risk is perceived by various types of tourists but more importantly how to manage it. In another recent study that focused on the perception of Israel as a risky destination, Fuchs and Reichel (2004) found that level and quality of perceived risk also diverge along cultural and nationality lines. They also discovered that risk reduction strategies by those who took the risk and visited Israel in times of security turmoil also vary significantly according to religious and nationality backgrounds. These findings and others mentioned above shed light on the complexity of the risk perception phenomenon and on its centrality in managing security induced tourism crises.

What actually constitutes the most frequent perceived travel risk? Apparently, it can be attributed to a wide array of uncertainties regarding tourists' ability to fulfill their travel motivations without being exposed to unfortunate situations. The main possible adverse situations documented in the literature are natural disasters, unexpected extreme weather conditions, outbreaks of epidemics, political unrest, crime, terror events, wars, and other possible security situations (Richter and Waugh, 1986; Enders et al., 1992; Gartner and Shen, 1992; Cossens and Gin, 1994; Bar-On, 1996; Mansfeld, 1996; Wall, 1996; Mansfeld, 1997; Carter, 1998; Sönmez and Graefe, 1998a; Sönmez and Graefe, 1998b; Sönmez et al., 1999; Brunt et al., 2000; Santana, 2001; McKercher and Hui, 2003; Fuchs and Reichel, 2004). In most cases, such events are not totally unforeseen (Faulkner, 2000). However, if tourists are unaware of a given probability that such events might occur, they could make a travel decision that would put them at some level of risk or might endanger their ability to fulfill their travel motivations and expectations (Thapa, 2003).

What is actually is expected from the consumer perspective in order to avoid, or at least to minimize, these risks to an acceptable level? The literature that covers destination choice and travel behavior points to the availability of (formal, informal, and experiential) travel information as one of the key risk reduction strategies (Mansfeld, 1992; Fodness and Murray, 1997; Sönmez et al., 1999, 1998b; Baloglu and MaCleary, 1999; Jenkins, 1999; Fuchs and Reichel, 2004). When purchasing a tourism product, a consumer cannot manipulate his or her basic senses in order to evaluate its value. Travel information collected by the individual actively or passively from various sources might, therefore, be used as a substitute. Obviously, the availability of travel information, the level of tourists' exposure to it, its nature, its quality, and its sources' characteristics all play a major role in shaping the consequent perceived image of the tourist product and, hence, the risk characteristics involved in purchasing it (Mansfeld, 1992; Jenkins, 1999).

Travel information is not only sought by travelers to support their risk-free destination choice. The tourist industry is also preoccupied with the attempt to establish a risk-free image to their destinations (Jenkins, 1999). Gunn (1972) was one of the first researchers of travel destination perceptions who noted that side-by-side with the "organic images" that are formed by non-tourism mass media sources, there exist "induced images" that are shaped by tourist oriented promotional agencies. In fact, once tour operators realized the relationship between image creation and actual travel patterns, they have been investing substantially in efforts to influence travel behavior through promotional material (Jenkins, 1999). However, despite the critical role of travel information in destination image formation, the majority of travel information that is produced for and conveyed to tourists is based on hard sell strategies. These stem from marketing segmentations reflecting pleasure preferences and travel motivations rather than risk perception levels and risk characteristics (Fodness and Murray, 1997; Pearce, 1988; Jenkins, 1999).

Communication of Security Information: A Vital Strategy in Tourism Crisis Management

Tourism crises are almost inevitable as they are generated in many cases by exogenous factors that are beyond the control of the tourism system itself (Brownell, 1990; Santana, 2003). However, the extent of the crisis, the level of resilience of the affected tourism sector, and the recovery phase can be controlled through an implementation of appropriate crisis management strategies. Effective management of a given tourism crisis is dependent on many factors such as the availability of reliable contingency plans, the availability of contingency funds, the level and type of public and private sector cooperation, and concerted efforts to change the situation by all tourism stakeholders (Mansfeld, 1999). Indeed, regardless of the theory used to explain the mechanism and the nature of tourism crises, there is common understanding that when they occur they force decision makers and managers to exert control over this sector in pursuit of balancing the tourism system through regenerated tourist flows to an affected destination (Coles, 2003; Prideaux, 2003; Santana, 2001).

One of the management strategies used to attain this end is security communication, which conveys security information to tourists and tour operators in the generating markets (Mansfeld, 1999; Ritchie et al., 2003). The aim of communication management is to limit, control, and balance the negative information conveyed through the media before and during the crisis and throughout the crisis recovery phase. Thus, through proper security communication management the negative effect of such crises on the receiving destination might be substantially mitigated and the destination's image may be safeguarded (Ritchie et al., 2003; Baral et al., 2004). However, the purpose of such a communication strategy is not always clear and well defined by those who advocate it. Previous studies of security induced tourism crisis have mentioned that advertising and PR campaigns have normally been integrated into crisis management plans. Affected destinations such as Croatia, Egypt, or Israel normally allocate a contingency budget to control image deterioration and to reactivate tourist flow once a security crisis seems to be over. However, the efficiency of these campaigns has never been thoroughly investigated. These communication campaigns were aimed at generating quantity of marketing information rather than quality security oriented information. In other words, the strategy has been to focus on the destination's touristic qualities, while ignoring the need to change tourists' perceived risk by conveying security information (Mansfeld, 1999). After all, personal security has become in recent years a leading factor in tourists' overall judgment of destinations' level of attraction (Hall and O'Sullivan, 1996; Poon and Adams, 2000; Cavlek, 2002).

In recent years, therefore, various studies concluded that the provision of information on the security situation and on the level of risk to travelers to the affected destination should be regarded as a prerequisite, and thus integrated into any crisis-management plan (Cavlek, 2002). This information should be communicated by the various information agents right after a security situation has commenced and must flow constantly, while providing updated information on the events accompanied by proactive risk assessments (Mansfeld, 1999; Thapa, 2003; Cavlek, 2002). Based on an analysis of many tourism crises in the past two decades, Cavlek (2002) highly recommends the improvement of accurate information flow between local agencies and the media. In fact, her information provision strategy is based on an effort to control the biased reports by the mass media by supplying the media with information on the security situation and on the steps taken to contain it (Cavlek, 2002, p. 488). Her postulate that the local tourism industry in affected destinations must adopt a proactive approach to security and risk information provision is logical and imperative. However, the assumption that the media will be more balanced just because it is provided with accurate information on the security situation seems to be naïve. After all, the mass media is known to prefer the negative, sensational mode of coverage even if this means conveying a distorted image of the actual security situation (Weinmann and Winn, 1994). An alternative, and what seems to be a more innovative and effective approach, is to establish new channels of security information that will reach consumers and tour operators alike through other modes of information dissemination, such as the Internet.

In any event, the recommendations on the use of security oriented information as a crisis-management strategy are still much too generalized and are not related to the situational constraints and to the choice behavior that characterize both tour operators and potential tourists. Therefore, there is a need to develop a more detailed and practical model that takes into account the situational characteristics and the constraints imposed on tourists and tour operators while engaged not only in the destination choice process but in all other stages of the travel behavior sequence. A conceptualization of this framework will be developed in this chapter after the current pattern of communicating security information has been characterized and critically analyzed.

Security Related Travel Information: Sources and Communication Patterns

Communication of official information on security and risk related issues that might affect tourists is currently distributed by four possible information sources:

- Security and risk information issued and communicated by governmental agencies in the generating markets (in the form of travel advisories);
- The global and local mass media;
- Governmental tourism organizations in the affected receiving destination; and
- The travel industry in the generating markets (Lepp and Gibson, 2003).

Following is an analysis of the way in which each information source functions and its advantages and limitations in conveying accurate, comprehensive, fair, dynamic, and up-to-date security related information to tourists, travel agents, and tour operators.

Travel Advisories

Governments in various generating markets issue travel warnings as part of their policy to protect their citizens and to save them from exposure to any sort of risk in their international travel. These travel warnings are issued to make potential travelers aware of the possible consequences when traveling to an insecure destination (Beirman, 2003). Practically, the information is compiled and communicated either by a central government ministry such as a ministry of foreign affairs or by embassies situated in an affected country. Travel warnings are in most cases ranked on a scale from "precautions are needed" to "strictly no-go destinations." Most travel warning providers will communicate their information through the mass media and by using the Internet. The security and risk information is dynamically updated following frequent security reassessments and as security events crop up. In most cases, travel warnings cannot legally prevent people from taking the trip and should be regarded as recommendations only. However, in the past decade they have practically become an important travel filter since insurance companies and large tour operators base their business policy on this mode of governmental risk assessment.

There is no doubt of the legitimacy of issuing governmental travel warnings in order to protect the life and health of a given country's citizens. However, a few researchers have raised some reservations regarding the accuracy and comprehensiveness of the security and risk information that is communicated to potential travelers. In their mind, governments prefer to convince people to avoid affected destinations altogether rather than deal with tourists trapped in the midst of some kind of security situation in a foreign and not always friendly country. Such a risk management strategy is more convenient from the government's perspective and protects governments from possible future liability lawsuits. However, if this preferred policy is implemented, it requires "massaging" of the objective security information and a certain level of distortion of the security reality on the ground. While this is a common mode of conduct by the mass media, if implemented by governments it could be perceived as violating the basic principles that drive international tourism, that is, reciprocity and the freedom of movement (Cavlek, 2002; Beirman, 2003).

Indeed, the academic literature has documented various cases where travel advisories caused an almost complete devastation of the local economy of the affected destination (Beirman, 2003). Thus, for example, Sönmez and Graefe (1998a) quoted Sharpley and Sharpley (1995) who investigated the impact of the British Foreign Office warnings to avoid any visit to Gambia. Apparently, the Gambian case is one of the worst examples, where travel warnings brought inbound tourist flow to a standstill, hence causing a domino effect that adversely affected the entire economy. However, researchers suggest that other countries have been badly affected by these warnings too (e.g., Mansfeld, 1999; Cavlek, 2002; Beirman, 2003).

Convincing or instructing tourists not to go to security affected destinations is the official task of travel advisories. However, unintentionally, they might end up functioning as catalysts for security upgrade and improvement in destinations hit by terrorism or other security events. Beirman (2003) observed, for example, that in the face of very explicit travel warnings issued by Western governments to avoid Egypt after the 1997 Luxor terror attack, the Egyptian government was very anxious to encourage their early revocation (Beirman, 2003, p. 81). Egypt immediately increased security measures in all major tourist attractions and invited foreign diplomats to see how the security situation had improved. This caused a quick reaction by those governments who rescinded their warnings, allowing Egypt a quick recovery from this crisis. This example supports Cavlek's (2000) postulate that travel advisories are perceived by affected countries as highly effective, and, therefore, should not be ignored by the receiving tourist industry and governmental tourism agencies. However, bearing in mind their occasional use of partial, biased, and/or intentionally blown-out-of-proportion information, the affected destinations must react in two possible directions. One is to provide tourists, travel agents, and tour operators in the generating markets an alternative, parallel, accurate, up-to-date, and genuine information platform of their security situation. The second measure would be to "push" its security assessments and risk evaluations proactively to those in charge of travel advisories in the generating markets. This concerted effort might help in defusing the destructive potential of travel advisories, while maintaining their reputation as a reliable risk-prevention mechanism among the generating markets.

The International and National Mass Media

The majority of security crises around the world are documented and brought to the awareness of tourists by the mass media (Wahab, 1996; Frisby, 2002; Thapa, 2003). Fodness and Murray (1997) suggested that what really influenced tourists' choice is not so much the type of the information source but rather its determinants and qualities. Indeed, the mass news media, in particular, has a major role not only in conveying information on security issues but also in interpreting it. Hence, it plays a crucial role in the creation of public opinion and the opinion of individuals. It is claimed that the opinion or image created in the wake of the news media is shaped by the frequency of its report and interpretation and by the type of media used (Wood and Peake, 1998). In fact, in the case of terror events, which are the most detrimental causes of security-generated tourism crises, the mass media is heavily utilized and manipulated to put across political and ideological manifestations (Buckley and Klemm, 1993; Weimann and Winn, 1994; Hall and O'Sullivan, 1996; Weimann, 2000; Pizam and Smith, 2000). When such terror events involve targeting tourists and tourism facilities, the mass media plays an even more influential role in shaping tourists' perception, awareness, and attitudes toward these affected destinations. Moreover, in cases of recurrent and frequent security situations, the media's continuous coverage and interpretation of the conflict deepens the fixation of a long-term negative image of such affected destinations (Iyengar and Kinder, 1987). Countries such as Israel, Egypt, Colombia, and others have been experiencing this outcome, namely, a strong and persistent negative image that maintains a very low tourist demand and lack of propensity to invest in their deteriorating tourism sector (Wahab, 1996; Mansfeld, 1999; Pizam and Smith, 2000). Further advancement of information technologies used by the global media means that security events will be continuously brought to potential tourists in real time and in the most graphic way (Pizam and Smith, 2000). These sensational live reports are often regarded by the travel industry as the cause for hysterical and disproportionate travel behavior—mainly in the way of trip cancellations (Schiebler et al., 1996; Lepp and Gibson, 2003). Such mass media coverage deepens the exclusion of affected destinations from the global tourist map unless other information and communication modes are to be used to balance the lead of the media as a security information provider (Baral et al., 2004).

Sönmez (1994) and Sonmez et al. (1999) also recognized the importance of mass media in affecting a destination's image, and called for further research examining the relationship between the two. However, Sönmez and Graefe (1998b), in their study, discovered that media is just one of the external factors that shape international tourists' risk perceptions. Therefore, since awareness and perception of risk are eventually interpreted and translated into some kind of behavior, it is vital to understand not just their relationship but to examine the relative and inter-relative role of the media vis-à-vis other factors that contribute to the overall destination perception.

National Tourist Organizations

The preparation of a priori contingency plans to handle security induced tourism crises is not, in most countries, part of the governmental management culture (Mansfeld, 1999; Prideaux, 2003; Santana, 2003). However, post priori and in an ad hoc manner, many national tourism organizations and tourism ministries around the world have played a major role in initiating and coordinating crisis management operations to mitigate the negative impacts of security situations in their respective territories (Mansfeld, 1999; Frisby, 2002). Their involvement in crisis operation focused on co-funding and coordinating marketing and PR campaigns to regenerate tourist demand once the security crisis was over (Wahab, 1996; Mansfeld, 1999; Baral et al., 2004). However, in most cases this organizational involvement in crisis management was reactive rather than proactive, ignoring the necessity to adopt a long-term crisis-management approach. Mansfeld (1999) argued that the policy of ignoring the cyclic nature of many security induced tourism crises could be attributed to the urge to save the cost involved in continuous marketing and security communication for potential markets. Evidently, such an ongoing crisis-management strategy might be costly but is imperative in security affected destinations that experience detrimental irregularity of tourist demand as a result of cycles of security situations (Sönmez et al., 1999).

In some cases, the involvement of national tourism organizations is problematic not only because of its shortsighted reactive characteristic. This involvement also tends to be very slow and cumbersome and tends to supply uncoordinated information on the crisis as well as on the level and risk involved in taking a trip to the affected destination while the security situation is still on (Richie et al., 2003). One of the constraints imposed on national tourism organizations is the effect of uncertainty as to how the crisis will develop and how long it will last. Due to budgetary limitations, the tendency of these organizations is to assume that the life span of a given security induced tourism crisis will be short and that they can count on tourists' short memories about such events (Mansfeld, 1996; Wall, 1996; Mansfeld, 1999; Prideaux, 2003; Richie et al., 2003). Therefore, the unanticipated additional budget needed to communicate with potential tourists and convey crisis oriented information to them in a comprehensive, constant, and accurate manner is not normally allocated (Baral et al., 2004). This strategy of "Let's wait, do nothing, and see if the situation improves" is counterproductive and contributes to further deterioration of affected destinations' safety and security image (Sönmez et al., 1999; Beirman, 2003).

It was noted earlier in this chapter that a security image is formed not only among potential tourists but also among tour operators in the generating markets. Since their decision to include an affected destination in their promotional material is critical to many receiving countries, it is imperative that national tourism organizations will convey their crisis communications on this level too. However, such contingency operations should, yet again, be carefully planned before a crisis situation evolves, and taken off the shelf for implementation immediately after a security event has begun (Prideaux, 2003). If such a policy is adopted, as was the case in Cyprus in the aftermath of the 1974 civil war; in Fiji after the 1987 and 2000 political coups, and in Egypt after the 1997 Luxor massacre of tourists, it could increase the chance of faster recovery. However, documentation and analysis of a previous tourism crisis made by Beirman (2003) shows that concerted marketing campaigns are not enough. Even if Tourism Action Groups (based on the Fijian strategic model) react quickly with carefully prepared marketing efforts, their actions cannot be efficient enough unless accompanied by risk and security communication systems provided for the generating markets by the tourism authorities of the affected destinations (Mansfeld and Kliot, 1996; Mansfeld, 1999; Sönmez et al., 1999; Berno and King, 2001; Bierman, 2003).

The Tourism Industry in the Generating Countries

Traditionally, it has been the role of the travel agents to act as a link between the tourist and the travel industry and to advise their clients on a variety of destination characteristics, including security issues (Hsieh and O'Leary, 1993). This is expected of them, since they are widely perceived not just as booking clerks, but predominantly as travel consultants who are supposed to help customers locate a quality and risk-free tourist product. Despite the recent growth of Internet usage as an alternative communication between customers and the industry, many studies still perceive the travel agent as an important opinion shaper who can provide accurate up-to-date and comprehensive information to support the destination choice process (Lawton and Page, 1997).

Lovelock (2003), in his study of New Zealand's travel agents practice of advising tourists on travel risk, developed an interesting model that explains how external and internal factors affect travel agents' perception of risk. Using this model he subsequently explained that travel agents' perception of risk is differentiated.

However, he concluded that personal or workplace characteristics could not be used as good predictors of such different risk perceptions. Also, his findings suggest that the exposure to media and past experience of travel agents are contributing factors for their understanding of the characteristics and level of risk perception (Lovelock, 2003, p. 274). Moreover, Lovelock claimed that travel agents are often caught between two sets of ethics: the humanitarian one that advocates not sending tourists to dangerous destinations, and the business ethic that calls for a hard sell regardless of the risk involved (Lovelock, 2003, p. 277). In times when the travel agency sector has been experiencing many business obstacles, mainly in the form of commission cuts and fierce competition, there is a high probability that travel agents will lean toward business survival over their customers' safety. Therefore, though it is clear that travel agents can act as a possible information channel of security related information, it is currently impossible to conclude to what extent they can be perceived as a reliable, updated, and comprehensive source by their customers. If this is the case, should the tourism industry still rely on travel agents as a reliable and effective information provider in times of security induced crisis? In light of the hitherto mentioned business and ethical limitations, some kind of more rigorous security information provider should be considered.

On the wholesale level it has also been expected that tour operators, who are in charge of designing and putting together the tourist product, will avoid selling high-security-risk destinations. Through their promotional material, they are also expected to make the customer aware of the potential of experiencing security situations and the ways to avoid them (Cavlek, 2002; Lovelock, 2003). In some countries, such as within the EU, tour operators bear a legal responsibility with regard to their clients' safety and security (Cavlek, 2002). Thus, their safety obligation to their customers is not a case of moral responsibility but rather a legal requirement. Consequently, tour operators must take any possible measures to avoid the exposure of their clients to any kind of safety and security risk. Hence, in times of security crises, they will be the first to react and will either evacuate their guests, exclude the affected destination from their travel brochures, stop operation in destinations already included in their products, or temporarily relocate their traveling clients (Cavlek, 2002). It is important to appreciate the consequences of such steps since tour operators are an important link in the tourism chain production. They are in charge of producing the opportunities for mass tourism to travel and reach destinations worldwide. As such, they have enormous economic power, influencing not only tourist experiences, but also the economic wealth of destinations. Therefore, the safety and security image they convey could have a major impact on affected destinations.

The fact is that generally the large international tour operators have always preferred the easier solution, namely, to call off operations in affected destinations until the security crisis is over. This kind of step was normally explained as a normative reaction to travel warnings issued by government agencies in the generating markets (Cavlek, 2002). Such a strategy protects tour operators from the legal consequences and in some cases prevents surging insurance premiums. But, at the same time, it might create major damage to its reputation when it faces frustrated clients who could not fulfill their travel plans just because their tour operators refused to be exposed to potential legal action. Moreover, the money lost as a result of hampering operations with a given affected destination can be detrimental to the tour operator. On top of these potential losses, Cavlek (2002) mentions another interesting constraint that might convince tour operators to maintain operation in affected destinations instead of canceling them. She claims that in some cases, tour operators feel more committed to affected destinations. This commitment most often stems from vertical integration that puts tour operators in the generating markets under the same umbrella of ownership as tour operations, hotels, and other tourist services in the receiving destinations. This fact, according to Calvek (2002), explains why Turkey and Spain were not dropped out altogether from the major European tour operators' brochures, despite being hit by recent terror activities (Calvek, 2002, p. 486).

To sum up this section, despite the general understanding of the role and function of the above analyzed security information, not much has been written on the way that security information is conveyed and to what extent it actually shapes tourists' travel behavior (Lawton and Page, 1997; Lovelock, 2003). Moreover, since there are no comparative studies examining tourists' use of security related travel information from different information sources, a few important research questions still remain open. Thus, it is still unclear how tourists' background, location, and situation shape their search and consumption of security information. Furthermore, it is important to further explore the relative importance of the available security information sources and to try to ascertain which of them is perceived to be more accurate and more reliable by tourists of different backgrounds. In order to lay the foundation for these research directions, there is first a need to conceptualize tourists' choices at each stage of the travel sequence, and to understand what security information they need in order to control their perceived risk and thus successfully pursue their travel plans. This conceptual framework will be discussed in the next section.

The Role of Security Information as a Risk Control Mechanism along the Travel Behavior Sequence

The theoretical review on the risk construct and risk reduction strategies among potential tourists, discussed above, proved that the majority of researchers confined their discussion on the perception of risk and its manifestations to the stage where tourists are engaged in destination choice. Sönmez (1998b) was one of the very few researchers who concluded that tourists might change their risk perception after a destination had been chosen, and attain levels that could cause drastic changes in their subsequent travel behavior (Sönmez, 1998b, p. 125). Following Sönmez's (1998b) observation, it is claimed here that the relationship between security induced risk perception and travel behavior should be studied on a travel behavior sequence basis, rather than on a single destination choice level. Moreover, it is argued here that tourists' risk perception and resultant risk reduction strategies are dynamic and change along the travel behavior sequence if risk factors have changed too. If these postulates are correct, the role of security oriented travel information and communication of this information should also be studied with special reference to these possible dynamics. The following discussion will define the concept of a travel behavior sequence and will discuss how tourists might alter their travel behavior should security risk factors change along this sequence. Once these ideas are formulated, the next section will explore how proactive, dynamic, and comprehensive security information, generated at the receiving destination, can be used in order to construct a substantially more balanced risk perception.

Tourists' dynamic risk assessment takes place in four basic stages of the travel behavior sequence. Figure 1 illustrates these stages, the possible risk levels involved, and the perceived risk directions as the tourist moves from one travel behavior stage to the next. The sequence of the four bases of tourist behavior stages starts with the *destination choice process,* which is triggered by travel motivation and ends with a choice of the preferred travel alternative. The second stage, the *in between period,* starts immediately after the tourist has acted upon the destination choice and booked the holiday, and ends when the tourist takes the actual trip. The third stage, the *on-site period,* starts when tourists commence their holiday at the destination until they travel back home. The last stage, the *pre-next trip period,* takes place upon returning home and just before starting to plan the next tourism experience (see Figure 1).

All along this travel behavior sequence, tourists use their subjective acceptable risk threshold (ART) and compare their dynamic perceived risk of a given destination against it. If circumstances such as the security situation deteriorate at a given destination that is under their consideration in the first stage (i.e., the *destination choice process*) their perceived risk of that destination might drop below their ART. If this is the outcome, this affected destination will evidently be eliminated from their destination search spectrum. In the same way, if the security situation in the chosen destination deteriorated after the choice had been made, it might drop tourists' perceived risk image below their ART. This will most probably compel these tourists to take some kind of drastic action such as cancellation of their booking if they are in the in between period stage, evacuation of their holiday destination if they are in the on-site period stage, or to avoid consideration of

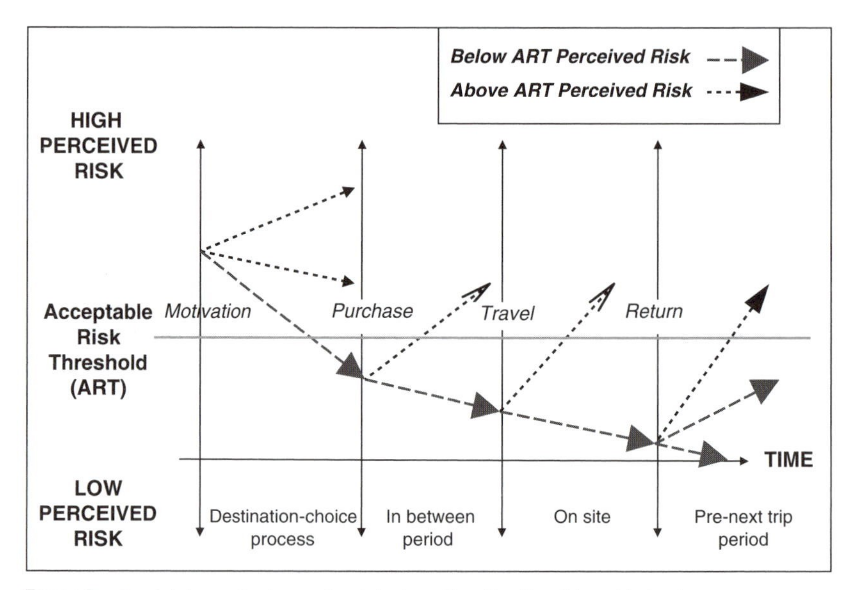

Figure 1 Tourists' perceived security risk along the situational travel behavior sequence.

this destination as a possible travel destination for their future holiday if they are at the pre-next trip period (see Figure 1).

It was discussed earlier in this chapter that many researchers came to the conclusion that security oriented information is one of the most important perceived risk reduction factors. In fact, security information helps the formulation not just of the perceived risk but also each tourist's ART. However, while perceived risk is highly influenced by dynamic changes in risk generating factors (such as security events), ART is less susceptible to occasional security events and is built into our cognition over a longer period than the tourist travel cycle. Thus, any intervention to keep tourists committed to their travel choice in each travel sequence stage should focus on efforts to maintain their risk perception below their ART.

What sources of security information are used at each stage of the travel behavior sequence? While global mass media and travel advisories provide continuous and situational security information during all stages of the travel behavior sequence, governmental tourism organizations in the affected receiving destinations and the travel industry in the generating markets provide only reactive situational information when security events take place and sometimes when the security crisis is over. However, the latter do not provide it on a continuous basis and do not operate an ongoing proactive information source that could balance the biased, partial, and distorted information communicated by travel advisories and the mass media.

A Dynamic Risk Management Framework Based on a Proactive Platform of Security Information

Following the theoretical discussion to this point, it is evident that destinations affected by security situations suffer from a major crisis-management flaw. Simply put, they are not engaging their public and private tourism sectors in providing any kind of balancing proactive security information and communication to tourists, travel agents, tour operators, the mass media, and travel advisories in the generating countries. Had they created such an information platform, and had they communicated security information to all the above stakeholders in the generating markets, they could have much more effectively balanced the risk perception among those generating markets—risk that at present is predominantly shaped by the mass media and by the travel advisories. Moreover, as was thoroughly explained in this chapter, risk perception does not stop influencing travel behavior once a destination is chosen by tourists. Therefore, this imperative security information source must be aimed at tourists at all four stages of the travel behavior sequence.

Figure 2 illustrates a proposed functional framework of such a security information platform. First, it depicts the components that should generate, process, and interpret the currently missing security information. Second, it differentiates between two security information clients: on the one hand, tourists in the four possible stages of the travel behavior sequence; and on the other hand, public and private tourism sectors, travel advisories, and the mass media in the generating markets.

One of the challenges of such security information platforms is to produce and deliver security information based on a variety of characteristics. It has to be balanced, objective, comprehensive, integrative, evaluative, interpretive, and, furthermore,

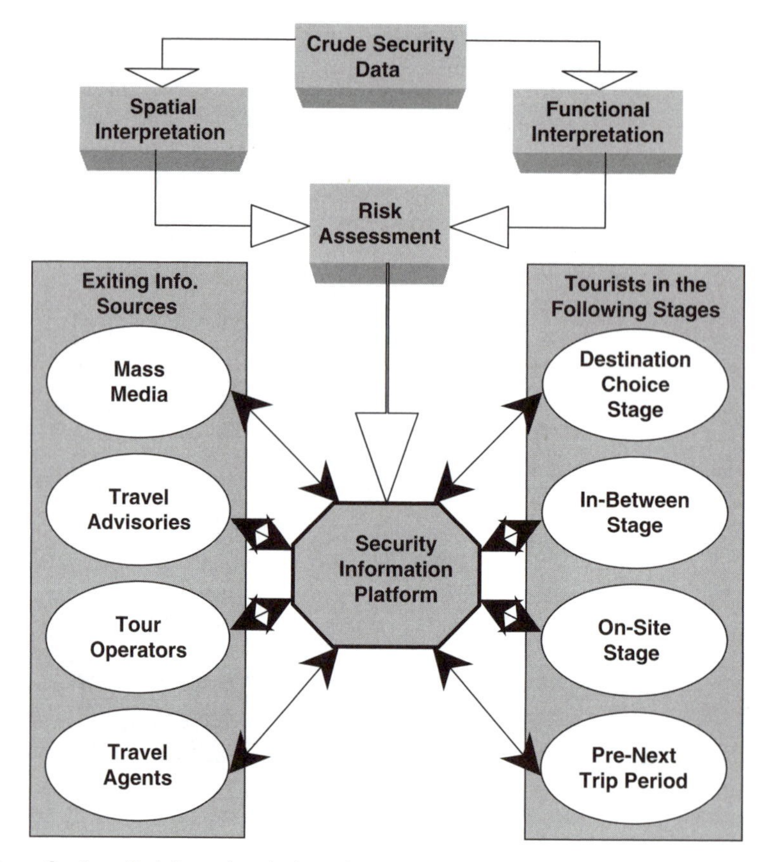

Figure 2 Security information platform: functional framework.

supply relevant information that can better enable tourists and prospective tourists on the one hand, and the tourist industry in the generating countries, on the other, to construct a balanced risk perception. The framework suggested here advocates filtering the crude security information through two prisms, each confronting the security data with its tourism as well as its spatial implications. Thus, it is important to evaluate to what extent security situations can influence the freedom and safety of movement in a given geographical region. Furthermore, to what extent do affected regions overlap spatially with the location of tourism attractions? By the same token, it is important to assess to what extent a given security situation can hamper the operation of tourism facilities and services in the affected regions and in their surroundings. Once these two evaluations are made, the filtered security information moves to the phase of constructing an overall risk assessment, which is then conveyed to the wide array of information clients (see Figure 2).

In terms of the *communication medium* preferred for this information platform, the Internet would seem to be the most appropriate option. This medium is an optimum solution since it can be updated frequently and kept up-to-date as security and risk assessments change on the ground. The Internet is highly accessible to the travel industry, governmental agencies that issue travel advisories, and the mass media. Tourists are also becoming more and more exposed to the Internet, and, in fact, already use it quite widely when consuming travel products. However, for those with limited or no access to the Internet, the security information produced by the receiving affected destinations could be alternatively obtained through their travel agent. It is proposed that this information platform on the Internet will reside on the affected country's national tourism web site and will be accessible through all the major Internet search engines. With the cooperation of all tourism providers in the affected country, all their promotional web sites will have links to this security information platform. This will improve its accessibility and make it a more widely used tool.

As far as *mode of operation* is concerned it is suggested that this security information source will be proactive. In other words, it should provide security information and produce risk assessments all year round regardless of level of risk involved in visiting a given affected country. A proactive approach assumes that even when a security event is over, the crisis is not, since recurrent security events are becoming more and more common. Also, when adopting a proactive approach, the information such a source is producing and communicating entails high levels of perceived reliability by all its potential customers. Another important feature of such security information sources is its dynamic operation. In other words, it must be updated on a daily basis; not just in terms of up-to-date security information but also in terms of its transformation into risk assessment for travelers. The information platform needs to be dynamic in the sense of being a two-way communication channel. Thus, it should allow prospective travelers, tourists in the in-between stage, on-site tourists, and returning travelers to consult it by sending their queries using e-mail and virtual discussion groups. By functioning also as a tailor-made information source that caters to individuals and not just for the public at large, this push-pull operational mode will enhance the credibility and reliability of tourists and the travel industry in the generating countries.

Another important feature of the security information platform is its proposed *ownership and operational responsibility*. It is suggested that the responsibility for obtaining, compiling, interpreting, and disseminating the security information will be shared equally by the receiving tourism industry and receiving governmental tourism agencies. Funding of this platform should also be equally shared by both parties. Again, such joint ownership and operational responsibility will prove that information is not misinterpreted and that it professionally translates security information into travel risk assessment. Joint ownership also conveys a clear message that commitment to tourists' safety comes before any urge to promote and sell affected tourist destinations at any cost.

Once the information on the current security situation is processed through all the components that feed the information platform, it is communicated, as explained earlier, to prospective and current tourists as well as to the mass media and travel advisories. By complementing and balancing the information issued by the mass media and by travel advisories it will allow all tourism stakeholders in the generating markets to make better decisions and to base their risk perception on a more solid informational ground. Establishing such an information source will also be an important manifestation of the fact that the public and private sectors in affected destinations *do* take responsibility over the fate of their tourism industry. Furthermore, it is a cross-sectoral declaration that both are cooperating to ensure survival of the tourism industry in times of security crisis. And on top of all the previous arguments, such a platform could convince the generating tourism markets that a security situation should not necessarily cause an immediate cancellation of trips. It will help convey a message that risk to travelers changes over time and space and if these phenomena are brought to the tourist and to the travel industry, they can still take the necessary precautions and tour these countries in a safe manner instead of avoiding them altogether. Thus, such a security information system can contribute to the survival of affected destinations and to avoidance of long-term economic hardship to host communities.

Conclusion

The study of security induced tourism crises has generated many case studies that successfully characterized and analyzed those crises. However, as indicated by Sönmez (1998a), Cavlek (2002), and others, much less theoretically developed is the discussion on how to effectively manage those crises. This chapter made an effort to explore one component of tourism crisis management, namely, the significance and role of crisis communication and security information.

Following an in-depth and critical theoretical discussion on the "risk construct," the chapter placed the role of security information and its communication in a crisis-management conceptual framework. The discussion was then directed toward a theoretical analysis of how the existing security information sources operate vis-à-vis tourists and the tourism industry in the generating markets. Subsequently, the discourse on the role of information in risk reduction strategies exposed the main current security information sources, and explained why they could not be left as the sole actors of crisis communication in times of a security crisis affecting a given tourism industry and local communities. That discussion was then related to the concept of travel behavior sequence. It was argued in the chapter that until recently risk reduction strategies concentrated mostly on potential travelers who are engaged in a destination choice process. This chapter advocated an alternative approach, calling for the extension of such strategies to also target would-be travelers who had already booked their holidays, tourists experiencing their holidays on-site, and those returning from their recent trip. It was claimed that security information should reach and become available to these travelers too. After all, they could also be susceptible to security situations and thus might take drastic measures and cancel their booked trip or evacuate destinations as their perceived risk levels cross their acceptable risk threshold. It is further claimed that, in many cases, such drastic measures by tourists can be avoided if they become exposed to accurate, balanced, comprehensive, up-to-date, interpretive, and evaluative information on the security situation and the ways it changes over time and space. A framework conceptualizing a proposed security information platform is exposed and explained in light of the need for such security information. This concept calls for the dynamic and proactive supply of security information by the affected receiving countries that have not explored this path in their crisis-management strategies so far. Thus the main conclusion of this chapter is that if such a risk reduction strategy is adopted by affected countries, the loss of income, clientele, jobs, and reputation as a result of security induced tourism crises will be substantially mitigated. Obviously, this conclusion is based on the theoretical review provided above and on the consequent conceptual development in its wake. In order to test the improved efficiency of crisis management strategies after such a security information platform is introduced, there is a need to implement this idea and to examine its practicalities.

Concept Definitions

Perceived tourist risk A cognitive state of mind that, if beyond an acceptable level, might affect travel behavior in various ways mainly through nonbooking, cancellations, and evacuation of affected destinations.

Security information Any information (formal or informal) on the security situation in a given affected destination that originates from either the generating markets or the receiving destination.

Security communication Conveying information on the security level and security situation in a given affected travel destination through various communication channels such as the mass media, travel advisories, or travel agents.

Acceptable risk threshold (ART) An individual level of perceived risk that marks a possible change in tourists' behavioral pattern. If one's perceived risk is higher than the ART then that person will either avoid booking a given affected destination; cancel the trip if the booking had already been made; or evacuate the destination if in the middle of the holiday.

Proactive security information platform A security information and risk evaluation system operated by the affected destinations that is supposed to balance the distorted security image generated by other information sources and thus contributes to an effective crisis-management strategy.

Review Questions

1. What are the relationships among travel information, travel behavior, and security situations?
2. How might perceived risk change tourists' behavior?
3. How do the various security information sources affect tourists' risk perception and what are the limitations and the advantages of each source?
4. How does security information communication fit into risk reduction strategies?
5. What is the main goal of a security information platform provided by the receiving destinations?
6. Why is the Internet such an important tool in operating the proposed security information platform?

References

Baloglu, S., and MaCleary, K. (1999). Model of destination image formation. *Annals of Tourism Research,* 26(4), 868–897.

Baral, A., Baral, S., and Morgan, N. (2004). Marketing Nepal in an uncertain climate: Confronting perceptions of risk and insecurity. *Journal of Vacation Marketing,* 10(2), 186–192.

Bar-On, R.R. (1996). Measuring the effects of tourism and violence and of promotion following violent acts, in A. Pizam and Y. Mansfeld (eds.), *Tourism, Crime and International Security Issues*. New York: John Wiley & Sons, pp. 159–174.

Beirman, D. (2003). *Restoring Tourism Destinations in Crisis—A Strategic Marketing Approach*. Wallingford, UK: CABI International.

Berno, T., and King, B. (2001). Tourism in Fiji after the coups. *Travel and Tourism Analyst,* 2, 75–92.

Brownell, J. (1990). The symbolic/culture approach: Managing transition in the service industry. *International Journal of Hospitality Management,* 9(3), 191–205.

Brunt, P., Mawby, R., and Hambly, Z. (2000). Tourist victimization and the fear of crime on holiday. *Tourism Management,* 21(4), 417–424.

Buckley, P. J., and Klemm, M. (1993). The decline of tourism in Northern Ireland: The causes. *Tourism Management,* 184–194.

Carter, S. (1998). Tourists and travelers' social construction of Africa and Asia as risky locations. *Tourism Management,* 19(4), 349–358.

Cavlek, N. (2002). Tour operators and destination safety. *Annals of Tourism Research,* 29(2), 478–496.

Coles, T. (2003). A local reading of a global disaster: Some lessons on tourism management from *Annus Horribilis* in South West England. *Journal of Travel and Tourism Marketing,* 15(2/3), 173–179.

Cossens, J., and Gin, S. (1994). Tourism and AIDS: The perceived risk of HIV infection and destination choice. *Journal of Travel and Tourism Marketing,* 3(4), 1–20.

Elsrud, T. (2001). Risk creation in traveling backpacker adventure narration. *Annals of Tourism Research,* 28(3), 597–617.

Enders, W., Sandler, T., and Parise, G. F. (1992). An econometric analysis of the impact of terrorism on tourism. *Kyklos,* 45, 351–354.

Faulkner, B. (2000). Towards a framework for disaster management. *Tourism Management,* 22(2), 135–147.

Fodness, D., and Murray, B. (1997). Tourist information search. *Annals of Tourism Research,* 24(3), 503–523.

Frisby, E. (2002). Communicating in a crisis: The British Tourist Authority's responses to the foot-and-mouth outbreak and 11th September, 2001. *Journal of Vacation Marketing,* 9(1), 89–100.

Fuchs, G., and Reichel, A. (2004). Cultural differences in tourist destination risk perception: An exploratory study. *Tourism,* 52(4), 7–20

Gartner, W. C., and Shen, J. (1992). The impact of Tiananmen Square in China's tourism image. *Journal of Travel Research,* 30(4), 47–52.

Gunn, C. (1972). *Vacationscapes,* Austin Bureau of Business Research, University of Texas.

Hall, C. M., and O'Sullivan, V. (1996). Toursim, political stability and violence, in A. Pizam and Y. Mansfeld (eds.), *Tourism, Crime and International Security Issues*. New York: John Wiley & Sons, pp. 105–122.

Hsieh, S., and O'Leary, T. (1993). Communication channels to segment pleasure travellers. *Journal of Travel and Tourism Marketing,* 2(2/3), 57–75.

Iyengar, S., and Kinder, D. K. (1987). *News That Matters: Television and American Opinion*. Chicago: University of Chicago Press.

Jenkins, O. H. (1999). Understanding and measuring tourist destination images. *International Journal of Tourism Research,* 1, 1–15.

Lawton, G., and Page, S. (1997). Evaluating travel agents' provision of health advice to travelers. *Tourism Management,* 18(2), 89–104.

Lepp, A., and Gibson, H. (2003). Tourist roles, perceived risk and international tourism. *Annals of Tourism Research,* 30(3), 606–624.

Lovelock, B. (2003). New Zealand travel agent practice in provision of advice for travel risky destinations. *Journal of Travel and Tourism Marketing,* 15(4), 259–279.

McKercher, B., and Hui, E. L. L. (2003). Terrorism, economic uncertainty and outbound travel from Hong Kong. *Journal of Travel and Tourism Marketing,* 15(2/3), 99–115.

Mansfeld, Y. (1992). From motivation to actual travel. *Annals of Tourism Research,* 19(3), 399–419.

Mansfeld, Y. (1996). Wars, tourism and the Middle East factor, in A. Pizam and Y. Mansfeld (eds.), *Tourism, Crime and International Security Issues.* New York: John Wiley & Sons, pp. 265–278.

Mansfeld, Y. (1997). Risk, in J. Jafari (ed.), *Encyclopedia of Tourism,* London: Routledge, p. 508.

Mansfeld, Y. (1999). Cycles of war, terror and peace: Determinants and management of crisis and recovery of the Israeli tourism industry. *Journal of Travel Research,* 38(1), 30–36.

Mansfeld, Y., and Kliot, N. (1996). The tourism industry in the partitioned island of Cyprus, in A. Pizam and Y. Mansfeld (eds.), *Tourism, Crime and International Security Issues.* New York: John Wiley & Sons, pp. 187–202.

Mayo, E. J. (1973). Regional images and regional travel behavior. *Research for Changing Travel Patterns: Interpretation and Utilization. Proceedings of the Fourth Annual Conference of the Travel and Tourism Research Association*, pp. 211–218.

Oppermann, M., and Chon, K. S. (1997). *Tourism in Developing Countries.* London: Thompson Business Press.

Pearce, P. L. (1988). *The Ulysses Factor: Evaluating Visitors in Tourist Settings.* New York: Springer-Verlag.

Pizam, A., and Smith, G. (2000). Tourism and terrorism: A quantitative analysis of major terrorist acts and their impact on tourism destinations. *Tourism Economics,* 6(2), 123–138.

Pizam, A., Jeong, G. H., Reichel, A., Van Boemmel, H., Lusson, J. M., Steynberg, L., State-Costache, O., Volo, S., Kroesbacher, C., Kucerova, J., and Montmany, N. (2004). The relationship between risk-taking, sensation-seeking, and the tourist behavior of young adults: A cross-cultural study. *Journal of Travel Research,* 42(3), 251–260.

Poon, A., and Adams, E. (2000). *How the British Will Travel—2005.* Bielefeld, Germany: Tourism Intelligence International.

Prideaux, B. (2003). The need to use disaster planning frameworks to respond to major tourism disasters: Analysis of Australia's response to tourism disasters in 2001. *Journal of Travel and Tourism Marketing,* 15(4), 281–298.

Ritchie, B. W., Dorrell, H., Miller, D., and Miller, G. A. (2003). Crisis communication and recovery for the tourism industry: Lessons from the 2001 foot and mouth disease outbreak in the United Kingdom. *Journal of Travel and Tourism Marketing,* 15(2/3), 199–216.

Richter, L. K., and Waugh, W. L., Jr. (1986). Terrorism and tourism as logical companions. *Tourism Management,* 7(4), 230–238.

Roehl, W. S., and Fesenmeier, D. R. (1992). Risk perception and pleasure travel: An explanatory analysis. *Journal of Travel Research,* 2, 17–26.

Santana, G. (2001). Globalization, safety and national security, in S. Wahab and C. Cooper (eds.), *Tourism in the Age of Globalization*. London: Routledge, pp. 213–241.

Santana, G. (2003). Crisis management and tourism: Beyond the rhetoric. *Journal of Travel and Tourism Marketing,* 15(4), 299–321.

Schiebler, S. A., Crotts, J. C., and Hollinger, R. C. (1996). Florida tourists' vulnerability to crime, in A. Pizam and Y. Mansfeld (eds.), *Tourism, Crime and International Security Issues*. New York: John Wiley & Sons, pp. 37–50.

Sharpley, R., and Sharpley, J. (1995). Travel advice—Security or politics? in *Security and Risks in Travel and Tourism. Proceedings of the Talk at the Top Conference.* Östersund, Sweden: Mid-Sweden University, pp. 168–182.

Sönmez, S. F., and Graefe, A. R. (1998a). Determining future travel behavior from past travel experiences and perceptions of risk and safety. *Journal of Travel Research,* 37(2), 171–177.

Sönmez, S. F., and Graefe, A. R. (1998b). Influence of terrorism risk on foreign tourism decisions. *Annals of Tourism Research,* 25(1), 112–144.

Sönmez, S. F. (1994). An exploratory analysis of the influence of personal factors on international vacation decisions within the context of terrorism and/or political instability risk. Doctoral dissertation, State College, Pennsylvania State University.

Sönmez, S. F., Apostolopoulos Y., and Tarlow, P. (1999). Tourism in crisis: Managing the effects of terrorism. *Journal of Travel Research,* 38(1), 13–21.

Thapa, B. (2003). Tourism in Nepal: Shangri-La's troubled times. *Journal of Travel and Tourism Marketing,* 15(2/3), 117–138.

Uriely, N., Yonay, Y., and Simchai, D. (2002). Backpacking experiences: A type and form analysis. *Annals of Tourism Research,* 29, 520–538.

Wahab, S. (1996). Tourism and terrorism: Synthesis of the problem with emphasis on Egypt, in A. Pizam and Y. Mansfeld (eds.), *Tourism, Crime and International Security Issues*. New York: John Wiley & Sons, pp. 175–186.

Wall, G. (1996). Terrorism and tourism: An overview and an Irish example, in A. Pizam and Y. Mansfeld (eds.), *Tourism, Crime and International Security Issues*. New York: John Wiley & Sons, pp. 143–158.

Weber, E. U., and Bottom, W. P. (1989). Axiomatic measures of perceived risk: Some tests and extensions. *Journal of Behavioral Decision Making,* 2(2), 113–131.

Weimann, G. (2000). *Communicating Unreality: Modern Media and the Reconstruction of Reality*. Thousand Oaks, CA: Sage Publications.

Weimann, G., and Winn, C. L. J. (1994). *The Theater of Terror: Mass Media and International Terrorism*. New York: Longman.

Wood, B. D., and Peake, J. S. (1998). The dynamics of foreign policy agenda setting. *American Political Science Review,* 92, 174–184.

16

Crisis Management and Recovery: How Washington, DC, Hotels Responded to Terrorism

Greg Stafford, Larry Yu, and Alex Kobina Armoo

Learning Objectives

- To define different types of business crises that will affect tourism operations.
- To analyze terrorism as a malevolent crisis and its impact on global tourism.
- To examine the impact of the September 11, 2001, terrorist attacks on the hotel industry in Washington, DC.
- To understand crisis-management planning and coordination by the hotel industry in Washington, DC.
- To study how to launch an effective recovery marketing campaign after a major crisis.
- To assess tourism recovery by evaluating hotel performance.
- To understand the response and recovery process for managing a terrorism crisis.

Introduction[1]

This chapter reviews studies on terrorist attacks on the tourism industry worldwide and examines crisis management for handling business crises caused by terrorist attacks. It analyzes the calamitous impact caused by the September 11, 2001, terrorist attacks, focusing on the human and financial effects on the hospitality industry in the Washington, DC, metropolitan area, and discusses the recovery efforts exerted by the local hospitality industry.

It first identifies terrorism as one form of malevolent crisis for business operations and discusses the negative impact of terrorism on hospitality operations by an extensive literature review of this subject. It then focuses on crisis-management practices that can effectively handle such a malevolent crisis. As one of the two September 11 terrorist attack targets, Washington, DC, is used as a case to illustrate how the local hospitality industry worked closely with other tourism partners to manage the unprecedented crisis and effectively developed recovery strategies to bring the city's hospitality business back to normalcy. The lessons and experience gained from managing this crisis can serve as recommendations for the hospitality industry in other destinations to prepare for and handle similar crises should they occur in the future.

Based on the review of the response and recovery efforts by the hotel managers in the Washington, DC, area, this study suggests a model for responding to and recovering from a malevolent crisis. At the heart of this model are two key recommendations: (a) crisis preparation and management expertise should be added to the qualifications of hotel managers at different levels; and (b) local clusters of hotels, brands, and stakeholders affected by a crisis should coordinate their efforts in a local manner to optimize the speed and completeness of the initial response and shorten the recovery time. A thoughtful and well-planned response, individually and collectively, will best mitigate the damage of crises by terrorist attack.

Type of Business Crisis

The September 11, 2001, terrorist attacks on the World Trade Center in New York City and the Pentagon in Washington, DC, touched off a myriad of events that had a devastating impact on the US economy (Pearlstein, 2001; Schneider, 2001). One of the business sectors that suffered massive losses from the terrorist attacks was the tourism and hospitality industry (Pressley, 2001; Higgins and Drucker, 2001; Business and terrorism, 2001). For the first time in US history, there was a complete shutdown of the US aviation system, creating such huge losses that a federal government bailout and guaranteed loans for airlines were warranted. Hotel occupancy rates in many major cities dropped precipitously after the September 11 terrorist attacks, as guests cancelled or postponed their travel plans. Since the tourism and hospitality industry mainly caters to visitors who are away from home enjoying attractions, visiting retail outlets, conducting business, or attending meetings and visitors use the transportation infrastructure in all these activities, the hospitality industry is particularly vulnerable to any disastrous event impacting transportation services. The immediate aftermath of the September 11 terrorist attacks was a sudden and severe disruption to normal operations.

The operation of any business involves risks. Sudden disruptive events, triggered by uncontrollable external forces such as nature, human elements, or internal organizational failure, are the most devastating. The occurrence of a sudden or unpredictable disruptive event is often depicted by the news media as a crisis. In the context of business, a crisis is defined as an event that can tarnish a company's reputation or is severely detrimental to a company's long-term profitability, growth, and even its survival (Lerbinger, 1997).

Managers of business operations may encounter various types of crises. Crises are conceptually defined in seven categories. Table 1 describes the nine types of crises that managers are likely to experience in operations. Crises of the physical environment include natural disasters and technology failure. Natural disasters have always been a threat to hospitality operations because natural hazards such as volcanoes, hurricanes, tornadoes, earthquakes, landslides, and floods menace life, property, and the environment. Hospitality operations on the Caribbean islands and Hawaii are often interrupted by hurricanes (called cyclones in the Pacific), such as Hurricane Luis and Hurricane Marylyn on St. Thomas, St. John, and St. Croix in September 1995. Technology failure relates to accidents caused by human application of science and technology in harnessing the physical environment. The nuclear power plant explosion at Chernobyl in 1986 was a good example of technological crisis because airborne radiation reached many parts of Europe, and

Table 1 Causes of Crisis for Business Organizations

Major Factors	*Specific Environment*	*Type of Crisis*	*Example of Crisis*
External factors	Physical environment	Natural disaster	Earthquake and tsunami damage coastal resort property; volcano eruption scares away tourists
		Technology failure	Oil spill contaminates a resort beach and prevents tourists from visiting the resort
	Human/social environment	Confrontation	Hotel union strike disrupts normal operations; special interest group boycotts fast food
		Epidemic	Mad cow disease and foot and mouth disease raise concerns of food safety and health problems; SARS epidemic spreads through human contact
		Malevolence	September 11 attack on NY and Washington, DC; product tampering by poisoning food; extortion by threatening to introduce a virus into computer reservation systems
		War/political conflict	Second Gulf War keeps many international tourists from the Middle East region; recent political upheaval diminishes tourism to Haiti, Venezuela, and some African countries
Internal factors	Management failure	Skewed values	Cruise ships dump waste oil into the ocean, placing economic value over concern for the environment
		Deception	Knowing the food item is contaminated, the restaurant continues to serve it to the customers
		Misconduct	Corporate CFO embezzles funds and receives kickbacks

Source: Adapted from Lerbinger, O. (1997). *The Crisis Manager: Facing Risk and Responsibility.* Mahwah, NJ: Lawrence Erlbaum, pp. 10–14.

many European countries had to destroy cash crops to prevent radiation from entering food supplies.

Crises of the human and social environment refer to confrontation, epidemics, malevolence, and war. Individuals or interest groups who fight business for certain issues purposely provoke confrontation crises, since they know the use of confrontation is most effective in attracting media, and eventually, management attention. Labor union strikes and boycotts of products and services are some commonly used confrontation tactics. Epidemics in livestock or the human population immediately halt travel to the infected areas, as clearly demonstrated by the SARS epidemics in 2003. Any political conflicts such as war will completely close the country or region at war to tourists and this is best illustrated by the Iraq war and the political upheaval in Sudan. Crises of malevolence refer to the criminal means or extreme tactics deployed by individuals or groups against a business organization or an entire industry. Malevolent acts include product tampering, extortion, corporate espionage, and terrorism. These extreme measures aim to destroy a company's business or a country's economic system.

Crises of management failure include skewed values, deception, and misconduct. Unreasonable financial expectations and failures of corporate governance are often at the root of these unethical or sometimes criminal behaviors by corporate leaders. Such crises can destroy owners' and shareholders' value in the company, as best illustrated by the accounting scandals exposed in Enron and WorldCom in the United States.

All these crises share three distinct characteristics: suddenness, uncertainty, and time compression (Lerbinger, 1997, pp. 6–9). A crisis normally occurs suddenly even though some early warning signs are usually detectable. A crisis that is unpredictable and erupts suddenly is always the most shocking (Mindszenthy, Watson, and Koch, 1988). Terrorist attacks, in any size, shape, or form, fit such a pattern of crisis. Lippman refers to this element of surprise as "shadowy, mobile, and unpredictable" (1999, p. A14). To the terrorists, the element of surprise helps them achieve their aim.

Due to the element of surprise associated with terrorist attacks, there is the problem of uncertainty—uncertainty of the kind of crisis an organization is likely to face or the kind of crisis for which one should prepare. Uncertainty is a major concern confronting business management especially when the organization's business is global and complex. Managers are normally tied to the daily operations of their organization. They have limited time or resources to scan the business environment for predictable external changes. Low-probability, high impact events, such as a terrorist attack on the United States homeland, were likely given little or no attention by management. Therefore, managers were generally poorly prepared for a terrorist-related crisis. When dealing with a crisis, management faces great pressure to make decisions rapidly and with incomplete information. Time compression adds to the enormous stress and anxiety that a crisis causes, not only for managers, but also for employees. Managers tend to look at aberrant events and determine the cause. They examine what is likely to happen next and determine how to best mobilize resources to deal with that next set of events. In the September 11 terrorist attacks the targets were symbolic and the victims were irrelevant to the perpetrators, and the perpetrators are unknown and apparently unconcerned about their own safety. Managers managing the effects of such events close to home will have grave difficulty if they have not at least thought carefully about the basic process of managing a crisis. After the initial shock and under uncertain conditions, management teams with well-defined crisis plans can most expediently and effectively limit the damage and regain control over their operations.

Terrorism and Crisis Management

Terrorist attack is therefore defined as one form of malevolent act against a government or a country's economic system, with an aim of disrupting the targeted country's social, political, and economic systems. Specific terrorist attacks on tourists have occurred often over the past few decades (Pizam and Mansfeld, 1996). Hijackings of commercial airplanes and cruise ships and the murder of innocent tourists have been perpetrated often over the past several years. D'Amore and Anunza conducted one of the earliest studies on terrorism and tourism in the mid-1980s. They reported that terrorism of major events increased from 206 in 1972 to 3,010 in 1985 (D Amore and Anunza, 1986). Some terrorist groups specifically attack international tourists and hotels as targets to achieve their cause or political purpose (Auerbach, 1998; Wahab, 1996; Wall, 1996). The Shining Path (*Sender Luminoso*) terrorist group in Peru deliberately targeted hospitality businesses and international tourists. Many international tourists were attacked, kidnapped, and killed by this terrorist group from the mid-1980s to the early 1990s (Ryan, 1991). The Al-Jihad terrorist group in Egypt attacked and killed sixteen Greek tourists in front of a hotel in April 1996. Another 58 international tourists were killed by Islamic militants at the 4,000-year-old Temple of Hatshepsut in Luxor in late 1997 (Lancaster, 1997). The Al-Qaeda movement headed by Osama bin Laden has been attacking US properties and American interests for the past few years. In 1998, they attacked the US embassies in Kenya and Tanzania, killing 263 people. Such acts of terrorism had a strong negative impact on tourism (Pizam and Smith, 2000; Tarlow, 2000).

The Al-Qaeda network, a sophisticated international terrorist organization, committed the September 11 terrorist attacks on the World Trade Center in New York City and the Pentagon in Washington, DC. The main objectives of the terrorists were apparently to paralyze the US economy and to spread panic and chaos among the American people. Prior to September 11, terrorist attacks involving airlines either involved foreign planes or took place outside the United States. The use of US planes, coupled with the attacks occurring on US sites, created a collective fear of flying. This condition was a paramount factor in the drastic decline of tourism in most parts of the United States immediately after the September 11 attacks.

The hotel industry was shocked and decimated by the impact of the crisis. Hotel managers were immediately under great stress to manage the crisis and thereby regain control of their operations. Effectively managing a crisis, through coordination of human resources, financial resources, and public relations, is vital for any organization that has been impacted by a terrorist act. Sönmez, Apostolopoulos, and Tarlow call for a carefully planned and implemented crisis management strategy and policy (Sönmez, Apostolopoulos, and Tarlow, 2000). Generally, hotel managers are trained to handle various types of crises in hotel operations: a natural disaster such as an earthquake, a safety issue such as a hotel fire or food poisoning, a security concern such as a robbery, or a legal charge such as racial discrimination or sexual harassment (Barton, 1994; Brewton, 1987; Stutts, 1990). Effective management of these crises can limit the widespread negative publicity of the property and minimize the damage to the company's image and its revenues. However, the effect of the September 11 terrorist attacks was widespread, impacting not just a business or a destination, but the entire hospitality industry and, indeed, the American way of life. In this environment, hotel managers had to collectively manage the impact of these attacks on their business and their staff while feeling the core of their own existence rocked like everyone else.

The following discussion illustrates how the Washington, DC, hospitality industry collectively responded to the September 11 crisis, and conducted recovery efforts to return business to normalcy. The actions taken by the DC hospitality industry can be emulated by other tourism destinations when facing similar crises.

The Impact of the September 11 Terrorist Attacks on the Hotel Industry in Washington, DC

Crisis Response

Internal Response

During the September 11 terrorist attacks and the moments immediately following, hotel managers were mostly concerned with conditions inside their hotels and gathering factual information about the attacks. Additional staff members were deployed to public areas to calm guests. Operating department managers focused on reassuring staff members and preparing them to react. Emergency action plans and crisis plans were accessed, although few hotel plans addressed a situation like the one that hotel managers were experiencing during the September 11 terrorist attacks. Emergency equipment was mobilized to prepare for the possibility of additional safety or security incidents related to these terrorist acts. Those guests or staff members who feared for safety of family members or friends were provided special attention. Televisions were placed in public areas and meeting rooms so that guests could monitor events. Command posts were set up to coordinate activities via radio communication, to take stock of conditions, and to communicate in a centralized manner with the outside world.

External Response

Once internal conditions were initially assessed, hotel managers communicated, consulted, and coordinated with external respondents. Formally, conditions at the Pentagon and throughout the area were identified by the Hotel Association of Washington, DC, through contact with Fire Department, Police Department, and DC Emergency Management Agency liaisons and then communicated to hotel general managers. The needs of relief personnel assisting at the site of the attacks or with victim relief were identified and communicated through the same channels. Hotel general managers in turn communicated requests for additional information and assistance through the Hotel Association. In most cases, managers consulted with brand, management company, and/or owner representatives for advice on handling a variety of service, safety, and financial issues. Additionally, managers reached out to each other to offer assistance wherever possible. Finally, media responses were coordinated with brands, management companies, and owners and especially with the Hotel Association of Washington, DC, and the

Washington Convention and Tourism Corporation to ensure that the media focus was directed towards the caring and responsible manner in which the crisis was being addressed internally and externally by hotels in the city.

The Damage

Though most tourist destinations in the United States were adversely affected by the September 11 terrorist attacks, the targets, New York City and Washington, DC, were particularly devastated by these events. Washington, DC's tourism industry weathered not only the initial blow of September 11, but also the closure of the Reagan National Airport for several weeks after the attack and reduced flight schedules for many months. Additionally, famous tourist attractions, such as the White House and the US Capitol Building, were shut down to tourists for several weeks. The situation was further exacerbated by anthrax scares that erupted in October 2001. As a result of these conditions, hotel occupancy plummeted to 41.8% the week following the attacks, a 52% decrease from the same period in 2000. REVPAR declined to $41.25 the first week after the attacks, a 62% plunge from the same period in 2000 (STAR Report, 2001). Hotel taxes decreased substantially in the September through November timeframe compared to the previous 2 years as recorded in Table 2.

Such sudden prolonged decreases in business created a business crisis for hotel owners and managers. Hotel managers were forced to make hard decisions to conserve cash by severely reducing operating costs and deferring capital improvements. As room sales remained dismal for several more weeks, hotel managers were forced to make the most difficult decision to severely reduce the work schedules or lay off most of the 75,000 employees directly supported by the hospitality industry in Washington, DC (Hedgpeth and Irwin, 2001; Twomey, 2001; Sheridan, 2001). Steep discounts on room rates and other sales promotions were offered by hotel operators to entice visitors to return to hotels (Lonnig and McCaffrey, 2001).

Table 2 Monthly Hotel Sales Tax Revenue in Washington, DC, 1999–2002 (in US: $)

	1999	*2000*	*2001*	*2002*
January	2,083,597	1,771,533	3,058,810	3,008,947
February	1,577,849	1,566,657	726,744	1,432,790
March	2,488,339	3,386,583	3,817,280	3,125,787
April	3,151,812	2,584,570	3,241,299	3,387,121
May	3,627,154	4,186,446	5,159,761	4,826,261
June	3,784,271	4,577,912	4,100,036	2,929,979
July	3,190,065	3,376,907	3,552,188	3,153,909
August	2,452,792	3,405,266	3,092,084	—
September	2,209,657	2,922,143	1,794,747	—
October	3,117,743	2,957,742	2,158,430	—
November	4,015,429	4,957,742	2,672,048	—
December	2,726,053	2,702932	2,354,936	—

Source: Washington Convention Center Authority, August 2002.

Clearly, the Washington, DC, hotel industry desperately needed to find a better way to address the market than the cutthroat, zero-sum rate war that appeared inevitable.

Recovering from the Crisis

The Hotel Association of Washington, DC (HAWDC), coordinated with the Washington, DC, Convention and Tourism Corporation (WCTC), the Washington, DC, Convention Center Authority (WCCA), the Restaurant Association of Metropolitan Washington (RAMW), and the District of Columbia government to promote tourism in the District. The early recovery initiatives of this group included (i) assuring a coordinated response by the Washington, DC, hospitality industry; (ii) campaigning to reopen Reagan National Airport and major tourist attractions; (iii) promoting government-related business travel to DC; and (iv) building a marketing fund and developing a marketing plan to reestablish Washington, DC, as a must visit destination (Irwin, 2001; WCTC, 2001a). The following discussion explains each of the crisis-management and recovery initiatives deployed by the DC hospitality industry in coordination with its partners.

Industry Coordinated Response

Led by the Washington, DC Convention and Tourism Corporation, the Washington, DC hospitality industry worked closely with DC government officials to ensure a coordinated response from the industry to the public and the media. The WCTC coordinated feedback from industry leaders and they were in contact on a daily basis for several weeks updating events and information regarding the September 11 terrorist attacks' impact and the recovery efforts. This communication occurred most frequently via email to WCTC members, although phone calls and meetings were used where more interaction was desirable or to convey more discreet or subtle information. The CEO of the WCTC, Bill Hanbury, was established as the spokesperson for the hospitality industry to further minimize conflicting information or statements to the media, partner organizations, or government officials. The WCTC retained the Burston-Marsteller Public Relations firm to assist the industry with public relations strategy. General managers were consulted about the impact of the terrorist attacks on their business and the strategy for recovery.

The initial recovery coordination efforts were focused on four distinct fronts. First, leaders of key tourism and travel stakeholder industries and organizations communicated on a frequent basis to coordinate the status of their industries and report on exactly how the crisis was impacting them on a moment to moment basis. Additionally, relationships were strengthened with the DC Government, the Greater Washington Board of Trade, the Greater Washington Initiative, the DC Chamber of Commerce, and many other regional economic development agencies to ensure that collective resources were used in a coordinated fashion to mitigate the damage inflicted on the image of the Nation's Capital. In this manner, relevant information became more accurate and comprehensive, problem solving became more collaborative rather than competitive, and solutions were more efficient on a regional basis.

Second, several hundred valued convention and group clients were contacted by the Mayor of Washington, DC, the CEOs of WCTC and WCCA, and other hospitality leaders. These important groups were reassured that Washington, DC was open for business. As a result, DC incurred very minimal major convention group attrition or cancellations. Additionally, the WCTC web site was updated daily to provide information for travelers regarding schedule changes and facility closings. Third, residents of the Washington, DC region were engaged regarding the crisis conditions confronting the city's hospitality industry and the depth of the local economic impact. Despite these extremely difficult circumstances, the local hospitality industry collaborated with the Pentagon Victim Assistance Center to aid local families directly affected by the September 11 terrorist attacks. Hotels provided complimentary hotel rooms and meals for immediate family members suffering losses in the terrorist attacks. The local hospitality industry also sent representatives to the Emergency Operations Center operated by the District of Columbia Emergency Management Agency. The industry received timely updates of information and fully participated in community-based decisions during the crisis. Fourth, the hospitality industry lobbied DC and federal government leaders to provide assistance in the form of extended unemployment compensation benefits, health care benefits, and income tax relief to displaced hospitality workers. Many hotels also initiated a variety of special assistance efforts for their displaced workers. The most critical aspect of these efforts was that they sent a clear message to hospitality workers that hotel companies and leaders were concerned about their workers' personal security and not just the survival of their businesses. In this manner, labor disaffection with the hospitality industry was minimized.

Reopening Reagan National Airport and Other Major Attractions

Three major airports serve the Washington, DC metro area: Ronald Reagan National Airport (DCA), Dulles International Airport (IAD), and Baltimore-Washington International Airport (BWI). DCA is conveniently located in Arlington, Virginia, just across the Potomac River from Washington, DC and serving over 16 million passengers annually. IAD and BWI are both located in suburban areas several miles from the District. About 23% of visitors to the Washington, DC area arrive through DCA.

DCA was immediately closed for several weeks after September 11, because of its proximity to the White House, the US Capitol Building, and other possible terrorist targets. At the time, it was unclear whether DCA would ever reopen. The closure of DCA had an immediate negative impact on all segments of the hospitality industry. Leaders of the hospitality industry realized that, if DCA remained closed, business travelers, meeting attendees, conventioneers, leisure tourists, and even members of Congress would spend much more time and endure more cost and inconvenience traveling in and out of DC. Indeed, some visitors would choose other destinations or not travel at all. Further, many travelers perceive DCA as the air gateway to the Nation's Capital. City and hospitality leaders feared that the public perception would be that DC was unsafe and under siege and effectively closed unless DCA was reopened. Reopening DCA was an economic and symbolic necessity. Frankly, most industry executives in DC and beyond deemed the reopening of DCA as another critical step in the broader issue of restoring the public's faith in flying.

A campaign to lobby Congress, key Cabinet members, and presidential advisors to reopen DCA was discreetly, but aggressively, waged by the local hospitality industry. The President of WCTC, District and local political leaders, and area Congress members led this charge. Despite the resistance of many key presidential military and security advisors, this political campaign and lobbying effort eventually fueled such a groundswell of political pressure on Congress and the President that DCA was reopened on October 4, 2001. The initial capacity was limited at less than 50% of the passenger capacity prior to September 11. Gradually, flights were increased so that the passenger capacity of DCA has returned to nearly the pre-September 11 level.

On a similar note, tourism to Washington, DC would remain crippled if major attractions such as the US Capitol building and the White House remained off limits to the public. The hospitality industry and local political leaders launched similar campaigns to reopen these notable tourist attractions, albeit with many safety precautions and modifications. Likewise, these efforts were successful after much political lobbying. Tours would only be available to guided groups in secured areas of these facilities. But at least they were open and free again for tourist viewing. The reopening of the White House, which took several weeks, was coordinated as a major press event welcoming an out of town school class and having a local school class be present as hospitality ambassadors. This step was crucial in the efforts of the DC hospitality industry's efforts to reconnect with the critical school trip and family markets.

Promoting Government-Related Business Travel

Hotels in Washington, DC serve diverse market segments: government business, leisure travel, meetings and conventions, international tourism, and many special-interest travel groups. Amid the drastic decline of all forms of travel to DC, industry leaders realized that federal government travel, including meetings, was either being curtailed or in some instances even relocated out of DC. Leaders of the hospitality industry mobilized an appeal to members of Congress and the President to promote travel broadly to Washington, DC, but particularly to ensure that government travel to Washington, DC was encouraged and that government meetings continued as scheduled in DC. In this regard, key government travel related issues were identified by the local hospitality industry and a list was communicated to the appropriate federal and congressional offices. The hospitality industry requested the following support activities be initiated:

- Issuing an executive order encouraging deferral agencies to continue to convene conferences, meetings, trade shows, and training sessions in Washington, DC.
- Urging the State Department to communicate with embassies to encourage international visitation to Washington, DC.
- Urging members of Congress and their staffs to communicate positively with constituents regarding travel to the nation's capital.
- Because of Washington's position within the global economy, encouraging allies to support the resumption of international travel.
- As the administration deals with this crisis, urging federal agencies to cooperate with DC authorities to minimize disturbance to local commerce and negative national images. The perception that Washington, DC is "under siege" was having highly negative results on local tourism (WCTC, 2001b).

The local hospitality industry worked very hard to seek moral and business support from the federal government in order to promote government travel to sustain its operations during the first few months. Not all these efforts were successful. However, federal government travel was very robust within DC in the fourth quarter of 2001 and the first quarter of 2002. Additionally, some of the inflammatory depictions of DC in a state of war (rather like London 60 years earlier) by government officials and staffers and foreign embassy staff were doused.

It should be noted that the DC hospitality industry did not seek a financial bailout. Rather it sought moral and business support from the federal government to overcome the most difficult period of time. The hospitality business community was determined to pull itself through this crisis with its own resources and initiative.

Launching the "City of Inspiration" Marketing Campaign

National and international media following the terrorist attacks on September 11, 2001, constantly publicized the images of a burning icon of American military strength and the horrific counts of victims. Washington, DC was powerfully portrayed as a "capital under siege" or "at the heart of an international crisis" throughout the United States and every part of the world. Subsequently, a series of bioterrorism events and frequent terrorist warnings issued by the Justice Department allowed the media to continue to project an image that it was unsafe to travel to the Washington, DC area. Facing devastating negative publicity at home and abroad, the hospitality industry was keenly aware that a marketing plan had to be developed and funded to restore and even capitalize on Washington, DC's character and tourism assets when travelers were ready to receive that message.

The planning for the marketing campaign was coordinated by WCTC, and Burston Marsteller was contracted to assist in developing this marketing plan. WCTC, the District of Columbia, Washington Convention Center, and the Greater Washington Initiative contributed an initial marketing campaign fund of $3.37 million. Those funds were subsequently supplemented by donations of private industry. The objective of the marketing campaign was to target local, regional, and national travel markets. The promotional strategy included a full range of marketing and advertising activities: newspaper, radio, television, Internet, public relations, collateral materials, and direct consumer promotions. In the months immediately after September 11, tourists were not only reluctant to fly, they were also reluctant to travel far from home, and many were specifically reluctant to visit Washington, DC because of continued bioterrorism events and a nagging fear that the Nation's Capital and its many symbolic landmarks were obvious targets for additional terrorist activities. Further, because of the cultural and historical attractions of Washington, DC—the White House, Washington Monument, National Mall, the US Capitol Building, Lincoln Memorial, Smithsonian Museums, Arlington National Cemetery, and so on—many of its most devoted tourists are families and school-age children. Parents were understandably even more reluctant to place their children's safety at risk. Most critically, even when it became reasonably safe and enjoyable to return to Washington, DC, people outside the region would remain very reluctant to visit the city as long as somber terror alerts and war updates were being issued by various political leaders based in Washington, DC. The marketing campaign would have to be developed and launched in a series of expanding circles around Washington, DC: small and then larger and larger as tourists' fears of visiting DC receded.

The marketing campaign for the local market was themed "Be Inspired in Your Hometown" and "Hometown Homecoming." It was launched in October less than five weeks after the initial terrorist attacks and just prior to the series of anthrax events, which shook Washington, DC with a second set of terrorist tremors. Many Americans were seeking ways to demonstrate their patriotism, and probably no other city in the world evokes such an atmosphere of courage and inspiration as Washington. The initial campaign sought to enable area residents within a short drive of Washington, DC to be inspired by their "hometown." The Washington Metropolitan Transit Authority offered free bus and rail passes to visitors. Restaurants and attractions offered special promotions during the weekend as added value for local residents to rediscover the monuments, museums, and spirit of the Nation's Capital.

Nationally, a comprehensive and large-scale marketing campaign was launched with the message: "Washington, DC is the City of Inspiration. Home of the American Experience." Television advertising was developed using a public service announcement format with the nationally recognized cast of *The West Wing* and the Mayor of Washington, DC. These ads were aired on select television programs. An additional ad was done featuring several notable Washington area personalities, some of America's most recognized political leaders, and the First Lady, Laura Bush. Cast appearances and most production costs were donated to the city's tourism recovery effort, as were many ad placements. Supporting print media ads were developed and placed in newspapers and magazines. Promotions were developed and launched on key Internet travel planning sites such as Travelocity and Orbitz. Tapes of the spots were incorporated in the WCTC convention sales efforts and supporting collateral materials were developed. An aggressive public relations program was initiated and its activities included:

- Managing media calls and requests;
- Developing positive stories, such as the reopening of Reagan National Airport;
- Collecting media clips from key daily papers to monitor consumer sentiment;
- Coordinating a letter writing campaign to the editor in order to generate more positive stories about DC;
- Holding editorial roundtable meetings with *The Washington Post, Washington Times,* and *Washington Business Journal;*
- Developing talk show strategy for programs such as *Good Morning America* and *The Today Show;*
- Communicating with school groups and school group organizations encouraging them to maintain their plans to visit Washington, DC; and
- Preparing news release and special event announcements (WCTC, 2001c).

Much focus was placed on developing the city's web site with rich, timely, and easy-to-access content and developing its Internet marketing capacity. An open letter of reassurance from Mayor Anthony Williams and Bill Hanbury, CEO of WCTC, was placed on the WCTC web site. Partner organizations were encouraged to distribute this letter to their customers and staff through all available channels in the interest of message penetration and consistency. Local hoteliers offered targeted "Be Inspired" rates through the online booking engines of WCTC: Hotel

Reservations Network and Capital Reservations. A digital version of the television broadcast spots was distributed to targeted lists purchased from Travelocity, AOL, Orbitz, and other travel service providers. The hotel booking engine sites were linked at the end of the broadcast spot. A special online section titled "Why Washington, DC Inspired Me" was developed to target school-age children. Well-written essays were scrolled on the WCTC home page.

A tremendous amount of work went into planning and executing an aggressive marketing plan. All the partners involved in developing the marketing plan were thoughtful about how to proceed in order to ensure long-term success rather than a quick fix. The marketing plan had to be comprehensive in order to communicate with its wide array of target audiences. The marketing plan had to be emotive in order to convey as much of the unique attitude and character of Washington, DC as possible. The city lacked the huge corporate and state-funded budgets of destinations such as New York, Las Vegas, or Orlando, and therefore could not afford to make any mistakes. Finally, the campaign had to be executed at the right time—the time when target audiences were ready to receive the message. This last aspect required enormous discipline on the part of the WCTC and the hospitality industry. Had the marketing campaign been launched with full force in early October, the investment would have been largely wasted, as Washington was soon besieged by a series of bioterrorism events and constant war and terrorism alerts. However, had the late winter timing of the marketing plan execution been more reticent, the spring tourism resurgence in DC would probably have been diluted significantly. In turn, it is quite likely that a softer spring would have led to residual tourism weakness in the summer.

The Recovery of the Washington, DC Hospitality Industry

Prior to the September 11 terrorist attacks, most hospitality leaders anticipated that 2002 would be a very challenging year for the hospitality industry in Washington, DC. It would be an election year in a narrowly controlled Congress. It would be the year before opening a major new convention center. Few significant new attractions were expected. Even Washington's insulated economy was feeling the pain of economic downturn prior to the September 11 terrorist attacks. Several declining demand factors seemed inevitable and a moderate increase in supply foreshadowed a disappointing year. Despite these conditions, Washington, DC has seen year after year occupancy increases in 16 out of the first 28 weeks of 2002 vs an average of only 3 weeks of increased occupancy nationally. More importantly, Table 3 shows that occupancies and REVPARS were close to stabilizing at prior year levels by spring 2002. Washington, DC rebounded better and faster than most other US markets. Many factors contributed to marketplace performance—overall marketing and sales effort of the destination, marketing budgets, changes in specific demand generators, changes in supply, adverse industry or company circumstances, and much more. The simple fact remains that every hospitality market had to decide how to respond to the impact of the September 11 terrorist attacks. As Table 3 further illustrates, all three lodging performance indicators, occupancy, average daily rate, and REVPAR, reached almost to their pre-September 11 levels by the end of 2003.

Table 3 Hotel Revenue Performance in DC and the DC Metro Area, 2000–2003

DC Occupancy Rate, 2000–2003

	2000	*2001*	*% Change*	*2002*	*% Change*	*2003*	*% Change*
January	51.0%	57.0%	11.9%	51.8%	−9.2%	54.3%	4.9%
February	65.6%	68.6%	4.6%	67.5%	−1.6%	64.6%	−4.3%
March	81.1%	82.8%	2.0%	76.9%	−7.1%	74.1%	−3.6%
April	84.8%	81.2%	−4.3%	86.7%	6.9%	73.9%	−14.8%
May	82.9%	78.7%	−5.1%	78.4%	−0.4%	76.1%	−2.9%
June	81.7%	80.0%	−2.2%	77.4%	−3.2%	79.5%	2.6%
July	76.6%	74.1%	−3.3%	71.2%	−3.9%	77.1%	8.3%
August	71.5%	66.6%	−6.8%	64.3%	−3.4%	63.9%	−0.7%
September	79.4%	45.8%	−42.3%	69.6%	52.1%	73.2%	5.1%
October	81.1%	59.3%	−26.9%	75.4%	27.2%	79.2%	5.1%
November	67.3%	56.5%	−15.9%	58.8%	3.9%	63.2%	7.5%
December	50.1%	46.1%	−8.1%	46.1%	0.0%	48.4%	5.0%

DC Average Daily Rate, 2000–2003

	2000	*2001*	*% Change*	*2002*	*% Change*	*2003*	*% Change*
January	134.21	177.32	32.1%	130.85	−26.2%	132.39	1.2%
February	143.01	156.17	9.2%	143.05	−8.4%	145.59	1.8%
March	153.64	167.81	9.2%	153.86	−8.3%	162.70	5.7%
April	162.42	168.91	4.0%	170.10	0.7%	155.56	−8.5%
May	161.12	164.88	2.3%	164.57	−0.2%	162.84	−1.0%
June	150.03	155.10	3.4%	149.12	−3.9%	151.77	1.8%
July	131.63	132.14	0.4%	128.35	−2.9%	131.13	2.2%
August	128.99	124.48	−3.5%	121.00	−2.8%	118.84	−1.8%
September	159.72	149.25	−6.6%	164.85	10.4%	158.00	−4.2%
October	168.03	149.95	−10.8%	165.12	10.1%	159.06	−3.7%
November	153.78	140.34	−8.7%	141.45	0.8%	143.49	1.4%
December	137.87	125.49	−9.0%	128.70	2.6%	131.62	2.3%

DC REVPAR, 2000–2003

	2000	*2001*	*% Change*	*2002*	*% Change*	*2003*	*% Change*
January	68.40	101.09	47.8%	67.76	−33.0%	71.93	6.2%
February	93.76	107.06	14.2%	96.51	−9.9%	94.02	−2.6%
March	124.62	138.89	11.5%	118.28	−14.8%	120.58	1.9%
April	137.70	137.09	−0.4%	147.54	7.6%	115.01	−22.0%
May	133.65	129.82	−2.9%	129.04	−0.6%	123.95	−3.9%
June	122.63	124.01	1.1%	115.45	−6.9%	120.59	4.5%
July	100.90	97.92	−2.9%	91.40	−6.7%	101.12	10.6%
August	92.17	82.92	−10.0%	77.85	−6.1%	75.91	−2.5%
September	126.77	68.31	−46.1%	114.77	68.0%	115.66	0.8%
October	136.20	88.91	−34.7%	124.53	41.0%	126.04	1.2%
November	103.42	79.35	−23.3%	83.12	4.8%	90.68	9.1%
December	69.11	57.81	−16.4%	59.27	2.5%	63.64	7.4%

Source: Compiled from monthly STAR Summary Reports, *Smith Travel Research*, 2000–2004.

Conclusion

Terrorism has always been a threat to the normal operations of tourism business worldwide. Such low-probability but high-impact attacks by terrorist groups can disrupt hospitality operations and cause heavy financial losses in a particular destination area. Subsequent media coverage of the terrorist attack tends to scare potential visitors from visiting the destination since they perceive the destination as unsafe or in continued turmoil. Managing a crisis caused by terrorist attacks requires thoughtful coordination of all stakeholders with interests in the hospitality industry and also requires well-developed crisis-management and recovery plans.

The September 11 terrorist attacks on the Pentagon and the ensuing bioterrorism events had an immediate and devastating impact on the hotel industry in the Washington, DC metropolitan area. Room sales decreased drastically due to the closure of Reagan National Airport and major tourist attractions, the series of anthrax scares and cleanup publicity, and tourists' perception of Washington, DC and its primary tourist assets as prime targets for terrorists. A malevolence crisis of such magnitude tested the local hotel industry's ability to manage the crisis and recover from the unexpected attack.

In light of the unprecedented nature of this event, the hotel industry in the Washington, DC metropolitan area worked closely with local tourism and convention authorities, related industries, and government officials to reopen tourist facilities and attractions and to stimulate travel activities. The four clearly defined objectives were quickly leveraged to stabilize the deteriorating business environment and to bring the business community back to normalcy. The coordinated industry response provided a single and reliable information source about the industry status to the valued repeat and potential customers, to the government, to the media, and to the local community. Therefore, any misinformation caused by internal communication mishaps or poorly coordinated communications among different partner groups could be minimized. The push for the reopening of Reagan National Airport and the promotion of government related travel business helped local hotels survive through the most difficult time in their business operations. The marketing campaign sent a reassuring and revitalizing message to the local, regional, national, and international travel markets that Washington, DC was open for business and more than ever a place that Americans and those who love America simply must visit. Perhaps Reagan National Airport and major attractions would have reopened promptly without the concerted, coordinated, and collective commitment and campaigning of Washington, DC's hospitality industry leaders. Perhaps government business in Washington, DC would have materialized regardless of the efforts of these leaders. Perhaps tourists would have been inspired to return to DC in massive numbers even if a large and expensive supplemental marketing effort had not been funded and executed or had been poorly conceived or executed. After all, few places in the world are as spectacular as Washington, DC when the cherry blossoms burst forth in early April. However, it is hard to imagine a better barometer of the success of the Washington, DC hospitality industry crisis response and recovery efforts than examining a period of time when business normally is running on all cylinders and observing it as fully recovered during this time frame. In fact, as Table 3 shows, those hotels closest to the charred rubble of the Pentagon, closest to the exposure to anthrax spores, and closest to the sweet scent of the cherry blossoms enjoyed a banner month in April 2002.

What can hospitality industry and tourism destination managers learn from the Washington, DC hospitality industry response and recovery efforts following the September 11 terrorist attacks? While many of the circumstances and factors described in this chapter are unique to the Washington, DC region, some generic lessons and even a rudimentary model for addressing future hospitality and tourism industry related malevolence crises emerge. The model for managing malevolence crises suggested by the DC hospitality industry September 11 terrorism experience is provided in Figure 1. Because malevolence attacks are sudden, uncertain, and characterized by time compression and the human, financial, and public image impact is calamitous, businesses should immediately access their crisis plans. These plans should be built on a solid knowledge of crisis management skills and refined through workplace practice.

When a malevolence crisis actually occurs, two response steps are vital. Management must first assess conditions internally with guests, staff, and property. This process of assessment also involves reassuring guests and staff, determining additional dangers, mobilizing emergency preparedness, protecting the safety of people and property, and establishing information outposts. When internal conditions have been established, these should be quickly coordinated externally with the local community through preestablished, efficient channels. In the case of the Washington, DC hospitality industry, the conduit to emergency personnel and between hotels was the office of the President of the Hotel Association of Washington, DC, Reba Pittman Walker; and the conduit to local and federal government, regional tourism organizations, and the media was the office of the CEO of the Washington Convention and Tourism Corporation, Bill Hanbury. Different arrangements may work better in different communities. However, a coordinated, orderly response in the early stages of a crisis will minimize disruption of emergency officials' primary focus of rendering assistance. Centralized response will also reduce misinformation. Naturally, internal crisis response must be coordinated with management company, owner, and brand representatives as well.

When the emergency conditions have subsided, management must transition quickly to the process of recovery. There are four steps that appear to be critical to ensuring a quick recovery. Elements of the first three steps may be intermingled and may occur concurrently. First, all regional tourism stakeholders must coordinate their activities. Forums for sharing information and ideas regionally must be rapidly established if they are not already in place. Regional assessments of the impact of the crisis on the area tourism infrastructure and assessments of how to communicate with the media, the local community, and customers should be conducted. At the same time, individual hotel managers should have a voice in this process and information and decisions should be broadly and quickly shared with hotel managers. Fortunately, e-mail communication facilitates such a process today. Second, the local community must be engaged. This effort should include assisting with aid to victims and relief workers, accessing local, state, and federal government agencies for business and worker assistance, steering local media toward positive stories about the industry and its contributions to alleviating the crisis, and encouraging local/regional visitation. Third, clients must be reassured. Hotel general managers should contact as many group and regular clients as possible in their respective hotels. At the destination level, political leaders and industry leaders should contact and reassure convention groups and major demand-generating organizations. Internet sites should be coordinated and updated with timely information. Press events should be staged and

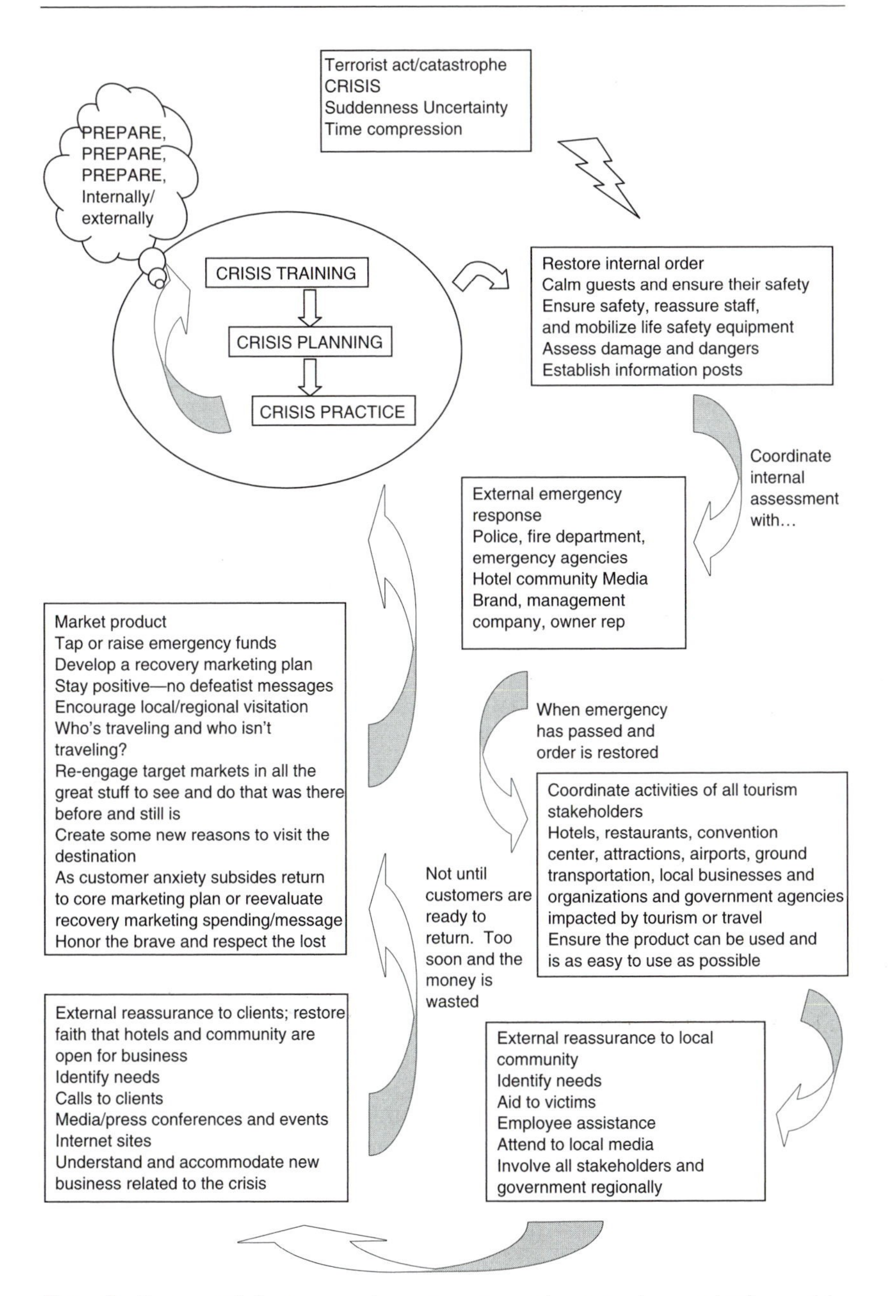

Figure 1 Recommended response and recovery process when managing a malevolence crisis.

coordinated to broadly reassure the public that the destination is open for business. If new business related to the crisis emerges, hotel managers and industry leaders should coordinate their response to this new demand factor to ensure that the destination maximizes its benefit from this new demand. The fourth step in recovery, funding, developing, and launching a recovery-marketing plan must not be taken until the first three steps have occurred and until target audiences have recovered from the initial shock. Special funding may be achieved through reallocating existing resources, generating contributions from private industry, appealing to foundations, or appealing to local, state, or federal government agencies. Normal market plan actions and funding will undoubtedly be insufficient. Target audiences may be emotionally prepared to visit a destination impacted by crisis in different stages. Typically, customers who have compelling reasons to visit, or who live closest to the destination, will be the earliest visitors to return. In all marketing and media communication it is vital that the hospitality industry project a positive, confident reassuring message and avoid any defeatist, self-pitying, or negative tone or language. The hospitality industry needs to identify and target markets that are most likely to travel to the destination, and remind prospective customers of the great reasons to visit and, if possible, create some new reasons/events to generate visitation. As the market recovers and as the anxiety of normal customer bases of the destination recedes, hospitality and tourism leaders should reduce marketing spending to normal levels and reevaluate marketing messages and the distribution of marketing expenses. Finally, destination hospitality and tourism leaders should find appropriate opportunities to honor those who bravely served on the front lines of the crisis response and recovery efforts and properly pay respects to those lost in the crisis. This model provides a framework of key steps in managing response to and recovery from a malevolence crisis as gleaned from the successful efforts of the Washington, DC hospitality industry. It is not meant as an exhaustive checklist of crisis response and recovery activities. Individual hotels and individual communities require more specific planning.

The September 11, 2001, terrorist attacks stunned the hospitality industry in the Washington, DC, metropolitan area. However, hospitality leaders gained valuable experience in handling the crisis and managing the recovery. Two lessons are clear from this experience. First, familiarity with crisis conditions and training in all aspects of crisis response and recovery will better enable hotel managers to handle future crises. It is highly recommended that crisis management expertise and the ability to plan and practice crisis response be added to the qualifications of all key hotel managers. Second, resiliency and a positive, cooperative attitude hasten recovery. Local clusters of hotels, brands, and all stakeholders affected by a crisis should coordinate their efforts in a local manner to optimize the speed and completeness of the initial response and shorten the recovery time. No matter how serious the crisis, the damage is reduced, and the recovery is hastened, when managers are prepared, and when all stakeholders in a crisis cooperate and coordinate response and recovery efforts.

Concept Definitions

Crisis management The planning, coordination, and execution of response and recovery strategies to tourism and hospitality business crises caused by natural disasters, epidemics, terrorist attacks, and management failures.

Malevolent crisis Tourism and hospitality business crisis caused by people with vicious ill intention to harm people and facilities, such as terrorist attacks, extortion, and product tampering.
Skewed values Placing unrealistic financial expectations over corporate values results in unethical or even illegal business conduct.
Time compression Due to the suddenness of a crisis occurrence, managers have limited time to respond to the crisis and experience enormous stress and anxiety.
Anthrax An acute infectious disease caused by the spore-forming bacterium *Bacillus anthracis* in wild and domestic lower vertebrates (cattle, sheep, goats, camels, antelopes, and other herbivores), but it can also occur in humans when they are exposed to infected animals or tissue from infected animals.

Review Questions

1. How can you define different types of business crises that pose a potential threat to normal tourism operations?
2. Why is terrorism described as a malevolent crisis to the tourism industry?
3. To what extent was the hotel industry in Washington, DC affected by the September 11, 2001, terrorist attack on the Pentagon?
4. How did the tourism and hospitality industry in Washington, DC plan and coordinate crisis management immediately after the terrorist attack?
5. Why was the timing for launching the marketing campaign so crucial for a successful recovery?
6. What are the major themes of the recovery marketing campaign and the targeted markets?
7. What do you think of the effectiveness of the recovery marketing campaign?
8. How can you develop a response and recovery process for managing a catastrophic crisis caused by terrorist attacks?

Note

[1] This chapter was previously published as: Stafford, G., Yu, L., Armoo, A. K. (2002). Crisis management and recovery: How Washington, DC hotels responded to terrorism. *Cornell Hotel and Restaurant Administration Quarterly,* 43(5), 27–40. Revision was made to the original article to reflect the latest hotel industry recovery in Washington, DC.

References

Auerbach, A. H. (1998, July 12). When travelers are targets: The growing threat of kidnapping abroad. *The Washington Post,* pp. C1, C2.
Barton, L. (1994). Crisis management: Preparing for and managing disasters. *Cornell Hotel and Restaurant Administration Quarterly,* 35(2), 59–65.
Brewton, C. (1987). Managing a crisis: A model for the lodging industry. *Cornell Hotel and Restaurant Administration Quarterly,* 28(3), 10–15.
Business and Terrorism: Taking Stock. (2001, September 20). *The Economist,* 53.

D'Amore, L. J., and Anunza, T. E. (1986). International terrorism: Implications and challenges to global tourism. *Business Quarterly,* November, 20–29.

Hedgpeth, D., and Irwin, N. (2001, September 28). Running on empty: With few guests, Hotel Washington and its staff feel the pinch. *The Washington Post,* pp. E1, E2.

Higgins, M., and Drucker, J. (2001, September 28). Facing crisis, US travel industry cuts rates. *The Wall Street Journal,* p. A1.

Irwin, N. (2001, October 2). Mayor backs aid for business: Travel-related firms may get loans. *The Washington Post,* pp. E1, E4.

Lancaster, J. (1997, November 17). Killers hunted down tourists in Egyptian temple's recesses. *The Washington Post*, pp. A1, A27.

Lerbinger, O. (1997). *The Crisis Manager: Facing Risk and Responsibility*. Mahwah, NJ: Lawrence Erlbaum.

Lippman, T. W. (1999, January 9). Reports on terrorism suggest closing some U.S. embassies. *The Washington Post,* p. A14.

Lonnig, C. D., and McCaffrey, R. (2001, October 14). Freebies fail to lure public to downtown DC. *The Washington Post,* pp. A1, A7.

Mindszenthy, B. J., Watson, T. A. G., and Koch, W. J. (1988). *No Surprise: The Crisis Communications Management System*. Toronto: Bedford House Communications Limited.

Pearlstein, S. (2001, November 4). Slump stirs specter of worldwide recession: Quick U.S. recovery appears unlikely. *The Washington Post,* pp. A1, A33.

Pizam, A., and Mansfeld, Y. (1996). Introduction, in A. Pizam and Y. Mansfeld (eds.), *Tourism, Crime and International Security Issues*. New York: John Wiley & Sons, pp. 1–7.

Pizam, A., and Smith, G. (2000). Tourism and terrorism: A quantitative analysis of major terrorist acts and their impact on tourism destination. *Tourism Economics,* 6(2), 123–138.

Pressley, S. A. (2001, October 10). The crowds that aren't: Dependent on air travel, Miami tourist industry takes beating. *The Washington Post,* pp. E1, E7.

Ryan, C. (1991). Tourism, terrorism and violence: The risks of wider world travel. *The Study of Conflict and Terrorism,* September, 1–30.

Schneider, G. (2001, September 19). Aftershocks rock local economy: Travel, tourism jobs bear brunt of cuts. *The Washington Post,* pp. A1, A25.

Sheridan, M. B. (2001, October 31). Wall Street to Washington, layoffs shatter lives. *The Washington Post,* pp. A1, A21.

Sönmez, S. F., Apostolopoulos, Y., and Tarlow, P. E. (2000). Tourism in crisis: Managing the effects of terrorism. *Journal of Travel Research,* 38(1), 13–21.

STAR Report: Washington, DC—September 2001. Smith Travel Research.

Stutts, A. T. (1990). *The Travel Safety Handbook.* New York: Van Nostrand Reinhold.

Tarlow, P. E. (2000). Creating safe and secure communities in economically challenging times. *Tourism Economics,* 6(2), 139–149.

Twomey, S. (2001, October 15). In the line of the fired, attack's cost is adding up: Hospitality workers find no jobs at the inn. *The Washington Post,* pp. B1, B7.

Wahab, S. (1996). Tourism and terrorism: Synthesis of the problem with emphasis on Egypt, in A. Pizam and Y. Mansfeld (eds.), *Tourism, Crime and International Security Issues*. New York: John Wiley & Sons, pp. 175–186.

Wall, G. (1996). Terrorism and tourism: An overview and an Irish example, in A. Pizam and Y. Mansfeld (eds.), *Tourism, Crime and International Security Issues*. New York: John Wiley & Sons, pp. 141–158.

Washington Convention and Tourism Corporation (WCTC). (2001a, September 18). News Release. Response to events of September 11.

WCTC. (2001b, September 23). White House Briefing Paper.

WCTC. (2001c, October 5). Summary of marketing, sales and communications activities for "City of Inspiration" campaign.

17

Hospitality Crisis Management Practices: The Israeli Case

Aviad A. Israeli and Arie Reichel

Learning Objectives

- To understand the turbulent nature of the hospitality industry in Israel.
- To understand how geopolitical development emerged into a demand crisis that crippled the hospitality industry.
- To examine factors considered as important by hospitality industry executives for dealing with geopolitical crises.
- To examine tactics actually utilized by executives in order to cope with crisis.
- To investigate the possible discrepancy between factors viewed as important in dealing with crisis in hospitality vs the actual utilization.

Introduction[1,2]

Countries throughout the world derive a large part of the Gross National Product from tourism and hospitality industries. Unfortunately, current times impose some threats on these industries. One of the most significant of these threats is terror, which has been experienced at diverse levels of intensity by different countries. For example, the Israeli hospitality and tourism industry historically has been subjected to cycles of various war and terror related crises (for a review of the conflicts from a hospitality and tourism industry perspective, see Mansfeld, 1999). The current crisis in the Israeli industry originates from the intensification of terror activities, and it provides an opportunity to study crises through a case example, to look at the impact crises have on the hospitality industry, and the tactics employed for coping with crises (in the short and long run). We believe that the present study can serve as an impetus for additional studies in other nations and locations that will enhance the understanding of hospitality management crisis practices and their effectiveness.

To achieve the purpose of this chapter, the study of crisis management practices, a tool dedicated to this purpose first must be developed. An analogy between crisis management and decision making is made in order to develop an evaluation tool that is based on the principles and logic of the multi-attribute decision-making approach (MADM). Next, the tool is validated by using it to investigate the Israeli industry and to describe, using a large sample, crisis management practices or tactics employed by managers in the Israeli industry.

The crisis in the Israeli hospitality industry (2000–2002), and in the Israeli economy as a whole, originates primarily from the instability of the peace process with the Palestinians. The most apparent outcome of this crisis is a sharp decline in the number of tourists arriving in Israel. In order to cope with this decline, the local hospitality industry increases its reliance on the domestic market by offering a variety of package deals. Consequently, the decrease in foreign visitors is compensated for by an increase in domestic tourism, resulting in a change of consumer proportions. As a result, while in 1995 only 40% of guests in Israeli hotels were from the domestic market (Israeli Hotel Association, 1999), the numbers from March 2002 suggest that this trend peaks to above 86% (Table 1 provides a summary of recent data). It should be noted that the domestic market is powerful and organized. Therefore, many consumers in this market understand that they have a powerful position and use it for extensive price negotiations. The most recent economic performance of the

Table 1 Guests in Israeli Hotels

Month	*Year*	*Total Guests (in thousands)*	*Israeli Guests (in thousands)*	*Percentage of Israeli Guests*
September	2000	636.00	336.00	53%
October	2000	437.80	249.90	57%
November	2000	405.20	297.10	73%
December	2000	473.30	377.90	80%
January	2001	417.20	299.00	72%
February	2001	443.40	331.00	75%
March	2001	542.80	414.60	76%
April	2001	461.70	338.90	73%
May	2001	516.40	400.80	78%
June	2001	530.20	430.50	81%
July	2001	619.40	535.10	86%
August	2001	735.50	654.90	89%
September	2001	395.90	347.30	88%
October	2001	459.60	407.60	89%
November	2001	447.40	397.50	89%
December	2001	469.90	405.40	86%
January	2002	388.60	311.30	80%
February	2002	443.00	384.50	87%
March	2002	467.80	402.40	86%

Source: Israeli Central Bureau of Statistics, 2001, 2002, 2003.

hospitality industry, as reported by secondary data and the news media, confirms that the industry is in a deep financial crisis (see a review of performance and yields in Israeli, 2002). Clearly, the decision to rely on domestic tourism is one tactic or practice that was publicly declared and yielded limited results. However, there were other more subtle practices that were (and still are) employed by managers in the industry in order to cope with the lingering crisis.

As mentioned before, this chapter offers a methodology for studying crisis management practices in the hospitality industry. In the next section, we present a tool for studying crisis management practices. Then, in the following section, we illustrate how this procedure is employed to study crisis management in the Israeli hospitality industry. In the conclusion, we provide interpretations to the findings and suggestions for future research.

A Tool for Studying and Evaluating Managerial Practices in Times of Crisis

The tool developed and used to study the way hospitality managers cope with a crisis should provide an evaluation of two related factors: first, recognition of the importance of measures that assist the organization in times of crisis; and second, the level of usage that managers demonstrate for each of these measures. In other words, we are looking for the weight of the measure and its level of use for managers trying to cope with a crisis. This is similar to the classical Simple Additive Weighting model (SAW). This model is probably the best-known, most widely used, compensatory model. In a decision problem with i alternatives, each characterized by j attributes, the value of each alternative can be expressed as

$$EQ=V(A_i)=\Sigma_j w_j v_j(X_{ij}),$$

where $V(A_i)$ is the value function of alternative A_i, and w_j and $v_j(x_{ij})$ are weight and value functions of attribute X_j, respectively.

Using the additive form in our case suggests that we can evaluate the value of management practices employed by the managers. If we could assign a weight to practices, we would be able to approximate the w_j for the additive function. Similarly, if we could evaluate how important each practice is to the manager, we would be able to approximate the value of the practice $v_j(x_{ij})$ for the manager.

For example, assume that managers have three practices that they consider for coping with a crisis in their industry: reducing labor force ($j = 1$), lowering prices ($j = 2$), and engaging in a marketing campaign for new markets ($j = 3$). If we want to gauge managers' actions with respect to this case, we can use an additive function. First, we assign a certain scale to the weights (techniques for assigning weights are discussed in Yoon and Hwang, 1995). In our case, we use a Likert scale where 5 is most important and 1 least important. Second, we use a similar scale (1–5) to find the level of use for each of the three practices.

Continuing with the example, Table 2 provides a hypothetical response from three different managers. The SAW formulation assists us in calculating the ordinal value index of each manager's practices. For example, for manager A, $V_A = (5)^*3 + (3)^*5 + (1)^*3$; therefore, $V_A = 33$. Similarly, $V_B = 43$ and $V_C = 23$. The example suggests that manager B acts in the most effective manner in employing the practices available to him. Managers A and C follow.

Table 2 Importance and Usage of Practices by Managers in Crisis Management

	Manager A		*Manager B*		*Manager C*	
Practice	*Importance*	*Level of Use*	*Importance*	*Level of Use*	*Importance*	*Level of Use*
Reducing the labor force ($j = 1$)	5	3	3	1	3	5
Lowering prices ($j = 2$)	3	5	5	5	3	3
Engaging in a marketing campaign for new markets ($j = 3$)	1	3	5	3	3	1

Another significant measure that results from this approach is an ordinal measure of importance and a measure of usage. Note that the sum of all possible weights is the maximum weight allowed (five in this case) multiplied by the number of practices (three in this case). The actual sum of weights assigned by the manager $\Sigma_j w_j$ is individual for each case. For example, manager A assigned 5, 3, and 1 to practices 1, 2, and 3, respectively. Therefore, dividing the actual weights assigned by the total possible weights generates a measure of achievement on importance. Similarly, we can also gauge the achievement on usage. The measure is computed by dividing the sum of the actual levels of usage by the maximum possible sum. Table 3 summarizes these measures for our example with managers A, B, and C.

The information in Table 3 provides some interesting insights. First, we can see that according to Table 3, although manager B identified the most valuable practices to face the crisis, their usage provides yet another facet to the analysis. Manager B was better than the others overall, but with respect to usage, manager B's performance is below that of manager A. These measures allow us to conclude that manager B is better in identifying the importance of practices but less effective in using the practices in comparison with manager A.

There may be some criticism on the abovementioned approach for crisis management. Critics may argue that the selection of practices is crucial. This is a valid comment, but we should note that there are established criteria for the

Table 3 Measure of Importance and Usage in Crisis Management

	Manager A	*Manager B*	*Manager C*
Measure of importance	60%	87%	60%
Measure of usage	73%	60%	60%

identification and selection of practices in general problems (Keeney and Raiffa, 1976; Yoon and Hwang, 1995). For example, Pardee (1969) suggested that the list of practices should be complete, exhaustive, mutually exclusive, and restricted to the performance degree of the highest degree of importance. Additionally, we try to illustrate in this chapter that a list of practices with general guidelines can also be assembled for specific problems, such as the case of the crisis in the hospitality industry. In the following sections of the chapter, we will demonstrate how to assemble a list of relevant practices for addressing and managing the crisis in the Israeli hospitality industry.

Another issue regards the relationship between managerial actions and performance. The approach presented here provides a measure of the managerial actions taken to manage a crisis. The approach does not and cannot address the long-term implications of these measures. Specifically, we do not know if, in the long run, a manager whose practices ranked high for importance and usage will truly outperform a manager with lower scores for these criteria. The relationship between good management and firm performance is still a subject of extensive research activity. There are two opposing views regarding firm performance: the first argues that structural characteristics of different industries are the primary determinants of performance (Porter 1980). The second view argues that firm-specific characteristics, such as superior management teams, can explain superior performance (Barney, 1991). In a recent paper, Hawawini, Subramanian, and Verdin (2003) summarized prior studies, analyzed current information, and demonstrated that industry-specific factors are more important for performance than firm-specific factors (such as qualified managers). However, with respect to management as a firm-specific factor, Hawawini et al. argue that their impact on performance is only in extreme cases, or what they title as managers who are value creators (winners) vs value destroyers (losers). Nevertheless, we will attempt to demonstrate that there is some evidence of a relationship between crisis management and performance, even though the crisis in the Israeli industry has not ended, and we do not yet have a long-term perspective. Despite the difficulties mentioned above, it should be noted that this chapter provides an innovative method for a multidimensional evaluation of managerial practices for crisis management that may, in the long run, lead to an above average performance for the firm.

Analysis of Crisis Management Using a Multi-Attribute Approach

The first step in studying crisis management practices from a multi-attribute perspective requires that the attributes (or practices) be identified. This can be achieved by assembling a list of practices based on an investigation of the literature, as well as by interviewing practitioners. Apparently, the literature on crisis management practices in the hospitality industry is relatively sparse. The academic research is based on a description of different occurrences of terror in hospitality and tourism contexts (Aziz, 1995; Pizam and Mansfeld, 1996; Leslie, 1996), the classification of violence activities relevant to the tourism industry (Pizam, 1999), the potential involved with the cessation of terror activities (Anson, 1999; Butler and Baum, 1999), and general recommendations of preparations for times of terror (Sönmez et al., 1999).

Mansfeld (1999) reviews the cycles of war and terror in Israel and provides some determinants of crisis management. He suggests that one method for coping with a crisis is by using extensive marketing, especially to local markets. Another determinant includes decisions concerning infrastructure. Specifically, in times of crisis the infrastructure should not be expanded and may even need to be limited. A final recommendation calls for the government to play a supporting role in times of crisis. While much of the aforementioned academic research revolves around classifications and general recommendations, the analysis does not reach the micro-level required to truly evaluate what managers actually do (or should do). Therefore, the second step in the current study involved in-depth interviews conducted with 13 managers of the leading hotel chains in Israel in order to receive their perspectives about crisis management and ensure a wide geographical representation.

The interviews, coupled with the literature review, assisted in constructing four categories of practices: marketing, infrastructure (or hotel) maintenance, human resources, and governmental assistance (see Table 4). In each of the categories, practices relevant to the crisis in the Israeli industry were listed. For example, in the marketing category, the managers suggested that marketing efforts towards foreign tourists, which were traditionally considered ineffective under the terror stricken reality, may become more relevant if they emphasize the fact that the specific geographical area promoted is remote from the areas under threat. Also, in the government category, requests for government support and tax relief were traditional practices. However, some managers added that noticeable protests, such as demonstrations and strikes, might also serve as a viable tool for achieving special attention from the government. Therefore, protest against the government was added in this context to our study of crisis management practices. With respect to maintenance, the managers distinguished between the postponement of scheduled "cosmetic" building maintenance and the maintenance of engineering systems. They also suggested that scheduled payments for maintenance might need to be postponed (the complete list of practices according to category is provided in Table 4).

The Hospitality Crisis Management Questionnaire

The practices were used to build a questionnaire that was made up of two parts. The first part examined the level of importance managers assigned to each of the 21 practices using a Likert scale of 1, least important, to 7, most important. The second part included questions about the level of actual use for each of the 21 practices also using a Likert scale ranging from 1, rarely used, to 7, extensively used. The questionnaire was pre-tested by four experienced executives and then was sent out to 328 general managers from all of the hotels in Israel registered with the Ministry of Tourism. To ensure an adequate response rate, an accompanying letter explained that the questionnaire was sponsored jointly by university researchers as well as by the Israeli Hotel Association. Additionally, the letter emphasized the significance of the issue under investigation and promised to release major findings to the respondents upon completion. Finally, a self-addressed stamped envelope was provided to participants. Out of 328 questionnaires, 116 usable questionnaires were returned, constituting a response rate of 35%. General descriptive statistics for each practice (according to importance and usage) are provided in Table 5.

Table 4 Practices in Crisis Management

Category	*Practice*	*Title*
Human resources	Firing employees to reduce labor force	Practice 1
	Using unpaid vacation to reduce labor force	Practice 2
	Decreasing number of working days per week	Practice 3
	Freezing pay rates	Practice 4
	Replacing high-tenure employees with new employees	Practice 5
	Increased reliance on outsourcing	Practice 6
Marketing	Marketing to domestic tourists in joint campaigns with local merchants (such as Visa, MasterCard)	Practice 7
	Marketing to domestic tourists with focus on specific attributes of the location	Practice 8
	Price drop on special offers	Practice 9
	Reducing list price	Practice 10
	Marketing to foreign tourists with specific focus on the location's distinctive features and relative safety	Practice 11
	Marketing and promoting new products or services (family events, catering)	Practice 12
	Marketing to new segments (such as ultraorthodox)	Practice 13
Maintenance	Cost cuts by limiting hotel services	Practice 14
	Cost cuts by postponing maintenance of the building (cosmetics)	Practice 15
	Cost cuts by postponing maintenance to the engineering systems	Practice 16
	Extending credit or postponing scheduled payments	Practice 17
Government	Organized protest against the lack of government support	Practice 18
	Industry-wide demand for governmental assistance with current expenses	Practice 19
	Industry-wide demand for a grace period on tax payments	Practice 20
	Industry-wide demand for a grace period on local tax (municipality) payments	Practice 21

Before we turn to data analysis, it is important to list the propositions that form the basis of this study. First, in Proposition 1 we assume that there will be a strong positive correlation between the importance one assigns to a certain practice and the level of usage of this practice. This is a necessary condition for rational and coherent crisis management, assuming that managers, through their actions, pursue the practices that they perceive to be important. This normative assumption forms the basis of the Performance-Importance Model (Martilla and James, 1977) that was considered to be an effective management tool but lost favor over the years when other, more quantitative methods became practical with computerization (Duke and Persia, 1996). The matching between performance and importance, or between usage and importance in the current study, is used as a general guideline for managerial decisions. This model was employed in the context of the hospitality and tourism industries in different variations (Hollenhorst, Olson, and Fortney, 1992; Evans and Chon, 1989; Mengak, Dottavio, and O'Leary, 1986).

Table 5 Descriptive Statistics for Practices in Crisis Management

	Importance			*Usage*		
	N	*Mean*	*Standard Deviation*	*N*	*Mean*	*Standard Deviation*
Practice 1	114	5.43	1.64	101	4.77	2.00
Practice 2	112	5.39	1.88	96	3.34	2.25
Practice 3	115	5.50	1.77	101	4.92	2.00
Practice 4	112	4.88	2.09	100	5.06	2.29
Practice 5	113	2.79	1.72	99	2.44	1.88
Practice 6	110	4.47	1.83	97	3.61	2.11
Practice 7	109	5.46	1.60	98	4.68	2.25
Practice 8	114	5.93	1.42	99	5.51	1.77
Practice 9	113	5.14	1.44	100	5.77	1.28
Practice 10	112	4.86	1.71	99	5.07	1.93
Practice 11	109	3.55	2.22	98	2.97	2.14
Practice 12	114	4.89	1.92	100	4.05	2.24
Practice 13	114	5.32	1.69	98	4.69	2.31
Practice 14	113	4.27	1.90	100	3.90	1.90
Practice 15	112	4.11	1.85	100	4.07	2.00
Practice 16	113	3.26	1.81	99	3.36	1.98
Practice 17	110	4.92	1.78	99	4.68	2.05
Practice 18	114	4.01	2.21	99	2.38	1.96
Practice 19	115	5.83	1.63	99	4.40	2.28
Practice 20	114	5.46	1.85	98	3.82	2.40
Practice 21	115	6.45	1.19	100	5.74	1.89

Second, in Proposition 2 we aim to identify the practices that can be grouped together for both importance and usage. As a reference point, we use the traditional categories found in the literature (i.e., human resources, marketing, maintenance, and government). The findings are compared and contrasted against this reference. Specifically, we assume that both importance and usage practices will follow the constructs of *human resources, marketing, maintenance,* and *government,* that is, there will be construct validity (see Table 4).

Results

In order to test Proposition 1, Pearson correlation tests were employed on all 116 questionnaires to find the correlation between the level of importance assigned to the practice and the level of usage of that practice. The results are listed in Table 6 and are presented in a descending order from the highest correlation to the lowest. First, it should be noted that the correlations are all positive, which suggests that, at a basic level, there is considerable correspondence between practices' importance and usage. The practice that received the highest rank on importance/usage was freezing pay rate from the human resources category (0.74), followed by postponing maintenance from the maintenance category (0.72). The next three were all practices from the marketing category: reducing list prices (0.70), price drops on

Table 6 Correlation among Practice's Importance and Use in Crisis Management

Practice	*Practice Details*	*Correlation between Importance and Usage*
4	Freezing pay rates	0.7444
16	Cost cuts by postponing maintenance to the engineering systems	0.7229
10	Reducing list price	0.6998
9	Price drop on special offers	0.6841
12	Marketing and promoting new products or services (family events, catering)	0.6637
6	Increased reliance on outsourcing	0.6625
13	Marketing to new segments (such as ultraorthodox)	0.6289
11	Marketing to foreign tourists with specific focus on the location	0.6184
8	Marketing to domestic tourists with focus on specific attributes of the location	0.6155
1	Firing employees to reduce labor force	0.6077
7	Marketing to domestic tourists in joint campaigns with local merchants (such as Visa, MasterCard)	0.5735
17	Extending credit or postponing scheduled payments	0.5527
14	Cost cuts by limiting hotel services	0.5478
20	Industry-wide demand for a grace period on tax payments	0.5160
18	Organized protest against the lack of government support	0.4562
3	Decreasing number of working days per week	0.4532
5	Replacing high-tenure employees with new employees	0.3920
21	Industry-wide demand for a grace period on local tax (municipality) payments	0.3871
19	Industry-wide demand for governmental assistance with current expenses	0.3700
2	Using unpaid vacation to reduce labor force	0.3287

special offers (0.68), and marketing and promotions for new products (0.66). The lowest correlation between importance of practice and usage was for using unpaid vacation time to reduce labor force (0.32).

In an effort to ascertain if Proposition 2 could be supported, the questionnaire was evaluated in terms of construct validity. An Orthogonal Varimax Rotated Factor Analysis was employed to detect which practices are clustered to factors for hospitality crisis management.

Dimensions of Practice Importance

The analysis was first applied to the importance of the various crisis management practices, and later to the use of these practices. The Factor Analysis of the importance of practices (Table 7) reveals that the 21 practices, grouped into four factors according to Principal Component Analysis and Varimax Rotation method, account for 49 percent of the variance. The minimum loading for each practice in a factor was 0.50.

Table 7 Rotated Component Matrix for Practice Importance

Practice	*Component*			
	1	*2*	*3*	*4*
19	0.814	−0.047	0.044	0.207
21	0.738	−0.107	0.240	0.271
20	0.690	0.245	0.237	0.067
7	0.606	−0.334	0.048	0.002
18	0.603	0.295	−0.096	0.001
8	0.564	−0.344	0.337	0.080
11	0.548	0.124	0.056	−0.369
2	0.498	0.058	−0.122	0.218
15	−0.072	0.766	0.216	−0.104
16	0.053	0.763	0.039	−0.310
14	−0.055	0.548	0.211	0.052
6	0.112	0.464	−0.073	0.070
5	−0.005	0.441	−0.102	0.190
10	0.032	0.053	0.773	−0.215
9	0.210	−0.067	0.772	−0.111
1	−0.175	−0.011	0.584	0.302
17	0.190	0.271	0.472	0.015
13	0.237	0.054	0.273	0.709
3	0.167	−0.026	−0.155	0.586
4	−0.010	−0.012	−0.107	0.547
12	0.277	0.198	0.340	0.471

Note: Extraction method: principal component analysis Rotation method: Varimax with kaiser normalization. A rotation converged in four iterations

The first factor includes practice 19: Industry-wide demand for governmental assistance with current expenses; practice 21: Industry-wide demand for a grace period on local tax (municipality) payments; practice 20: Industry-wide demand for a grace period on tax payments; practice 7: Marketing to domestic tourists in joint campaigns with local merchants (such as Visa, MasterCard); practice 18: Organized protest against the lack of government support; practice 8: Marketing to domestic tourists with focus on specific attributes of the location; and practice 11: Marketing to foreign tourists with specific focus on the location's distinctive features and relative safety. This factor accounts for 17% of the variance and was titled "reliance on government and marketing."

The second factor includes practice 15: Cost cuts by postponing maintenance of the building (cosmetics); practice 16: Cost cuts by postponing maintenance to the engineering systems; and practice 14: Cost cuts by limiting hotel services. This factor accounts for 11.5 percent of the variance explained and was titled "maintenance cost cuts."

The third factor consists of practice 10: Reducing list price; practice 9: Price drop on special offers; and practice 1: Firing employees to reduce labor force. This factor accounts for 11 percent of the explained variance and was titled "lowering prices through labor cutbacks."

Finally, the fourth factor of importance includes practice 13: Marketing to new segments (such as the ultraorthodox Jewish segment); practice 3: Decreasing number of working days per week; and practice 4: Freezing pay rates. The fourth factor explains 9.5 percent of the variance and was titled "finding neglected segments and tightening employment terms."

Dimensions of Practice Usage

The Factor Analysis was employed again to analyze the usage of practices. Table 8 reveals that the 21 practices, grouped into 4 factors for usage according to Principal Component Analysis and Varimax Rotation method, accounted for 48.5 percent of the variance. Again, the factor analysis used the principal component extraction method and Varimax rotation.

The first factor includes practice 16: Cost cuts by postponing maintenance to the engineering systems; practice 15: Cost cuts by postponing maintenance of the building (cosmetics); practice 14: Cost cuts by limiting hotel services, and practice 6: Increased reliance on outsourcing. The factor explains 13.5% of the variance and was titled "cost cutting practices."

Table 8 Rotated Component Matrix for Practice Usage

Practice	*Component*			
	1	*2*	*3*	*4*
16	0.833	0.137	−0.058	0.046
15	0.814	0.021	0.084	−0.075
14	0.709	0.085	0.161	0.011
6	0.537	−0.051	0.023	0.304
5	0.413	−0.155	0.164	0.009
4	0.199	0.140	−0.101	−0.007
19	0.010	0.751	0.323	0.068
18	0.019	0.670	0.041	0.007
20	0.160	0.634	0.268	0.317
2	−0.323	0.617	−0.201	0.131
21	0.006	0.609	0.476	−0.121
1	0.215	0.406	0.212	−0.264
10	0.205	0.307	0.273	−0.100
9	−0.005	0.120	0.779	0.120
12	0.105	0.066	0.638	0.106
17	0.248	0.187	0.618	−0.098
8	−0.164	−0.232	0.608	0.528
13	0.000	0.159	0.554	0.051
7	−0.092	−0.041	0.143	0.767
11	0.384	0.000	0.037	0.612
3	0.071	0.319	−0.005	0.548

Note: Extraction method: principal component analysis. Rotation method: Varimax with Kaiser normalization. A rotation converged in ten iterations.

Factor 2 consisted of practice 19: Industry-wide demand for governmental assistance with current expenses; practice 18: Organized protest against the lack of government support; practice 20: Industry-wide demand for a grace period on tax payments; practice 2: Using unpaid vacation to reduce labor force; and practice 21: Industry-wide demand for a grace period on local tax (municipality) payments. Factor 2 accounts for 13% of the variance and was titled "recruiting government support."

Factor 3 comprised practice 9: Price drop on special offers; practice 12: Marketing and promoting new products or services (family events, catering); practice 17: Extending credit or postponing scheduled payments; practice 8: Marketing to domestic tourists with focus on specific attributes of the location; and practice 13: Marketing to new segments (such as the ultraorthodox). Factor 3 explains 13% of the variance, similar to Factor 2, and was titled "massive marketing."

Finally, Factor 4 included practice 7: Marketing to domestic tourists in joint campaigns with local merchants (such as Visa, MasterCard); practice 11: Marketing to foreign tourists with specific focus on the location's distinctive features and relative safety; and practice 3: Decrease the number of working days per week. This factor accounts for 9% of the variance and was titled "focused marketing and shorter workweek."

Several practices were not included in the factors for importance or usage. First, practice 5: Replacing high-tenure employees with new employees was not included in both importance and usage. Practices that were not included in the factors of importance were: practice 2: Using unpaid vacation time to reduce labor force; practice 6: Increased reliance on outsourcing; practice 17: Extending credit or postponing scheduled payments; and practice 12: Marketing and promoting new products or services (family events, catering). Three practices were not included in the factors of usage. They were: practice 4: Freezing pay rates; practice 1: Firing employees to reduce labor force; and practice 10: Reducing list price.

Dimensions of Combined Importance and Usage Practice

A third factor analysis was utilized to assess the dimensions of hospitality crisis management practices when the importance assigned to each practice was multiplied by its corresponding level of importance. This multiplication is important because it provides the terms that are added to provide the grade of value function $V(A_i)$, as mentioned before. Again, a Principal Component Analysis and Varimax Rotation method was carried out, yielding four components that accounted for 47.54% of the variance.

Table 9 presents the loading of each practice. The first factor includes the following weighted practices: practice 19: Industry-wide demand for governmental assistance with current expenses; practice 21: Industry-wide demand for a grace period on local tax (municipality) payments; practice 20: Industry-wide demand for a grace period on tax payments; practice 18: Organized protest against the lack of government support; and practice 13: Marketing to new segments (such as the ultraorthodox), which is often done in an organized manner. This factor explained 18.8% of the variance and was titled "organized industry-wide efforts."

Factor 2 comprised the following weighted practices: practice 16: Cost cuts by postponing maintenance to the engineering systems; practice 15: Cost cuts by

Table 9 Rotated Component Matrix for Product of Practice Importance and Practice Usage

Practice	*Component*			
	1	*2*	*3*	*4*
19	0.809	0.006	−0.021	0.145
21	0.782	0.014	0.009	0.184
20	0.640	0.205	0.110	0.149
18	0.572	0.113	−0.091	0.023
13	0.538	−0.056	0.276	0.052
2	0.469	−0.312	0.069	−0.070
12	0.459	0.197	0.279	−0.003
4	0.328	−0.040	0.216	−0.247
16	0.064	0.852	−0.047	−0.075
15	0.131	0.840	−0.082	0.113
14	0.129	0.675	−0.057	0.120
6	0.011	0.479	0.277	−0.142
17	0.312	0.420	0.074	0.262
5	−0.122	0.381	0.037	0.070
8	0.164	−0.050	0.750	0.218
7	−0.015	−0.036	0.675	0.079
11	−0.035	0.358	0.615	−0.159
3	0.202	−0.021	0.456	0.004
9	0.113	0.026	0.319	0.834
10	0.061	0.093	0.065	0.817
1	0.268	0.084	−0.306	0.472

Note: Extraction method: principal component analysis. Rotation method: Varimax with Kaiser normalization. A rotation converged in five iterations.

postponing maintenance of the building (cosmetics); and practice 14: Cost cuts by limiting hotel services. Clearly, this factor deals with "maintenance and cost cuts" and was thus titled. The factor accounts for about 12% of the variance.

The third factor included practice 8: Marketing to domestic tourists with focus on specific attributes of the location; practice 7: Marketing to domestic tourists in joint campaigns with local merchants (such as Visa, MasterCard); and practice 11: Marketing to foreign tourists with specific focus on the location's distinctive features and relative safety. The factor was titled "marketing" and accounted for 9% of the variance.

The fourth and last factor comprised practice 9: Price drop on special offers; and practice 10: Reducing list price. It was titled "reducing prices" and accounted for 7.6% of the variance.

Analysis of the Crisis Management Questionnaire Findings

In our attempt to design a tool for measuring hospitality crisis management practices, we offer a list comprising items that were generated from the literature as well as from in-depth interviews. The 21 items (or practices), measured first in terms of importance and later in terms of actual usage, were grouped into four major aspects: human resources, marketing, maintenance, and government. The 21 practices were used as the basis for a questionnaire that was completed by 116 Israeli executives. From the analysis, we can see that the most important practice for managing hospitality crises is an industry-wide demand for a grace period on local (municipality) payments. This practice is also the most widely used. The practice that ranked lowest in terms of importance is the replacement of high-tenure employees with new employees. Also in term of usage, this practice scored second lowest. Clearly, the Israeli hospitality executives feel more comfortable turning, in an organized way, to the local government for help rather than replacing their workforce. This probably reflects both the dominant managerial culture as well as some past experiences regarding the effectiveness of various practices.

It is also interesting to note that in almost all cases, the average ranks assigned to the importance of practices are higher than the average ranks for usage. Only in four cases, freezing pay rates, price drop on special offers, reducing list price, and the practice of cost cuts by postponing maintenance to the engineering systems (practices 4, 9, 10, and 16, respectively), the average means for usage were slightly higher than the means for importance. We can therefore speculate that in most practices there is a hidden agenda that possibly more could be done. On the other hand, the four practices that are more used than important may reflect some ambivalent feelings about their effectiveness (or importance). This speculation is interesting, especially in light of the fact that practice 9: Freezing pay rates, scored the highest correlation between importance and usage; followed by practice 4: Cost cuts by postponing maintenance to the engineering systems; practice 16: Reducing list price: and practice 10: Price drop on special offers. One possible explanation is that these four practices have been extensively used in past crises and are almost automatically considered when a new crisis arises. Given the criticism of Mansfeld (1999) about the lack of learning from past experience, it is doubtful that Israeli hospitality executives actually examined the effectiveness of these four practices when the crises were over.

Proposition 1 stated that there will be a strong positive correlation between the importance executives assign to a certain crisis management practice and the level of usage of this practice. Clearly, Table 5 illustrates Pearson correlations ranging from 0.32 (the lowest) to 0.74 (the highest). All correlations were positive, supporting Proposition 1. Thus it may be argued that this basic correspondence between importance and usage constitutes a necessary condition for rational and coherent crisis management.

Interpretations to Dimensions of Practice Importance

Proposition 2 assumed that both importance and usage practices will follow the constructs generated from the literature and from the in-depth interviews with hospitality industry executives. Four major categories were presented, serving as the foundations of the questionnaire developed in this study: human resources, marketing, maintenance, and government. The results of the factor analysis partially supported Proposition 2. For example, the factors generated from the importance of crisis management practices formed somewhat different dimensions. Importance factor 1, “reliance on government and marketing,” encompasses all four practices of the *government* category, plus three practices related to the *marketing* category. The addition of marketing practices to the practices of turning to central and local governments for support can be explained in light of the aforementioned Israeli hospitality industry's past experience of approaching the government for joint marketing campaigns both domestically and internationally. The Israel Ministry of Tourism has long been involved in marketing campaigns planned and implemented jointly with the hospitality industry, usually through the Israel Hotel Association.

The second factor in the analysis of practice importance, "maintenance cost cuts," corresponds to the *maintenance* category. Cost cuts such as limiting hotel services and postponing maintenance of buildings and engineering systems all correspond to a very similar dimension. Consequently, the exclusion of practice 17: Extending credit or postponing scheduled payments, appears to be logical.

The third factor of importance, "lowering prices through labor cutbacks," includes the two practices of lowering prices, either list price or price for special offers (practices 9 and 10). These two practices were originally included in the category of *marketing*. Their grouping seems logical. However, the inclusion of the last practice in this factor, practice 1: Firing employees to reduce labor force, is hard to interpret.

Similarly, factor four "finding neglected segments and tightening employment terms," included practices from two categories: practice 13: Marketing to new segments (such as the ultraorthodox Jewish segment) from the *marketing* category and practices 3: Decreasing number of working days per week; and 4: Freezing pay rates, from the *human resources* category. In sum, factors three and four only partially support Proposition 2. Finally, the fact that all of the categories are represented in the factors and five practices are excluded enables us to suggest a shorter version of the original questionnaire, aimed only toward measuring the importance of hospitality crisis management practices.

Interpretations to Dimensions of Practice Usage

The results of the factor analysis of usage also partially support Proposition 2. Examining factor 1, titled "cost cutting practices" indicates that three practices from the original category of maintenance were included: practice 16: Cost cuts by postponing maintenance to the engineering systems; practice 15: Cost cuts by postponing maintenance of the building (cosmetic); and practice 14: Cost cuts by limiting hotel services. These three practices were also included in factor 2 of importance. However, practice 6: Increased reliance on outsourcing, was added in the usage factor. It can be speculated that under the harsh conditions of the 2000–2002 hospitality industry crisis in Israel, these practices were the first to be applied as an emergency measure.

Factor 2 of usage was titled "recruiting government support" and includes all of the practices in the *government* category: practice 19: Industry-wide demand for governmental assistance with current expenses; practice 18: Organized protest against the lack of government support; practice 20: Industry-wide demand for a grace period on tax payments; and practice 21: Industry-wide demand for a grace period on local tax (municipality) payments; as well as one practice related to human resources, practice 2: Using unpaid vacation to reduce labor force. The four *government* practices correspond to factor 1 of importance. The addition of unpaid vacation to reduce labor is conceptually similar to the addition of the marketing practices to factor 1 of importance: namely, the reliance on government support. Since "unpaid vacation" is actually financed by the Israel Ministry of Labor, it is another dimension of organized governmental support of the industry.

Factor 3 of usage, titled "massive marketing," is dominated by practices from the *marketing* category. Specifically, four items from the *marketing* category are included: practice 9: Price drop on special offers; practice 12: Marketing and promoting new products or services (family events, catering); practice 8: Marketing to domestic tourists with focus on specific attributes of the location; and practice 13: Marketing to new segments (such as the ultraorthodox). In addition, one practice from the *maintenance* category appears in factor 3: practice 17: Extending credit or postponing scheduled payments. Again, the grouping of the four *marketing* items is clear, but the issue included in practice 17 discolors the clarity of the content validity of the factor. Attempts to explain this pattern remained futile.

The fourth factor of usage, titled "focused marketing and shorter workweek," also only partially confirms Proposition 2. This factor includes two aspects of *marketing*, practice 7: Marketing to domestic tourists in joint campaigns with local merchants (such as Visa, MasterCard); and practice 11: Marketing to foreign tourists with specific focus on the location's distinctive features and relative safety; and one aspect of *human resources,* practice 3: Decreasing number of working days per week.

The mixed patterns of usage seem to be more difficult to interpret than the more solid patterns of importance. One possible explanation is that importance domains are more "ideal" than actual use, thus rendering more coherent patterns. This speculation is worth studying in the future.

Interpretations to Dimensions of Practice Combined Importance and Usage

The results of factor analysis conducted on weighted usage reflect higher construct validity. The inclusion of practices in each factor is much more easily understood and gives much more credence to Proposition 2.

Factor 1, titled "organized industry-wide effort," included all of the practices of the *government* category: practice 19: Industry-wide demand for governmental assistance with current expenses; practice 21: Industry-wide demand for a grace period on local tax (municipality) payments; practice 20: Industry-wide demand for a grace period on tax payments; and practice 18: Organized protest against the lack of government support; plus practice 13: Marketing to new segments (such as the ultraorthodox), from the *marketing* category. As mentioned earlier, the addition of the marketing practice here may reflect the Israeli hospitality industry's custom of sharing governmental resources for marketing campaigns. It should be noted that one of the most recent tactics of the Israel Ministry of Tourism, in collaboration with the Israel Hotel Association, was to convince both ultraorthodox Jews and devoted Southern Baptists in the United States to visit Israel, in spite of the unpleasant and distressing situation.

Factor 2 of weighted usage, titled "maintenance and cost cuts," falls within the domain of *maintenance*. The factor includes the following practices: practice 16:

Cost cuts by postponing maintenance to the engineering systems; practice 15: Cost cuts by postponing maintenance of the building (cosmetics); and practice 14: Cost cuts by limiting hotel services. Apparently, the issue of extending credit and postponing payments is not necessarily an integral part of this domain, and consequently is not included in this factor.

While factors 3 and 4 of weighted usage, titled "marketing" and "reducing price," respectively, fall strictly within the *marketing* category, they represent two different aspects of marketing. Factor 3 included practice 8: Marketing to domestic tourists with focus on specific attributes of the location; practice 7: Marketing to domestic tourists in joint campaigns with local merchants (such as Visa, MasterCard); and practice 11: Marketing to foreign tourists with specific focus on the location's distinctive features and relative safety. This factor consists of marketing to either domestic or foreign tourists with an emphasis upon unique features; specific attributes of the location (or destination), distinctive features and relative safety, and joint campaigns with local merchants. The fourth factor, on the other hand, deals with pricing issues of marketing and includes practice 9: Price drop on special offers; and practice 10: Reducing list price.

Examining the four factors of weighted usage indicates that in essence, the study's initial crisis management categories are supported to a large degree. It is interesting, however, that *human resource* practices were not included either as a unique factor or within the four factors extracted.

At this point there seem to be three major options for future utilization of the hospitality crisis management questionnaire developed in this study: a) using the original 21-item questionnaire with the four categories (*human resources, marketing, maintenance,* and *government*). This will enable future verification of the current study results. b) Utilizing two different questionnaires, one that includes practices that appear in the four factors of importance and a second one comprising practices included in the four factors of usage. This construct of two questionnaires is not synchronized in terms of the ability to multiply practice importance with its corresponding usage. Therefore, the weighted grade may not be available for all practices. Finally, c) use the condensed 13-item weighted usage question naire. Utilizing either the complete 21-item questionnaire or the condensed 13-item weighted usage questionnaire will allow us to employ the additive function and to calculate a certain grade for crisis management practices.

Suggestions for Future Research Using the Crisis Management Questionnaire

The above comment leads us to one interesting and significant question that has yet to be answered. Specifically, do crisis management practices improve the position and performance of the firm in the long run? At this time, we lack the long-term perspective needed to test the method proposed in this chapter and to comment on the above question. Therefore, this question remains unanswered as a challenge for future research. Nevertheless, we provide an anecdotal example to demonstrate how crisis management practices and firm performance may be associated. We caution that this example should not be used to draw conclusions.

Two different locations in Israel are considered to be significant tourist destinations: Tel Aviv and Eilat. There are approximately 35 hotels with about 6,600 rooms in Eilat; in Tel Aviv there are 41 hotels with about 4,700 rooms. Eilat is primarily a resort and vacation destination, and Tel Aviv is primarily a business center. The performance of the local industries in terms of yield was approximated in the past by Israeli (2002) for the purposes of evaluating the economic performance. We borrow the yield proxies from the abovementioned analysis to evaluate its relation to crisis management practices. Another possible aspect of performance is average revenue per employee. This measure was received from the Israeli Bureau of Statistics. Table 10 combines the data from our crisis management results and from the performance results mentioned above. In the development of the crisis management evaluation model, we presented the concept of achievement on importance and on usage. The average achievement for practice importance and usage (with its standard deviations) was calculated for the 24 hotels in Eilat that participated in the study and for 25 hotels from Tel Aviv. A grade was then provided for crisis management by employing the additive function: for each hotel we summed the products of importance and usage grades for each practice. The average grades and the standard deviations for Eilat and Tel Aviv are also provided in the table. The annual yield of hotels in the location is provided next, followed by the average revenue per employee (in New Israeli Shekels).

The findings serve as an example of evaluation crisis management and their relation to performance. There are only small differences between hotels in Eilat and Tel Aviv with respect to the importance assigned to crisis management practices. However, there are significant differences in the level of usage, which is higher in Tel Aviv than in Eilat. As a result, the average grade given to crisis management in Tel Aviv is also higher. The data can support a possible argument that managers in Tel Aviv are better at managing the crisis compared to managers in Eilat, but this improved performance is derived from usage and not from the recognition of the importance of practices. With respect to performance, hotels in Tel Aviv outperform hotels in Eilat with respect to both average yield and average revenue per employee. As we mentioned before, we do not know at this time if a cause-and-effect relationship exists between crisis management and short-term performance, but we present evidence that could and should direct future research in this vein.

Table 10 Crisis Management and Performance: Anecdotes from Eilat and Tel Aviv

	Eilat (n = 24)	*Tel Aviv (n = 25)*
Average of importance achievement measure	67%	68%
Standard deviation of importance achievement measure	14%	15%
Average of usage achievement measure	42%	51%
Standard deviation of usage achievement measure	30%	25%
Average value function grade	203	259
Standard deviation of value grade	153	143
Proxy-for-yield (Israeli, 2002)	29%	57%
Average revenues per employee (Central Bureau of Statistics, 2002)	16,702	18,567

The suggested hospitality crisis management practice questionnaire presented here is only the first step in a series of steps needed to further check its reliability and validity dimensions. It is quite possible that in different countries there will be a need to add or delete practices. For example, the items that deal with central and local governmental forms of support may reflect the Israeli norms and past experience and might have to be adjusted for different countries. On the other hand, it may be argued that given the magnitude of the crisis in Israel, in almost every other country hospitality executives might turn to the government for support. The case of the US aviation industry after September 11, 2001, may support this argument.

Conclusion

The current crisis in the Israeli industry originates from the intensification of terror activities, and it provides an opportunity to study crises through a case example, their impact on the hospitality industry, and the tactics employed for coping with them. The tool developed and used to study the way hospitality managers cope with a crisis provides an evaluation of two related factors: first, recognition of the importance of measures that assist the organization in times of crisis; and second, the level of usage that managers demonstrate for each of these measures.

Based on a literature review and interviews with experts in the Israeli hospitality industry, four categories of crisis management practices were identified: *marketing,* infrastructure (or hotel) *maintenance, human resources,* and *government* assistance. In each of the categories, practices relevant to the crisis in the Israeli industry were listed.

The questionnaire was sent out to 328 general managers from all of the hotels in Israel registered with the Ministry of Tourism. One hundred sixteen usable questionnaires were returned, constituting a response rate of 35%. There were two leading propositions to the study. Proposition 1 assumed that there will be a strong positive correlation between the importance managers assign to a certain practice and the level of usage of this practice. Proposition 2 assumed that practices can be grouped together for both importance and usage to the traditional categories found in the literature (i.e., human resources, marketing, maintenance, and government).

To test Proposition 1, Pearson correlation tests were employed to find the correlation between the level of importance assigned to the practice and the level of usage of that practice. The correlations were all positive, which suggests that at a basic level, there is considerable correspondence between practices' importance and usage. The practices that received the highest rank on importance/usage were freezing pay rate from the human resources category (0.74), postponing maintenance from the maintenance category (0.72), reducing list prices (0.70), price drops and special offers (0.68), and marketing and promotions for new products (0.66).

The results of the factor analysis partially supported Proposition 2. The factors generated from the importance of crisis management practices were as follows: reliance on government and marketing, maintenance cost cuts, lowering prices through labor cutbacks, and finding neglected segments and tightening employment terms. The results of the factor analysis of usage also partially support Proposition 2 forming the following factors: cost cutting practices, recruiting government support, massive marketing and focused marketing, and shorter workweek.

The questionnaire is only the first step in characterizing hospitality crisis management. It is possible that in different countries there will be a need to add or delete such practices. For example, the items that deal with central and local governmental forms of support may reflect the Israeli norms and past experience and might have to be adjusted for different countries. On the other hand, it may be argued that given the magnitude of the crisis in Israel, in almost every other country hospitality executives might turn to the government for support. The case of the US aviation industry after September 11, 2001, may support this argument.

Concept Definitions

Hospitality crisis A sharp decline in demand of incoming tourists, due to negative geopolitical developments and act of terror.

Important crisis management tactics A means for dealing with the hospitality crisis that is considered by hotel managers as important. Such tactics may include special marketing campaign efforts geared toward neglected foreign segments or toward domestic tourists.

Utilized crisis management tactics Tactics employed by hotel executives in order to actually deal with the declining number of guests. The purpose of these tactics is to control the damages of the crisis by cutting costs on the one hand and attempting to attract alternative segments on the other hand. The utilized tactics are expected to correspond to the "important" standard of coherent crisis management.

Coherent crisis management Defined in this chapter as a high level of compatibility between crisis management tactics that were reported as "important" and tactics reported as "actually utilized." The rational model of crisis management assumes a high degree of congruence between importance and usage.

Review Questions

1. Please explain the assumption of the rational coherent model of management tactics. How is this model applied to the crisis management study in Israel?
2. What are the crisis management tactics that were deemed as "important" by hospitality executives?
3. Please explain congruent tactics vs the incongruent tactics in dealing with the hospitality crisis.
4. Can we conclude from this chapter that a high level of congruence between "important" and "utilized" crisis management tactics will result in superior performance? Please explain.

Notes

[1] This chapter was previously published as: Israeli, A. A., and Reichel, A. (2003). Hospitality crisis management practices. International Journal of Hospitality Management, 22, 353–372.

[2] This chapter was supported by the Dean's Fund for Support of Research at the School of Management, Ben Gurion University of the Negev and by the Israeli Hotel Association.

References

Anson, C. (1999). Planning for peace: The role of tourism in the aftermath of violence. *Journal of Travel Research,* 38(August), 57–61.

Aziz, H. (1995). Understanding terrorist attacks on tourists in Egypt. *Tourism Management,* 16, 91–95.

Barney, J. (1991). Firm resources and sustained competitive advantage. *Journal of Management,* 17, (1), 99–120.

Butler, R. W., and Baum, T. (1999). The tourism potential of the peace dividend. *Journal of Travel Research,* 38(August), 24–29.

Duke, C. R., and Persia, M. A. (1996). Performance-importance analysis of escorted tour evaluations, in D. R. Fesenmaier, J. T. O'Leary, and M. Uysal (eds.), *Recent Advances in Tourism Marketing Research*. London: The Haworth Press.

Evans, M. R., and Chon, K. S. (1989). Formulating and evaluating tourism policy using performance importance analysis. *Hospitality, Education and Research Journal*, 29, 176–188.

Hawawini, G., Subramanian, V., and Verdin, P. (2003). Is performance driven by industry—Or by firm-specific factors? A new look at the evidence. *Strategic Management Journal*, 24(1), 1–16.

Hollenhorst, S., Olson, D., and Fortney, R. (1992). Use of performance-importance analysis to evaluate state park cabins: The case of West Virginia state park system. *Journal of Park and Recreation Administration,* 10, 1–11.

Israeli, A. (2002). Star rating and corporate affiliation: Their influence on pricing hotel rooms in Israel. *International Journal of Hospitality Management,* 21(4), 47–64.

Israeli Hotel Association (1999). The Israeli hospitality industry—National conference. Tel Aviv (in Hebrew).

Keeney, R. L., and Raiffa, H. (1976). *Decisions with Multiple Objectives*. New York: John Wiley & Sons.

Leslie, D. (1996). Northern Ireland, tourism and peace. *Tourism Management*, 17, (1), 51–55.

Mansfeld, Y. (1999). Cycles of war, terror and peace: Determinants and management of crisis and recovery of the Israeli tourism industry. *Journal of Travel Research*, 38(August), 30–36.

Martilla, J., and James, J. (1977). Importance-performance analysis. *Journal of Marketing*, 4, 77–79.

Mengak, K. K., Dottavio, F. D., and O'Leary, J. T. (1986). Use of importance-performance analysis to evaluate a visitor center. *Journal of Interpretation*, 11, 1–13.

Pardee, E. S. (1969). Measurement and evaluation of transportation system effectiveness. RAND Memorandum RM-5869-DOT.

Pizam, A. (1999). A comparative approach to classifying acts of crime and violence at tourism destinations. *Journal of Travel Research*, 38(August), 5–12.

Pizam, A., and Mansfeld, Y., (eds.) (1996). *Tourism, Crime and International Security Issues*. Chichester, UK: John Wiley & Sons.

Porter, M. E. (1980). *Competitive Strategy*. New York: Free Press.

Sönmez, S. F., Apostolopoulos, Y., and Tarlow, P. (1999). Tourism in crisis: managing the effect of terrorism. *Journal of Travel Research,* 38(August), 13–18.

Yoon, K. P., and Hwang, C. L. (1995). *Multiple Attribute Decision Making—An Introduction*. Sage University Paper.

18

Tour Operators and Destination Safety

Nevenka Cavlek

Learning Objectives

- To discuss the importance of tour operators in international tourism development.
- To be able to explain the behavior of tour operators towards destinations in a time of crisis.
- To identify the role and significance of tour operators in the recovery of destinations affected by a crisis on the international tourism market.
- To be able to describe the role of tour operators in creating an image of a tourism destination in the minds of both their potential clients and individual tourists.
- To understand why the image created by tour operators may not always reflect the real situation in the destination.

Introduction[1]

Peace, safety, and security are the primary conditions for the normal tourism development of a destination, region, or country and thus are the basic determinants of its growth. Without them, destinations cannot successfully compete on the generating markets, even if they present in their marketing campaigns the most attractive and best quality natural and built attractions. Tourism contributes to peace as much as it benefits from it (Savignac, 1994). According to Pizam (1999, p. 5), every minute of every day a crime or a violent act occurs at a destination somewhere in the world. At the same time, it would be difficult to deny that many types of safety risks coexist in everyone's daily life and within tourism as well. However, an important difference exists: people are rarely in a position to change their place of living, but nothing can force them to spend a holiday in a place that they perceive as insecure. The basic requirement of contemporary demand is a higher quality of supply and services, and that quality has become the most important factor in the existing climate and development of that demand. But usually not mentioned are factors that are the conditio sine qua non—peace, safety, and security, now generally taken for granted. Any threats to the safety of tourists causes a decrease or total absence of activity, not only in a particular destination, but also very often in neighboring regions or countries as well. Consequently, since tourism is an important contributor to national economies, host countries will find it necessary to take substantial measures to bring the country hit by crisis back onto the market as quickly as possible.

From 1950 to 1998 the number of international tourist arrivals in the world increased from 25 million to 635 million (WTO, 1999b), with an average annual increase of 6.97%. Over the past 15 years, international tourism receipts have grown 1.5 times faster than world GDP, with no signs of slowing down. In 1998 international tourism accounted for an estimated 8% of the world's total earnings and 37% of exports in the service sector (WTO, 1999a). According to data from the International Monetary Fund, in 1998 international tourism receipts and passenger transport amounted to more than US$504 billion, putting it ahead of all other categories of international trade (automotive products, chemicals, food, petroleum and other fuels, computer and office equipment, textiles and clothing, telecommunications equipment, mining products, iron and steel, etc.).

This rapid development of international tourism can partly be explained by the tremendous surge in international air traffic and the favorable package holidays promoted nationally and internationally. Indeed, tour operators represent one of the most powerful and influential entities in the tourism industry. They have a strong influence on international flows from main generating markets to various destinations. According to World Tourism Organization estimates, tour operators nowadays have a share of about 25% in the total international tourism market. This means that in 2000 tour operators organized at least 175 million international tourism trips. Therefore, the success of many destinations depends on whether foreign tour operators include them in their programs.

Crises have become an integral part of business activity, and tourism is no exception. Probably no other industry in the world can suffer more from crisis than tourism (Leaf, 1995). This can happen in many different forms, including natural disasters, human-caused disasters, and catastrophes caused by human or technical error. All disasters can turn tourism flows away from impacted destinations, but war, terrorism, or political instability has a much greater psychologically negative effect on potential tourists when planning their vacations. This applies not only to the time of crisis, but also to the period following it. In the case of human-caused disasters, usually three phases of crisis can be distinguished: the period of pre-disaster (with clear signals of a possible outbreak of a crisis), the period of real crisis, and the period of post-crisis—that is, the period up to the full recovery of tourism in the receiving county. All three phases (periods) can be marked as *tourism crisis cycles* during which the appeal of a particular receiving country suffers. The negative effects of human-caused crises at destinations have been described and analyzed by many (Bar-On, 1996; Bloom, 1996; Gartner and Shen, 1992; Leslie, 1999; Mansfeld, 1999; Pizam and Mansfeld, 1996; Radnic and Ivandic, 1999; Sönmez, 1998; Sönmez, Apostolopoulos, and Tarlow, 1999), with Pizam (1999) developing a comprehensive typology of acts of crimes and violence at destinations.

The risks that lie within tourism are not only related to the individual tourist and the host society, but also to the company that organizes the trip, namely tour operators (Steene, 1999). They play a very important role in creating the image of destinations and can significantly influence international flows toward a country hit by safety and security risks. This chapter examines the behavior of tour operators toward destinations in a time of crisis and analyzes their significance in the recovery of impacted destinations on the international market. Special consideration is given to the case of Croatia.

Tour Operators in a Time of Crisis

According to the *EC Directives on Package Travel, Package Holidays, and Package Tours* (Perez and East, 1991), tour operators are considered liable not only for the non-performance or improper performance of the services involved, but also for the physical injury of their clients if this could be in any way linked to negligence due to them, or even to their service providers. This means that tour operators are liable for all aspects of the contract with the client and can be relieved of such liabilities only if they result from force majeure. Therefore, it is natural that they take certain measures to secure the safety of their clients during the journey and while on holiday. In this way, tour operators try to diminish the safety and security risks that their clients could face. They assess destination safety even more critically than an individual would. For example, tourists are very often not aware of the quality of sanitation or health care at a destination (Steene, 1999). In order to avoid risks, tour operators decide whether or not to include in their programs destinations with different kinds of risk, to stop operations to certain resorts already included in their program, to reduce capacities at a destination, or to take certain measures to protect their clients on the spot.

Influencing Factors

The behavior of tour operators toward a destination hit by a crisis depends on many factors, and they make a final decision on their attitude toward it after analyzing all possible aspects of safety and security risks there. Nevertheless, their business practice shows that they primarily concentrate on the following issues, including type of crisis (human-caused crises like war, civil unrest, riots, regional tensions, terrorism, political instability, violence of any kind, crime); natural catastrophes (earthquakes, floods, hurricanes, volcanic eruptions, outbreaks of an epidemic disease, fire); catastrophes caused by human or technical error (nuclear pollution, nuclear tests, oil spills); dimensions of crisis (limited to just a certain place in a country, a certain region of a country, or the whole country); predicted length of crisis (long-term disruption, ongoing uncertainty, short-term/single event disruption, as categorized by EIU, 1994, and Pizam, 1999); consequences of the crisis (level of damage on tourism facilities); tour operator's own business interests at the destination (direct or indirect investments in the country and ownership of tourism establishments); and government decisions of generating country (travel advisory information and warnings). Pottorf and Neal (1984) note that all these factors and issues are interrelated; even if only one component of a destination's tourism or infrastructure is hit by crisis, other components will also experience consequences.

Tour operators have to follow government travel advisory information. For example, a warning by a foreign office is usually soon followed by the governments of other countries alerting their tourists to avoid the country at risk of war. Such was the case in Croatia in July 1991 when all tourists on tour packages had to be evacuated and returned to their home countries. This forced withdrawal resulted in tour operators immediately stopping the promotion and sale of holidays to the country, and led to the sudden halt of inbound tourism for the entire season. In contrast to terrorist attacks, whose time and place of action cannot be predicted, the borders between the regions affected by war and those nearby can be quite precisely drawn (as in the case of Croatia). Still, government warnings to potential tourists always have very strong psychological effects on them and are the main impediment in selling holidays even to parts of the country still safe for holidays.

Very often, the effect of a crisis also spreads to other parts of the country—or even to the neighboring countries where no such problems exist. Safety and security are inextricably connected to issues of international law and political relationships, and so have a ripple effect that goes far beyond the destinations and parties directly involved in the incident (Drabek, 2000, p. 352). Proof of this can be found in the NATO bombardment of Serbia in late March 1999. The war in Kosovo lasted three months, but its impact on tourism in neighboring countries was felt for the whole summer season. When the bombardment began, the repercussions on tourism were felt from Rome to Athens and Istanbul, and from Prague to Budapest, but nowhere more than on Croatia's Adriatic coast. Therefore, the country's tourism in 1999 suffered a 15% drop in international arrivals compared to 1998, while it had actually been estimated that it would achieve a rise of about 30% in 1999. When the bombardment started, tour operators of the main generating markets, like those in Germany and the United Kingdom, immediately cancelled their charter programs to Split and Dubrovnik. A specialist British tour operator to Croatia, beginning in early April, experienced a heavy drop in bookings from a previous level of 250 a day down to only 30 a week. Although bookings picked up slowly after the war, the operator lost two thirds of its Croatia trade, as 1999 client numbers fell from 18,000 to just 6,000 (Richards, 1999). Worse still, some tour operators did not restore Croatia to their summer program for 2000. Those offering packages to Eastern and Southeast Europe also felt this Kosovo effect (Richards, 1999). Many tourists were afraid to travel to Budapest, Prague, and Vienna. Romania was also hit by the crisis, although it had recovered by early August, thanks to the thousands of visitors who traveled to Bucharest to witness the eclipse of the sun on August 11, 1999.

Alongside the serious practical difficulties that confront receiving countries in times of crises is the lack of sound geographical knowledge of many tourists. This problem, for example, depleted Greece's US market, but for the oddest of reasons. Many Americans apparently confused the Greek island of Kos with the capital of Kosovo. But the lack of this kind of knowledge can sometimes also have a positive effect. This was the case with the island of Bali where tourism flourished during the crisis in Indonesia, because most of the tourists did not know that Bali belonged to Indonesia.

It has to be borne in mind that tourists nowadays have such a wide holiday choice that they usually do not even consider traveling near places where they might be at risk. One destination can easily be substituted by a similar or even a completely different one elsewhere. Tourists prefer to wait until the situation in the respective country becomes normal again. This is simply the basic attitude of tourists. Such a situation leads to negative economic effects from international tourism flows in respective destinations. Since the crisis usually outlives the physical damage, the tourism community/industry has to find ways to manage the disaster's aftereffects (Sönmez, Apostolopoulos, and Tarlow, 1999).

Attitude towards Destinations in Times of Crisis

By analyzing the problems that tour operators have had in the last decade with safety and security issues in many receiving countries, certain factors come to light. Travel agencies in the generating markets do not want to sell holidays to risky destinations, even if they have a client who is interested in them, as was the case in Croatia. Many clients who intended to spend their holidays there during the time of crisis were discouraged by their travel agencies from purchasing package trips from Croatian tour specialists. Some of these clients decided to book the holiday directly from the specialists. But the question arises, how many clients accepted the advice and instead went to other destinations? This clearly suggests that travel agencies are not willing to risk the problems their customers might face at unsafe destinations, a situation that also applies to foreign tour operators. To protect the interests of their clients, as well as their own interests, they withdraw from the country at risk and delete it from their program.

Media and journalists in such situations will not recommend to their readers or viewers that it is safe to travel to a destination their government considers to be a risky zone, even when they themselves might have been to the destination and are convinced that some parts of the country are safe. It should also not be forgotten that media bombard readers and viewers with news during the time of crisis and people find negative reports far more interesting than positive. The media and journalists are very much aware of this tendency. Further, insurance companies do not cover damage that is the result of violence, military actions, and the like. Insurance premiums for aircraft and buses operating to an area of high risk (which do not cover total risk) are dramatically higher than premiums in normal circumstances.

Once imposed, travel restrictions to a particular country under risk of war cannot easily be changed. Foreign governments prefer to lift restrictions for the whole country when all is safe. When the governments of generating countries do not issue special warnings to tourists to avoid particular countries or destinations, it is up to tour operators to decide how to react to each case. The problem arises when clients want to cancel their holidays themselves, even though tour operators do not intend to stop their program there. Although they have contractual rights with their clients to ensure payment is made for the journey, the final decision usually depends on the particular situation. Tour operators will try to consider what is in the interest of their clients and also what is in their own interest for the protection of their own good image.

The growth of terrorism, crime, and violence in the world has given rise to a new problem: tourists as targets of terrorist and criminal acts (Cavlek, 1998). As pointed out by Richter and Waugh (1986), tourists are targeted for their symbolic value. For terrorists, the symbolism, high profile, and newsworthiness of the international tourist are too valuable to be left unexploited (Sönmez, Apostolopoulos, and Tarlow, 1999, p. 15). The responsibility of tour operators toward their clients means that they cannot neglect this very fact. In cases of uncertain disruptions, they will usually stop their operation to the particular country for a period of time until local governments undertake adequate measures to diminish safety risks. Alternatively, tour operators will significantly reduce their capacity to the destination. In this way they keep the destination in their program, but at the same time try to minimize their risk of operation. The canceling or omission of a particular destination from a tour operator's program is also a signal to individual tourists that the destination is not safe (Cavlek, 1998). Thus, tour operators influence the way a particular destination is viewed, because their behavior toward it affects the attitude of all potential tourists. In this way, they influence the image of a destination and thus have a direct impact on the tourism income of a particular country. The business interests of tour operators and their partners in the receiving country in such situations lose their common ground, and marketing a destination in generating markets becomes their main problem. The respective destination thus stays without direct and adequate support on foreign markets, since their tourism mostly depends on outside tour operators. There have been many examples of such behavior in tour operators. A few cases of the reactions of foreign tour operators toward countries hit by safety and security risks support the points raised.

China 1989

After Beijing's Tiananmen Square incident, Silk Cut Travel cancelled the whole summer program for the country; China Travel Service cancelled all travel until September; and Voyages Jules Verne cancelled all travel until July. In total about 300 groups cancelled their travel plans, and the country's tourism earnings declined by $430 million in 1989 (Gartner and Shen, 1992).

Egypt 1992–1994

Over 120 tourists were targeted by Islamic terrorist attacks during this period. Some European tour operators struck the country from their program entirely, but most of them reduced their capacities significantly. In 1992–1993, the biggest tour operator in Europe, TUI, sold 30% fewer holidays to Egypt than the previous year, while Tjaereborg sold 77% fewer. In 1993–1994, the former further reduced its sales to Egypt by 22% and the latter by 27% (FVW-International, 1993). During this time, Egypt recorded a significant drop in arrivals (22%), in tourist nights (30%), and even more so in tourism income (43%) (Wahab, 1996). With signs of recovery from the crisis, TUI immediately increased its sales to the country in 1994–1995 by 20% and in 1995–1996 went back to the country completely (158% increase in sales compared to 1994/95) (FVW-International 1995; 1996).

Egypt 1997–1998

Most tour operators pulled out immediately after the terrorist attack in Luxor, in which 64 tourists were killed. They repatriated their clients from Egypt or organized other holidays for them in another country. Following travel advisory information, foreign tour operators abandoned the destination in the winter season. In the summer of 1998, tour operators slowly started to reintroduce the country in their programs, but with only one tenth of their earlier number of clients.

Florida 1992–1994

After several killings of foreign tourists in different locations in Florida, European tour operators still decided to keep Florida in their programs, but gave their clients the opportunity of changing their previously booked holiday to some other destination in their program, with an additional payment if their choice was more expensive. Many of their clients favored this opportunity. Tour operators themselves took some special security measures for their clients. As the targets of the attacks were tourists using rental cars to sleep in during the night, tour operators booked a hotel near the airport for their clients who landed in Florida for late arrivals, so that a car could be hired the next morning. The drop in tourists in 1993 from the British market alone was around 10%.

Russia 1993–1994

After the news was broken on Moscow television in August 1993 that more than 4,000 people were affected by diphtheria in Russia, with 900 in Moscow alone, the biggest British tour operator, Thomson, cancelled its whole Moscow and St. Petersburg program for the whole year. Many other European operators did the same. The clients who were supposed to travel there in August were informed of the operator's decision and were told that they could still travel, but that the tour operator would not bear any risks in case of illness.

Turkey 1992–1994

Since 1991, the Kurdistan Worker's Party had been directing their terrorist attacks against tourists by bombing sites and hotels and even kidnapping tourists. Still, major European tour operators did not cut Turkey from their program, but safety problems resulted in a drop-off in tourism demand. For example, TUI had 10% fewer clients than the year before, the second largest German tour operator, NUR, 14% less, etc. In 1993–1994, the safety situation did not change and TUI had a decrease in sales of 47%, but in 1994–1995 quickly returned with a 67% increase and in 1995–1996 with a further increase of 13%.

Turkey 1997–1998

The threat of the Kurdistan Worker's Party terrorist attacks resulted in 11.1% fewer German passengers flying to destinations in Turkey (ADV, 1998).

Commitment to Destinations

The actions that tour operators are ready to take themselves in order to allow their clients to visit a destination at risk depend greatly on the interest in a particular country. The more important the destination is in their entire business program, the more attached to it they are. If a tour operator is a shareholder in some hotels in the respective destination, or if its owners are a company that has made some direct foreign investments there, it is only natural that the tour operator would try to return to that country more quickly than to another without such business interests. Probably the most obvious example, from many that can be found to support this supposition, is the development of tourism in Spain and Turkey.

It is estimated that Spain today absorbs about 18% of the total world package market and thus represents the largest receiving destination for this form of tourism in the world. The most influential German tour operators, such as TUI, C&N, LTU, and FTI, are often owners or shareholders of hotel chains in Spanish resorts to which they send their clients, but they also have exclusive fixed contracts with many hotel companies. In this way, tour operators are able to ensure accommodation capacities at a particular destination for their clients, as well as to have a stronger influence on the quality control of the services provided and to achieve optimal standards. One also has to bear in mind that TUI and C&N are at the same time leading tour operators in the Dutch, Belgian, and Austrian markets, and have a very significant role in those and some other European generating markets from which Spain receives a significant number of tourists.

In the development of international tourism in Turkey, which started relatively late compared with traditional Mediterranean destinations, a spectacular growth took place over a relatively short period of time. According to WTO data, in 1979 Turkey had some 600,000 international arrivals, in 1989 it registered 2,855,500, and in 1998, 9,200,000. The most important role in the development of tourism in Turkey was played by tour operators organizing favorable air package holidays to this destination. They have made Turkey the bestseller on the Mediterranean. Of course, this would not have been possible if German capital had not been invested in infrastructure development. In 1998 Turkey was visited by 2.23 million tourists from Germany and almost a million from the United Kingdom. The majority of tourists from both countries go to Turkey on organized package tours. The most powerful and influential European tour operators are owned by different entities whose main field of activity often does not have any other close connection with tourism, such as the industrial concern Preussag, which owns the largest European travel concern, TUI and also the largest tour operator in the British market, Thomson Travel. Very often powerful tour operators are only segments of huge multinational concerns that control direct foreign investments in tourism as well. Hence, it is understandable that such companies push the tour operators owned by them to direct flows toward the destinations and countries in which the investors are interested. TUI and C&N have invested money in hotels in Turkey in which they are the main shareholders or complete owners (TUI owns the brand name Iberotels-RIU, and C&N has a 95.5% share in Paradise Hotels). Moreover, German tour operators are tightly linked to smaller or larger hotel chains elsewhere (like Tunisia, Cyprus, Egypt, Greece, and the Caribbean). For example, TUI partly or totally owns over 170 hotels in 19 receiving countries and in 42 resorts, with a total capacity of over 85,000 beds. In 5 years' time, TUI intends to raise its hotel portfolio by an additional 40,000 beds. Business logic requires a tour operator to care more about a destination in which it is financially involved.

Following these examples, it becomes clear why the safety risks experienced both in Spain and Turkey, coming specifically from terrorist groups, have so far not had such major repercussions as in some other countries. The strong commitment to destinations is even more marked by tour operators who specialize in only one or a few countries. They cannot simply abandon the only destination they may have in their program, and so they tend to stick to it fiercely. In a time of crisis, they are sometimes the only connection between the receiving and generating markets and are thus very important for the future of the destination. However, vertical integration with the accommodation sector in the respective receiving country represents a substantial risk for the tour operator, particularly when the country involved falls into a crisis. Since tour operators work with very low net profit margins, it is important for them to share in the profit of different components of the whole product that they supply. They are ready to take such risks, but only in those countries that are attractive for foreign capital and that have legal conditions that support foreign investors. The more the tour operator is vertically integrated with companies in the receiving countries, the more attention it pays to the speedy recovery of tourism in the respective country.

Necessary Actions of Destination Countries

With an increase in different types of crises in the world, there is a growing number of case studies with examples of crisis management strategies. Tourism theoreticians analyzing these cases try to make a framework for crisis management (Barton, 1994; Drabek, 1994, 1995; Ioannides and Apostolopoulos, 1999; Mansfeld, 1999; Pizam and Mansfeld, 1996; Radnic and Ivandic, 1999; Santana, 1999; Sönmez, Apostolopoulos, and Tarlow, 1999; Sönmez, Backman and Allen, 1994; Steene, 1999). The researchers point out that a tourism crisis must be controlled, and this can be done successfully through comprehensive crisis management. Therefore, the main task of the receiving country hit by some kind of crisis is to take all necessary steps to handle the crisis in the best possible way. But, although tourism is vulnerable to catastrophe, the risk is not fully recognized by those within it (Drabek, 1995, p. 14). The industry within the country can manage the crisis efficiently only by being prepared to take some actions in advance. Planning is an essential element of control. Without it, an organization is at the mercy of events (Barton, 1994). However, emergency planning specifically for tourism has only lately become an area of research and has not been widely discussed and theoretically developed. Therefore, most literature is based on case studies that illustrate actions usually taken by hotels when the crisis has occurred. Interestingly enough, the best studies until now have been made by academics who are not primarily involved in the tourism field, such as Drabek (1994, 1995, 2000), a sociologist. The recovery of this industry from a crisis is far more complicated than for others. As proven in practice and stressed by Sönmez, Apostolopoulos, and Tarlow (1999, p. 15), although tourism is quite adept at using established marketing principles, setbacks due to negative occurrences call for something more than traditional efforts. The industry must conduct recovery marketing that is integrated fully with crisis management activities. Its complexity requires a proactive role from all tourism officials in the public and private management of marketing activities. Very strong partnership and coordinated work among the government, national tourism organizations, foreign tour operators, local travel organizers, and local hospitality officials is essential. Each needs to participate to an important degree in order to secure the fulfillment of several important actions. These include successful rebuilding of the destination image, overcoming any adverse publicity resulting from the crisis, short-term restoration and long-term reconstruction of the damaged tourism facilities and infrastructure, and effective management of media coverage (Drabek, 2000).

The very fact that so many parties need to be involved in handling the crisis in a proper way requires intergovernmental and public-private partnership and harmonized and coordinated actions among them. This means that the ministry of tourism is not the only government body that needs to undertake measures to overcome the problem. Its policies and actions have to be supported by necessary measures taken by the ministry of foreign affairs and the ministries of transport, culture, economic affairs, and the like. The government actions form the groundwork on which national tourism organizations, tour operators, and local hospitality officials will continue to build. Therefore, it is essential to establish a governmental agency to handle public relation activities coordinated with the public and private sectors within the industry.

In times of crisis, local government bodies do not always act quickly enough, but journalists do. Using the fact that people are more interested in negative than positive news, and that they have an appetite for spectacular events, journalists sometimes exaggerate some incidents that happen to tourists abroad. Repeated negative news in the media creates a negative image of a particular destination, consequently causing the image of tourists toward this destination to change. Most scholars would agree that a destination image is a key factor that influences tourists' buying behavior. Gunn (1972) was among the first to examine the image formation process and made a distinction between the component parts of its "induced" and "organic level." Induced images are created from the strategic promotion by a destination and/or businesses (mostly national tourism organizations and tour operators). They tend to create images in the mind of potential tourists by producing and distributing promotional materials through different channels of distribution or even directly to selected individuals. Organic images, on the other hand, are not created directly by the organizations that depend on the tourism development of a particular destination, but are created by mass media like the television, newspapers, films, and other ostensibly unbiased sources of information. Such images consequently tend to be more influential. Still, in both cases, decision processes in choosing a particular holiday are based on tourists' mental images of a destination. Therefore, poor control of relations with the media during a crisis and afterward can severely damage the long-term viability of the industry, just as skillful relations can greatly enhance it. To minimize effects of bad publicity on the tourism of a receiving country hit by crisis it is necessary to coordinate all activities with the Ministry of Tourism, National Tourist Organizations, foreign tour operators, local travel organizers, airline companies (national and international), hoteliers, and other related organizations; to give the media accurate information about the crisis, because credibility is critical to this process; to inform the media about the steps taken in the country to solve the problem; to make use of media interest in the country to broadcast positive facts; to invite journalists to the destination to show them the real situation (McGuckin and Demick, 2000, p. 339); to work closely with tour operators and give them all the support they need to keep the destination in their program; to organize study trips to the country for tour operators and travel agents (sales staff); and to promote the destination in target markets, so that it stays present in the minds of potential tourists.

Although not all tourism experts will agree with the last recommendation, since they might find marketing campaigns during severe and ongoing security crises to be a waste of resources (Mansfeld, 1999, p. 35), the practice in the case of Croatia or Northern Ireland proved the opposite.

One could argue that the local tour operators and hospitality groups of receiving countries may be more willing to put tourists at risk than foreign counterparts. It might also be more difficult for foreign tour operators to evaluate accurately and to understand the real conditions in the respective destinations, and to use that understanding to enable them to restore the flow of tourists as soon as possible without putting their clients at risk. Although this might be true to some extent, it should be remembered that local tour operators and hospitality groups cannot change the destination image sufficiently well by themselves. Potential clients rely heavily on organic images that are created independent of local officials. Nevertheless, when it comes to induced images, clients again will trust their own compatriot tour operators more than site-based tour operators or hospitality officials of a receiving country. Therefore, the successful rebuilding of the destination image should first start with organic images, because this is the foundation on which induced images can be created with the help of foreign tour operators. Once they return to the country, the local counterparts can really begin to restore their business too.

Although nobody can be prepared for every conceivable type of crisis, every company should have some general procedures in place to effectively deal with various situations. However, many industries, especially in less developed parts of the world, usually do not see why they should invest in something that they might never need. This corresponds to Drabek's (1994) research documenting the reasons for many companies not having written plans. Therefore, according to him, a community partnership comprising local emergency managers and tourism representatives should be set up to stimulate greater awareness of current vulnerability and to encourage the implementation of a preparedness plan (1995, p. 16). Experience shows that crises have become a part of daily lives, and neglecting this fact can cost the company much more than the cost of establishing an action plan to handle a general crisis. That is why more and more companies do realize the necessity of having a crisis management plan prepared in advance. The necessary steps for every tourism organization can include activating a crisis management team made up of parties directly and indirectly involved with the crisis; following a prepared manual with necessary steps for certain situations and appointing persons with defined tasks and responsibilities; defining new working relationships needed in the event of a crisis; creating up-to-date lists of all important media (foreign and local press); drawing up a list of important people in the tourism business (tour operators, leading travel agents, credit-card companies, major hotel chains); and organizing internal exercises for the staff after the crisis to train them how to better handle a similar crisis in the future (based on Leaf, 1995).

The whole industry should be aware that the successful crisis management actually depends on constant education—not only internal education, as practiced by tourism companies that have experienced a crisis, but even more, long-term educational activities in receiving countries. Indeed, Drabek suggests (1995, p. 16) that educational initiatives should be implemented to ensure that university curricula in tourism and hospitality administration include more emphasis on disaster management, including mitigation, preparedness, response, and recovery. This suggestion might also be extended to other levels in addition to university education.

Croatian Experience

During the whole period of war in Croatia international tourism flows to the country did not stop. Although it was thought at first that the war in Croatia would be a short-term disruption, it actually became a long-term one. Soon after the governments of European countries issued warnings to tourists to avoid the country (in the middle of the summer season 1991), all tour operators withdrew from Croatia.

Before the war, the main generating countries for Croatia were Germany, Italy, Austria, the United Kingdom, and the Netherlands. Being near the receiving market, tourists from Austria and Italy have always preferred to travel to Croatia individually. In the case of Germany, more than one third of tourists used to come in organized package tours, as was mostly the case with the Dutch tourists. The majority of British vacationers also used to travel in an organized fashion, using the services of tour operators. But experience shows that even when tourists travel individually, they usually contact their travel agents for some kind of information about their intended destination and also take brochures from tour operators who include their choice in their program. Tour operators have always had quite a significant influence on the total generating market because, through their brochures, they create a kind of "travel fashion," so that even individual itineraries of tourists are prepared according to the suggestions of tour operators. The fact that in 1992 and 1993 no foreign tour operator included Croatia in its summer program leads to the conclusion that this situation caused enormous negative consequences for Croatian tourism.

In 1990, Croatia registered just over 7 million foreign tourists and in 1992 only 1.3 million. From its main generating market (Germany), Croatia in 1990 registered 1.5 million tourists, while in 1992 only 10% of this volume. The consequences of the absence of British tour operators were even more noticeable because Croatia is a typical package market for their holiday-makers. So, while in 1990 Croatia was visited by some 600,000 British tourists, in 1992 this number dropped to 9,000 individuals. Even these cannot be considered "real tourists," since they were mostly family members who came to visit British soldiers in UN peacekeeping forces.

The war of aggression on Croatia never affected its most developed tourism resorts in the Istrian region, on the northern part of the Adriatic. Therefore, the country tried to persuade foreign tour operators to come back to the places not affected by the war which thus were safe throughout the whole period of the crisis. Such initiatives demand a great deal of work and investment in order to reestablish business connections with tour operators, travel agents, and the media. They together play a major role in persuading potential tourists to spend their holidays in a particular country. During the whole crisis period, Croatia was present in all major generating markets through its national tour operator: Bemextours. Because channels of information for the new country had not been formed, it initially performed the task of the national tourist organization, promoting the country on different markets, and was the only place where information could be obtained. Being the specialist for Croatia, Bemextours, compared to all other tour operators that listed their programs, was the most severely hit by this crisis. If Bemextours had not had the support of the Croatian government to continue its activities in German, British, Dutch, Austrian, Italian, French, Swiss, Belgian, Czech, and Slovakian markets, its position under these long-term circumstances of crisis would certainly have come into question. But it was in the best interest of the Croatian government to back this national tour operator, which had a dual role during the period. This entailed national tourist promotion (at the beginning of the crisis, this tour operator was the only connection between the main generating markets), as well as political promotion of the new, independent country whose new name was unknown to the majority of potential clients. By declaring its independence, Croatia, once part of the former Yugoslavia but with a tourism tradition of over 150 years, was perceived on the foreign market as a new receiving destination. The national tour operator fulfilled this dual role by joint actions with the Croatian Ministry of Tourism, hoteliers, and the newly established airline company, Croatia Airlines, which offered the country's only air connection with the outside world (all other air companies had stopped their operations to Croatia).

The main thrust in promoting Croatian tourism in foreign markets was to be present at the main tourism fairs in Europe, to work with tour operators and travel agencies to reestablish their operations in Croatia, to organize press conferences abroad, to hold permanent contacts with the foreign media, to distribute promotional material to travel agencies and directly to potential clients, to organize study trips for journalists and travel agents, to prepare accurate press releases, and to organize individual talks and interviews with foreign journalists.

The work of Bemextours with German journalists in 1992 led to the organization of 10 press conferences and over 400 journalists and TV crews being invited to Croatia, which resulted in 207 published articles in newspapers, 63 radio interviews, and 24 TV interviews. This all had a very positive effect in promoting Croatia on the tourism market of Europe. The commercial value of the publicity that Bemextours achieved for Croatia in foreign markets was immense. Just one program on Dutch TV about the safety of the region of Istria, which lasted 35 minutes during prime time, according to a promotional tariff, would have cost $700,000. Not long after this publicity about safety issues, the Dutch Government lifted its travel alert to this Croatian region. Not only did Bemextours acquaint the foreign market with a new name for a known destination, but they also helped to keep this destination in the minds of those who used to go to Croatia for their holidays. Bemextours did not succeed in making a profit for itself with this promotional campaign, since a very small number of tourists decided to buy a package tour for a holiday in Croatia. However, the overall benefit of this promotion was very significant because, as the data in Table 1 show, this message had a much stronger influence on individual tourists.

At the fairs where Bemextours was present with its brochures and holiday offers to Croatia, it attracted the interest of travel agencies, other tour operators, tourists, and journalists. Color brochures presented 42 resorts, 130 hotels, and tourist villages in Croatia where it was completely safe to spend a holiday. This was a surprise

Table 1 Main Promotional Activities Organized by Bemextours in Europe in 1992

Bemextours Organization	*Brochures Issued*	*Travel Agents with Bemex Brochures*	*Presentation at Fairs*	*Journalists Invited to Croatia*	*TV Crews Sent to Croatia*	*Bemex Clients in Croatia*	*Tourists in Croatia*
Austria	270,000	1,247	5	9	—	4,500	164,000
Belgium	100,000	834	5	3	—	0	6,000
France	120,000	1,580	3	2	—	375	16,000
Italy	250,000	1,770	4	15	4	1,185	239,000
Holland	300,000	1,900	4	7	2	0	12,000
Germany	450,000	3,900	10	38	2	2,842	148,000
Switzerland	90,000	730	5	12	—	0	7,000
TOTAL	1,580,000	11,961	36	86	8	8,902	1,270,855
Total cost of promotion in dollars	965,172						

Source: Internal company data and Statisticki godisnjak Hrvatske (Republic Bureau of Statistics).

for all at the fairs, because other tour operators did not offer package holidays to Croatia. But at the same time, some foreign tour operators started to consider including Croatia in their programs, or to issue offprints. In this way, the distribution of the first Croatian brochure with package holidays and its presentation at all major fairs in Europe gave a clear sign to other tour operators not to exclude the whole of Croatia from their program. One of the aims of the brochure was also to stimulate other tour operators to issue programs for the country. The line of thinking was simple and clear: if a national specialist, who knows the situation in the country best, invests considerably in marketing a Croatian product, other tour operators should perhaps not be left behind (either for business or for competitive reasons). Therefore, this particular operator played the role of catalyst in the market and very much helped to "break the fear" of traveling to Croatia.

Although the crisis in Croatia has long passed, tourists from the main generating markets are returning only slowly. In 1998, Croatia achieved not more than 50% of the overnights spent by German tourists in 1990, about 68% of Italians compared with 1990, less than 30% of Dutch tourists, and the situation with the overnights of British holidaymakers has become marginal (9.1%). The reason could be that during this long period of negative media coverage, tourists and tour operators turned toward other destinations. Consequently, Croatia, which has emerged from a war, has to create its tourism identity from scratch, and develop its own image. But image has its time dimension and it changes slowly. It has to be built, developed, consolidatedm and monitored (Gartner, 1996; Pirjevec, 1998).

Today Croatia depends heavily on foreign tour operators that control European tourism flows. These can significantly help the country to return more quickly to international markets. Although the largest European operators have returned to the country, the accommodation capacities that they have taken are very small, and can hardly be compared with the period before the war. To be able to comprehend this situation completely, the ownership structure of the leading European tour operators should also be analyzed and understood (Cavlek, 2000). As foreign investors have still not found this destination worth investing in, Croatia has problems in positioning itself better on the international market. This statement can be supported by comparing the development of tourism in Turkey, which, for example, registered 36,000 tourists from the United Kingdom in 1988 when Croatia at the same time was visited by almost 400,000. Just 10 years later, the number of British tourists to Turkey had risen to 1.6 million, and Croatia registered 56,000. Another example is the engagement of the leader TUI on the Croatian market. In 1987, it had 124,000 clients in Croatia and the numbers for Turkey were marginal. In 1996 TUI had 17,800 clients in Croatia and almost ten times more in Turkey (172,000). But, as already noted, TUI and some other major European tour operators, or their parent companies, are financially engaged in Turkey, but none of them is in Croatia.

Conclusion

Peace, safety, and security are the conditio sine qua non for development of tourism. Any threats to the safety of tourists cause a decrease or total absence of activity in an affected destination, which in turn can negatively influence inbound tourism to neighboring destinations as well. The movement of demand from the most significant generating markets to foreign countries is greatly impacted by the leading tour operators, since they organize the largest number of trips abroad from these markets. They have become very powerful and influential players, even shaping the image of destinations. The crisis behavior of tour operators toward the destination primarily depends on the type of crisis, its dimensions, its predicted length, its consequences in the receiving country, the tour operator's own business interests in a country, and decisions by the governments of the generating counties. Tour operators always try to diminish the safety and security hazards that their clients could face. Therefore, they have to decide whether to include destinations with different kinds of risks in their programs. As such, they influence the way a particular destination is viewed, because their practice affects the attitude of all potential tourists.

The Croatian experience supports the findings on vacation decisions by Sönmez and Graefe (1998), which show that past tourism experience has only an indirect effect on future behavior. Therefore, it is possible to claim that in order to bring tourists back to a country following a crisis, it is not enough to rely on the fact that they know the destination from the time before the crisis. The situation requires commitment and considerable investment to reestablish the business with tour operators and travel agencies to restore traffic and regain the earlier position in the international market. The more the tour operator is vertically integrated with companies in the receiving countries, the more attention it pays to the quick recovery. But, not until the main and respected tour operators who dominate certain generating markets return to the receiving market after the crisis, can the country count on full recognition in the main generating markets. This suggests that the largest tour operators enjoy the trust of potential tourists and thus have crucial influence on them and the resulting demand. As long as the "big players," with real or even symbolic capacities do not return to a particular receiving market, tourists continue to question safety and security in vacationing there. Potential tourists view tour operators as strong signals of travel safety or risk. Thus, they create an image in the minds of both their potential clients and individual tourists. Interestingly, this image may not always reflect the real situation at the destination.

Concept Definitions

Tour operator An economic entity which, by uniting the services of different providers, creates and organizes inclusive tours in its own name and its own account, for yet unknown buyers, and by doing so continuously realizes its main source of income.
Induced image An image created from the strategic promotion directly by a destination and/or businesses.
Organic image An image created by mass media and other ostensibly unbiased sources of information.

Review Questions

1. Explain why tour operators assess destination safety more critically than an individual would.
2. Explain the factors influencing the crisis behavior of tour operators toward tourism destinations.

3. Who should be involved in handling the crisis in a tourism destination and why?
4. Explain why rebuilding of the destination image should first start with organic images.
5. Why does the study claim that a country affected by crises cannot count on full recognition in the main generating markets until the main and respected tour operators dominating these markets return to the receiving market?

Note

[1] This chapter was previously published as: Cavlek, N. (2002). Tour operators and destination safety. *Annals of Tourism Research,* 29(2), 478–496.

References

Bar-On, R. (1996). Measuring the effects on tourism of violence and of promotion following violent acts, in A. Pizam and Y. Mansfeld (eds.), *Tourism, Crime and International Security Issues*. New York: John Wiley & Sons, pp. 159–174.
Barton, L. (1994). Crisis management: Preparing for and managing disaster. *The Cornell Hotel and Restaurant Administration Quarterly,* 35, 59–65.
Bloom, J. (1996). A South African perspective of the effects of crime and violence on the tourism industry, in A. Pizam and Y. Mansfeld (eds.), *Tourism, Crime and International Security Issues*. New York: John Wiley & Sons, pp. 91–102.
Cavlek, N. (1998). Turoperatori i svjetski turizam. Zagreb, Croatia: Golden Marketing.
Cavlek, N. (2000). The role of tour operators in the travel distribution system, in W. Gartner and D. W. Lime (eds.), *Trends in Outdoor Recreation, Leisure and Tourism*. Wallingford, UK: CABI, pp. 325–334.
Drabek, T. (1994). *Disaster Evacuation and the Tourist Industry*. Boulder, CO: Institute of Behavioral Science, University of Colorado.
Drabek, T. (1995). Disaster responses within the tourist industry. *International Journal of Mass Emergencies and Disasters,* 13, 7–23.
Drabek, T. (2000). *Emergency Management, Principles and Applications for Tourism, Hospitality, and Travel Management*. Washington, DC: FEMA (Federal Emergency Management Agency). www.fema.gov/emi/edu/higher.htm.
EIU. (1994). The impact of political unrest and security concerns on international tourism. *Travel & Tourism Analyst,* 2, 69–82.
FV-W International. (1993–1998). *Deutsche Veranstalter in Zahlen*. Hamburg, Germany: Verlag Dieter Niedecken GmbH.
Gartner, W. (1996). *Tourism Development—Principles, Processes, and Policies*. New York: Van Nostrand Reinhold.
Gartner, W., and Shen, H. (1992). The impact of Tiananmen Square on China's tourism image. *Journal of Travel Research,* 30(4), 47–52.
Gunn, C. (1972). *Vacationscape: Designing Tourist Regions*. Austin, TX: Bureau of Business Research. University of Texas.
Ioannides, D., and Apostolopoulos, Y. (1999). Political instability, war, and tourism in Cyprus: Effects, management, and prospects for recovery. *Journal of Travel Research,* 38(1), 51–56.
Leaf, R. (1995). Presentation at the General Assembly of the WTO in Cairo.

Leslie, D. (1999). Terrorism and tourism: The Northern Ireland situation—A look behind the veil of certainty. *Journal of Travel Research,* 38(1), 37–40.
Mansfeld, Y. (1999) Cycles of war, terror, and peace: Determinants and management of crisis and recovery of the Israeli tourism industry. *Journal of Travel Research,* 38(1), 30–36.
McGuckin, M., and Demick, D. (2000). Northern Ireland's image—Platform or pitfall for gaining the competitive edge, in J. Ruddy and S. Flanagan (eds.), *Tourism Destination Marketing—Gaining the Competitive Edge.* Dublin Institute of Technology, Dublin: Tourism Research Center, pp. 335–343.
Perez, S., and East, M. (1991). *The EC Directive—An Analysis.* London: Travel Industry Digests.
Pirjevec, B. (1998). Creating a post-war tourist destination image. *Acta Turistica,* 10(2), 95–109.
Pizam, A. (1999). A comprehensive approach to classifying acts of crime and violence at tourism destinations. *Journal of Travel Research,* 38(1), 5–12.
Pizam, A. and Mansfeld, Y. (eds.). (1996). *Tourism, Crime and International Security Issues.* New York: John Wiley & Sons.
Radnic, A. and Ivandic, N. (1999). War and tourism in Croatia – Consequences and the road to recovery. *Turizam*, 47, (1), 43–54.
Richards, B. (1999). Special report—The Kosovo effect. *ABTA Magazine,* 36–38.
Richter, L. K., and Waugh, W. L., Jr. (1986). Terrorism and tourism as logical companions. *Tourism Management,* 7(4), 230–238.
Santana, G. (1999). Tourism: Toward a model for crisis management. *Turizam,* 47(1), 4–12.
Savignac, A. R. (1994). *WTO News,* 3, 1.
Sönmez, S. F. (1998). Tourism, terrorism and political instability. *Annals of Tourism Research,* 25(2), 416–455.
Sönmez, S. F., Apostolopoulos, T., and Tarlow, P. (1999) Tourism in crisis: Managing the effects of terrorism. *Journal of Travel Research,* 38(1), 13–18.
Sönmez, S. F., Backman, S. J., and Allen, L. R. (1994). *Managing Tourism Crisis: A Guidebook.* Clemson, SC: Clemson University.
Sönmez, S. F., and Graefe, A. R. (1998). International vacation decisions and terrorism risk. *Annals of Tourism Research,* 25(1), 112–144.
Statistički godišnjak Republike Hrvatske (Republic Bureau of Statistics). (1990). Zagreb, Croatia: Drzavni zavod za statistiku.
Statistički ljetopis. (1995–1996). Zagreb, Croatia: Drzavni zavod za statistiku.
Steene, A. (1999). Risk management within tourism and travel. *Turizam,* 47(1), 13–18.
Wahab, S. (1996). Tourism and terrorism: Synthesis of the problem with emphasis on Egypt, in A. Pizam and Y. Mansfeld (eds.), *Tourism, Crime and International Security Issues.* New York: John Wiley and Sons, pp. 175–186.
World Tourism Organization (WTO). (1999a). *The Economic Impact of Tourism.* Madrid: WTO.
WTO. (1999b). *Yearbook of Tourism Statistics.* Vol. II, 51st ed., Madrid: WTO.
WTO. (1999c). *Tourism Highlights 1999—Revised Preliminary Estimates.* Madrid: WTO.

Summary and Conclusions

Yoel Mansfeld and Abraham Pizam

What are the lessons that we have learned from the theoretical chapters and the case studies described in this book? By now it is possible to conclude that:

- Safety and security incidents will continue to occur at tourist destinations regardless of the efforts made by the private and public sector to prevent them. Since the majority of safety and security incidents, be they natural (such as hurricanes, tornadoes, floods, earthquakes, tsunamis, etc.), man-made (such as crime, terrorism, war, etc.), or health hazards, are caused by factors outside the control of tourists destinations, it is practically impossible to totally eradicate them.
- An absolute majority of safety and security incidents occurring at tourist destinations will have some degree of negative impact on affected destinations and their tourism industry, though the severity of the impact is not uniform and is influenced by numerous factors.
- Though it is impossible to totally prevent the occurrence of safety and security incidents, destinations can prepare themselves for the occurrence of such incidents through crisis planning and thus reduce their negative impacts on the community, its economy, the tourism industry, and the tourists themselves.
- The private sector (i.e., the tourism industry) bears the major responsibility for preventing or reducing the number of safety and security incidents occurring on their own properties. The public sector and, most importantly, governments of host destinations, bear the major responsibility for ensuring a high level of security for visiting tourists.
- There are presently numerous available methods—both hardware and policies—for preventing and/or reducing the number of safety and security incidents at tourist destinations, though their level of effectiveness varies widely.
- Generally speaking, the most effective way to prevent and/or reduce the number of safety and security incidents occurring at tourist destinations is by strong cooperation between the tourism industry, the local community, governmental authorities, and the tourists themselves.
- Despite the severe and devastating impacts that some safety and security incidents have on affected destinations, with few exceptions—such as total and continuous warfare—it is actually possible to fully recover from these impacts. The pace of

recovery will be rapid if there is a strong, latent demand for a tourist product, even though it might mean waiting until the security situation in the affected destination is over before being able to fully utilize that tourist product.

- To recover completely from severe and recurring safety and security incidents, tourism enterprises need the full technical, financial, and moral support of the local, regional, and national public sectors.
- The negative impacts of tourism security and safety incidents cannot usually be confined to the geographical area where they occur and usually spills over well beyond the location of the incident. In order to avoid this unnecessary damage there is a need to convey geographical messages defining the spatial dimensions of the affected area.
- Tourists need to be constantly educated about the safety and security hazards present during their trips and given practical tips to avoid them. They should be made to understand that in numerous instances they themselves through their own actions bear part of the responsibility for incidents of safety and security affecting them.
- The media and travel advisories issued by governments in generating markets play a major role in the formation of the image of a tourist destination following one or several incidents of safety and security. Therefore, to fully recover from such incidents, destination decision makers have to work closely with the media and governments of generating markets to reduce panic and bring back normalcy.

Above all it is possible to conclude that with each additional unfortunate safety and security incident occurring at a tourist destination, our knowledge base increases. Thus, as students and researchers of tourism safety and security it is our duty to continue to collect the data, process, and analyze it, so that we may draw the proper lessons and use them for reducing the occurrence of such incidents and mitigating their impacts.

Index

T